EDUARD BRÜCKNER – THE SOURCES AND CONSEQUENCES OF CLIMATE CHANGE AND CLIMATE VARIABILITY IN HISTORICAL TIMES

Eduard Brückner – The Sources and Consequences of Climate Change and Climate Variability in Historical Times

Edited by

Nico Stehr
Sustainable Research Development Institute,
University of British Columbia,
Vancouver, BC, Canada

and

Hans von Storch
Institute of Hydrophysics,
GKSS Research Centre,
Geesthacht, Germany

translations by
Barbara Stehr and Gordon Gamlin

KLUWER ACADEMIC PUBLISHERS
DORDRECHT / BOSTON / LONDON

Library of Congress Cataloging-in-Publication Data

Brückner, Eduard, 1862-1927.
Eduard Brückner : the sources and consequences of climate change and climate variability in historical times / edited by Nico Stehr and Hans von Storch.
p. cm.
Includes bibliographical references and index.

1. Climatic changes. 2. Paleoclimatology. 3. Brèckner, Eduard, 1862-1927. I. Title: Sources and consequences of climate change and climate variability in historical times. II. Stehr, Nico. III. Storch, H. v. (Hans von), 1949- IV. Title.

QC981.8.C5 B75 1999
551.6'0903--dc21

ISBN 978-90-481-5381-7 99-058834

Published by Kluwer Academic Publishers,
P.O. Box 17, 3300 AA Dordrecht, The Netherlands.

Sold and distributed in North, Central and South America
by Kluwer Academic Publishers,
101 Philip Drive, Norwell, MA 02061, U.S.A.

In all other countries, sold and distributed
by Kluwer Academic Publishers,
P.O. Box 322, 3300 AH Dordrecht, The Netherlands.

Printed on acid-free paper

Contents

About the Editors

Nico Stehr is Senior Research Associate in the Sustainable Research Development Institute of the University of British Columbia, Vancouver, British Columbia, Canada, and a visiting scientist at the Max-Planck Institut für Meteorologie, Hamburg, Germany. He is a Fellow of the Royal Society of Canada and editor of the Canadian Journal of Sociology. His research interests center on the transformation of modern society into a knowledge society, global change and public policy, the interrelation between climate and society and the uses of social and natural science knowledge. Among his recent publications are *Practical Knowledge* (1992), *Knowledge Societies* (1994) and *The Culture and Power of Knowledge: Inquiries into Contemporary Societies* (with Richard V. Ericson, 1992). His *The Fragility of Modern Societies* is forthcoming.

From 1987 to 1995, Hans von Storch was Senior Scientist and leader of the "Statistical Analysis and Modelling" group at the Max Planck-Institut for Meteorology. In 1996, he became director of the Institute of Hydrophysics at the GKSS Research Centre and professor at the Meteorological Department of the University of Hamburg. He published several books, among others "Statistical Analysis in Climate Research" (1999) and edited "Anthropogenic Climate Change" (1999) and "Analysis of Climate Variability" (1995).

His scientific interests are statistical analysis (especially transfer functions relating large-scale climate to local features, identification of modal structures in geophysical fields; data driven simulations), simulation of regional climates and budgets of matter, paleoclimatic modelling, dynamics and statistics of low-frequency climate variability, and transfer of knowledge from natural sciences to the public arena (in cooperation with economists and sociologists).

Acknowledgements

Before we acknowledge the crucial assistance and support of colleagues and organizations we want to affirm our belief that the transmission of intellectual traditions or the history of ideas and innovation in science are not simply opposing activities. The transfer of ideas from the past should not be conflated with the notion that the preservation and acquaintance with knowledge of the past equals mere repetition and preservation of these traditions. The transmission of ideas from the past is always a mediation of such ideas in the light of new circumstances and therefore present problems and issues. In short, familiarity with past ideas can be instrumental in the construction of new knowledge and not so much an obstacle to scientific discovery, as the practice of science today often appears to imply. We would anticipate that this also is the case with the formidable ideas Eduard Brückner developed about the dynamics of the climate system as well as its impact on society.

We are grateful to a number of individuals and institutions that have made the English publication of this anthology of the writings of the climate scientist and geographer Eduard Brückner possible. Gordon Gamlin has provided us with a competent translation of Brückner's writing. Barbara Stehr has spent much imaginary energy and countless hours vetting and improving upon the initial translation. We are most grateful to both. Skilful and responsible editorial assistance was provided by Robin Taylor and Ilona Liesner.

Some of the financial support in the form of research assistance was provided by the Canadian Social Sciences and Humanities Research Council. The institutional support of Green College of the University of British Columbia, Vancouver, Canada as well as the GKSS Forschungszentrum, Geesthacht, Germany also was of considerable help in preparing this volume.

Eduard Brückner's Ideas - Relevant in His Time and Today
Nico Stehr and Hans von Storch

1 INTRODUCTION

1.1 Temporal Flow of Ideas and the Failure of Diffusion

For a natural scientist, scientific discourse develops like the trunk of a tree. Each year, a new tree ring is formed based on the most recent findings incorporating previous results—from the most recent tree ring, so to speak—and newly established facts and interpretations. Knowledge obtained from earlier research is either encoded or obliterated in present knowledge—continuously transferred from tree ring to tree ring—or forgotten. If something has not been incorporated from cohort to cohort of scientists, it is considered to be irrelevant and of little interest. This approach in natural science to its own history is clearly manifest in practically all-contemporary articles in scientific journals. Most of the citations refer to work not older than 5 years. Sometimes casual reference is made to a handful of "classical" papers or books but the authors have likely never closely examined these classic works but know about them only indirectly.

This mode of operation is undoubtedly an efficient way of coping with the sheer amount of publications scientists face daily. It is simply not possible to digest all new results—even in a field as relatively narrow as climate science—let alone critically read many of the potentially relevant original documents of past research. For example, for the process of understanding a map displaying the global temperature by means of isotherms, it is not important to know that the technique of isotherms was invented by Alexander von Humboldt, or what his ideas about the technique were at the time.

In almost all cases, this "diffusive"[1] transfer of knowledge from cohort to cohort and generation to generation, or from "tree ring" to "tree ring", works effectively and is robust enough to filter out what are considered to be irrelevant constructions from the flow of knowledge. At the same time, during

[1] We use concepts geared towards the thinking of physicists. The relevant background is the transport of heat in a fluid. Heat can be transported either by *diffusion*, which is maintained by the collisions of the individual molecules within the fluid. This transport is relatively inefficient as any transport is made up of many little steps over the small distances between two molecules. In our metaphor, this means that knowledge is transferred through personal contacts among scientists and through recent publications. A more efficient way takes place when a current transports a parcel consisting of many molecules as a whole over a longer distance. This process is called *convective transport*. In our context, this refers to the introduction of forgotten concepts and results into contemporary thinking.

this process of consensus building all knowledge claims are continually and critically examined with respect to recent insights. Today, no one in the natural science community would advance claims based on old authoritative sources, as was common, for example, with the work of Aristotle during medieval times. However, this process is not effective when a line of inquiry in science is displaced for some reasons—and interest then reappears after a longer period of time has passed and the collective memory about past intellectual perspectives is no longer available in present-day journals and scientists. In such a case, the transfer of knowledge needs more than "diffusion" but outright "convective transport" from deeper placed tree rings to the surface.

This "convective" transfer of ideas from the past should not be misunderstood as an attempt of merely repeating and preserving cognitive traditions. Hand-in-hand with the transmission of ideas from the past goes a mediation and interpretation of such ideas in the light of new circumstances and therefore present problems and issues. Thus, familiarity with past ideas can be instrumental in the construction of new knowledge and is not so much an obstacle to scientific discovery, as the practices of the scientific community today often appears to imply but an intellectual asset in efforts to advance science.

We believe that climate science is a case for which the "convective" influx of past ideas is a compelling necessity. After having been the bookkeeper for geography and meteorology in the 19th century, climatology developed into a science of the physics and chemistry of the atmosphere and the ocean; the early view that climatology is foremost a field of study that deals with the impact of climate on people and society was virtually forgotten. In the 1980s and 1990s, climate science underwent another paradigmatic change: after the discovery that humankind is about to change climate, the old problem of anthropogenic climatic change and the influence of climate on individual and society re-emerged.

During our own work on the interrelations between climate and social conduct we came across a number of early climate scientists who had a significant impact on both their peers and the general public. One of them was the eminent geographer Eduard Brückner, who is today forgotten in climate science, and is considered by geographers to represent but a closed episode in their disciplinary history.[2] At the turn of the twentieth century, he was one of the central protagonists in a vigorous debate in science and society about global climate variability and its political and economic significance. We believe that his formidable ideas could have a significant impact on our present view of climate, climate variability and climate impact. It is

[2] An informative overview about Brückner's scientific career and achievements can be found in Grosjean (1991).

for this reason that we have assembled this anthology of Brückner's main work on climate variability and climate impact.

2 ORGANIZATION OF THIS BOOK

In this introductory chapter[3] we present information about Eduard Brückner and his scientific work, compare his approach with that of his contemporary Julius von Hann, and relate his views to the present-day discussion.

The main part of this book consists of reprints of Brückner's original work in climate science. As most of his publications were in German, they were translated. These translations of Brückner's texts conform strictly to the original. Only in the case of completely irrelevant notes have we decided to delete such references. Additions we have made are inserted in square brackets. All diagrams have been redrawn. Some native city names used may be less familiar than the English names: München is Munich; Wien, Vienna; Praha, Prague.

The following is a list of the material presented. These eleven items were chosen as they demonstrate well Brückner's interest in climate variability, his assessment of contemporary analyses and thinking about anthropogenic climate change (such as the widespread concern about desiccation), and how he has dealt with the transfer of knowledge into society.

1. *Groundwater and Typhus* [*Grundwasser und Typhus*. Mittheilungen der Geographischen Gesellschaft in Hamburg], Volume III, 1887–1888.
2. *Fluctuations of Water Levels in the Caspian Sea, the Black Sea, and the Baltic Sea Relative to Weather* [*Die Schwankungen des Wasserstandes im Kaspischen Meer, dem Schwarzen Meer und der Ostsee in ihrer Beziehung zur Witterung*], Annalen der Hydrographie und Maritimen Meteorologie, Volume II, 1888.
3. *How Constant is Today's Climate?* [*In wie weit ist das heutige Klima konstant?*], Verhandlungen des VIII Deutschen Geographentages, 1889.
4. *Climate Change Since 1700.* [*Klimaschwankungen seit 1700.* Excerpts from *Klimaschwankungen seit 1700.*] Wien; E.D. Hölzel, 1890;
   Chapter 1: *The Current Status of the Inquiry into Climate Changes* [*Der gegenwärtige Stand der Frage nach den Klimaänderungen.*]
   Chapter 8: *Periodicity of Climatic Variations derived from observations of ice conditions on rivers, the date of grape harvest and the frequency of severe winters* [*Die Periodizität der Klimaschwankungen, abgeleitet auf Grund der Beobachtungen über die Eisverhältnisse der Flüsse, über das*

[3] This introductory chapter incorporates some materials first published by Stehr et al. (1995)

Datum der Weinernte und die Häufigkeit strenger Winter] Chapter 9: *The Significance of Climatic Variations in Theory and Practice* [*Die Bedeutung der Klimaschwankungen für Theorie und Praxis*]

5. *About the Influence of Snow Cover on the Climate of the Alps* [*Über den Einfluß der Schneedecke auf das Klima der Alpen*], Zeitschrift des Deutschen und Österreichischen Alpenvereins, 1893.
6. *Influence of Climate Variability on Harvest and Grain Prices in Europe* [*Der Einfluß der Klimaschwankungen auf die Ernteerträge und Getreidepreise in Europa*]. Geographische Zeitschrift, 1895.
7. *Weather Prophets* [*Wetterpropheten*], Jahresbericht der Berner Geographischen Gesellschaft, 1886.
8. *An Inquiry About the 35-Year-Period Climatic Variations* [*Zur Frage der 35jährigen Klimaschwankungen*] Petermann's Mittheilungen, 1902.
9. *About Climate Variability* [*Über Klimaschwankungen*]. Mittheilungen der Deutschen Landwirtschaftsgesellschaft, 1909.
10. *Climate Variability and Mass Migration* [*Klimaschwankungen und Völkerwanderungen*]. Talk at Kaiserliche Akademie der Wissenschaften, Wien 1912.
11. *The Settlement of the United States as Controlled by Climate and Climatic Oscillations*. Memorial Volume of Transcontinental Excursion of 1912 of the American Geographical Society of New York, 1915.

Eduard Brückner and Albrecht Penck in the summer of 1893 on an excursion near Flims (Graubünden, Switzerland). Taken from Büdel, (1977).

They say it is observed in the Low Countries, that every five and thirty years the same kind and suit of years and weathers comes about again; as great frosts, great wet, great drought, warm winters, summers with little heat, and the like, and they call it the prime; it is a thing I do rather mention, because, computing backwards, I have found some concurrence.

Francis Bacon

3 THE CLIMATE SCIENTIST EDUARD BRÜCKNER

3.1 The Life of Eduard Brückner

Eduard Brückner was born on July 29, 1863, in Jena, Germany.[4] He lived for a while in Odessa, Russia, before moving with his parents to Dorpat (now Tartu, Estonia), where he spent most of his childhood. In 1879 he was sent to school in Karlsruhe, Germany. After graduating from high school, he studied at the universities of Dorpat, Dresden, and München. He attended lectures and seminars in geography, geology, paleontology, physics, meteorology, and history. In 1885, he completed his doctorate under the supervision of Albrecht Penck in München with a dissertation on the *Glaciation of the Salzach area* (*Die Vergletscherung des Salzachgebietes*) in Austria. In 1886, he moved to the Office for Marine Weather (*Seewarte*) in Hamburg to work with Wladimir Köppen. The first two of our translated articles originate from this early period of his scientific career. They pertain to the possible link between groundwater levels and the incidence of typhus, and the relationship between sea water level variations and weather conditions.

On the strength of his dissertation, Brückner was appointed professor of geography at the University of Bern in 1988. He stayed in Bern for 16 years, and became Rector of University of Bern in 1899/1900. During his stay in Bern, he lectured on various aspects of geography but also regularly offered public lectures. In 1904 he accepted an offer from the University of Halle in Germany and, in 1906, finally moved, as the successor of his former teacher Albrecht Penck, to the University of Vienna. Brückner died in Vienna in

[4] Cf. Grosjean (1991) and Oberhummer (1927).

1927 at the age of 64. While in Vienna, he was, as in Bern, engaged in the transfer of academic knowledge to the general public. He was chairman of a series of Public University Lectures (*Volksthümliche Universitätskurse*).

In 1890 he published the first extensive book-length discussion of recent climate fluctuations, that is, of climatic fluctuations in "historical times". Brückner (1894:1) credits the head of the Bavarian meteorological services, C. Lang, with the discovery of decadal scale climate variability in a study of the climate of the Alps. We have selected chapters 1, 8, and 9 of this monograph (item 4).

After 1890, Brückner published only a few smaller articles on the observational evidence of climate variability (Brückner, 1895, 1902; reprinted as items 6 and 8). He explains the small number of articles on the observational evidence as the result of a lack of new and appropriate meteorological data on the issue. In the present-day context of particular importance, though, are his articles in which he speculated about the geographical and socio-economic impact of climate change, i.e., the social consequences arising from the climate fluctuations, such as emigration, immigration, and migration patterns (Brückner, 1912; [1912] 1915; items 10 and 11), or on harvests, the balance of trade of countries and shifts in the political predominance of nations (Brückner, 1894, 1895, 1909; the last two reprinted as items 6 and 9) .

He was convinced that the issue of climate change and its impact was both of considerable scientific merit and that future climate changes are of great importance to the well-being of society as well as for the strategic and economic balance of political and economic powers. He therefore presented his conclusions about serious repercussions associated with climate change anticipated for the end of the past century in the form of oral presentations addressing the general public and especially affected segments of the public, such as farmers. As a result, Brückner presented his initial findings on climate change not only to a congress of professional geographers in Berlin in 1889 (our item 3), but also a year earlier in a public lecture entitled *Is our climate changing?* at the University of Dorpat that was duly noted in the local press (Brückner, 1888). Later Brückner (1894, 1909; the last one is reproduced as item 9) published newspaper articles about the general issue of climate change as well as about its specific economic and social consequences. His work on climate variability was discussed at length in the contemporary press (e.g., *Neue Freie Presse*, Vienna, February 11, 1891).

As a result of these activities and the response they generated, Brückner's work on climate variability found a considerable echo among the scientific community of climate researchers (e.g., DeCourny Ward, [1908] 1918), sociologists (e.g., Sorokin, 1928: 120–124), geographers (e.g., Huntington, 1915: 172–173; [1915] 1924:25), historians (e.g., Le Roy Ladurie, [1971]

1988:217, 220) and physicists (e.g., Arrhenius, 1903: 570–571), but to some extent also among the public at large, as is exemplified by the fact that he was often mentioned as an influential climate scientist in various encyclopaedia until the 1950s. Huntington (1915:172) elevates Eduard Brückner to "*one of the chief European authorities on climate*" and credits him for having initiated a kind of paradigm shift in climate research: "*Since the publication of Brückner's widely known book on 'Climatic Changes Since 1700' there has been a strong and growing tendency to treat climate as a dynamic instead of a static geographical force*" (Huntington, 1916:192).

3.2 Eduard Brückner's Analysis of Climate Variability

In the following section we summarise Brückner's attempt to synthesize the observational evidence for global-scale synchronous climate variability from his limited data and limited computing power. Most of this synthesis is described in his 1890 monograph.

Brückner (1889:2) indicates that he was first alerted to the possibility of climate change, aside from information about shrinking glaciers in the Alps, as the result of observations about changing water levels in the Baltic, the Caspian and the Black Sea (item 2). The changes in the water levels appeared to follow a specific pattern. The rhythm of the changes resembled changes in the glaciers of the Alps.

In his detailed discussion of "recent" climate fluctuations Brückner (1890) justified his approach by referring to the studies of E. Richter, C. Lang and A. Swarowsky. Richter concluded that the causes for the secular variations of one specific glacier (*Obersulzbachgletscher* in Austria) are wet and dry periods lasting for several years in that particular region. Lang showed that this result is valid for the entire Alpine region. Swarowsky stated a striking correlation between the variation of the water level of the Neusiedler See, a lake without any outlet near the Austrian–Hungarian border, and the secular variations of the glaciers in the Alps, thereby demonstrating that lakes without an outlet are excellent indicators of secular climate variability.

In his 1890 monograph on climate variability, Brückner started his analysis with a careful investigation of the world-wide largest "lake" with no outlet, the Caspian Sea. Brückner drew the conclusion that Lang's results not only hold for the Alps but may be extended to the vast catchment area of the Caspian Sea (Brückner 1890:86). He found that the climatic variation followed a characteristic 35-year pattern, with wet and cool conditions alternating with dry and warm conditions.

This inductive method of extending results from a smaller region to a larger one is, by the way, typical for Brückner's approach and consequently he searches in data available from several other lakes without an outlet all over the world for signals of secular variations. Brückner states that the mere existence of water variations in the lakes allows to the presumption that secular climate fluctuations take place in the corresponding catchments (Brückner 1890:115). In a further step, Brückner applies the concept of linking water levels of lakes to the rainfall in the corresponding regions also to lakes with an outlet (*Flusseen*) and even rivers thereby stating the existence of a more or less synchronous climate fluctuation over the entire land mass of the world (Brückner 1890:132).

The record of instrumental observations available to Brückner reached back for about 100 years. In these data he identified a rhythm of 35-year alternating wet/cool and dry/warm episodes. In order to trace these characteristic climate fluctuations further back, Brückner also studied the observed data of the ice conditions of the rivers, the grape harvest and the abundance of strong winters. According to his data, Brückner was able to identify 25 quasi-periodic cycles of about 35 years length during the last 1000 years (Brückner 1890:286).

He emphasized the fact that his mode of variability was not strictly periodic but that the alternating wet and dry periods lasted about 35 years on average. This fact is insofar noteworthy as in Brückner's years, the fashion of decomposing time series of all sorts into its Fourier components in an attempt to describe the time series as a sum of predictable components developed. Obviously Brückner stayed away from this fashion, which later was shown to be based on a simplistic misunderstanding of the mathematics of statistical time series.[5]

[5] The fascination with the notion of periodic cycles as a description and an explanation for the rise and fall of geological phenomena, of plants and animals as well as social and economic processes, was still a vibrant enterprise in science during Brückner's career. Sir N. Shaw's Manual of Meteorology from the mid-1930s featured a list of several pages length of various periods found in meteorological data. The conviction that "*the whole history of life is a record of cycles*" (Huntington, 1945:453) was widespread. The fascination arises from the fact that a process made up of a superposition of a finite number of periodic sub-processes makes the process predictable: "*It will be a vast boon to mankind when we learn to prophesy the precise dates when cycles of various kinds will reach definite stages*" (Huntington, 1945:458). In the 1920s and 30s, the Russian mathematician Slutsky showed that the Fourier analysis of a statistical time series always reveals some periodicities, even if the time series is constructed free of such periodicities. If different chunks of such time series are analyzed, different periodicities pop up and vanish. In spite of this finding, which today is completely understood, in some circles, and in particular among lay scientists, the interest did not cease. On the contrary, in 1941 the interest in the study of cycles led to the formation of a "Foundation for the Study of

He speculated that the dynamical mechanism behind his quasi-oscillation would be related to some unknown solar forcing mechanism (Brückner 1890: 240, 242) but was aware that no observational evidence for such an oscillation exists. In this context Brückner denied any connection between secular climate fluctuations and variations of sunspot activity (Brückner 1890:242).

Based on this 35-year period oscillation, Brückner prognosed a dry period at the turn of the century (Brückner 1890:286,287) with severe negative consequences in crops for continental regions, like Northern America, Siberia and Australia. It is noteworthy that this predictive scheme would have enabled Brückner to predict the "dust bowl" in the central part of the United States, which actually took place during the Thirties of this century.[6]

Brückner's methods were mainly limited to the exploratory statistical analysis of time series since confirmatory tools such as confidence intervals or hypothesis testing were not developed in combination with what might be called common sense. He was unfamiliar with dynamically arguments (for instance, concerning the geostrophic wind, which was well known among meteorologists of those days) and he was unaware of theories concerning the general circulation of the atmosphere (he failed to acknowledge the different dynamic character of the tropics as opposed to the extratropical westerly regime).

A fact that is impressive for modern climate researchers, who are used to being supported by computers and digital data files, is the amount of computational work done by Brückner. It seems that he did all the calculations himself. He computed 5-year totals, called *lustrum*, and checked their consistence by comparing records from neighboring stations. When data at neighboring stations at some time begin to diverge, he concluded that one of the two records is contaminated by some artificial effects, such as displacement of the instrument (such as a water level meter). He tried to correct for such *inhomogeneities*, and calculated correlations between his various time

---

Cycles" by Edward R. Dewey. The Foundation exists to this day and claims to have more than 3000 members.

6 In 1915 Brückner predicted that by 1920 "*we may expect a maximum of humidity*" in the United States (Brückner, 1915: 132). This prediction exploited two pieces of information: first, the dynamical insight about the existence of a 35-year oscillation and second, Brückner's finding that precipitation was at its minimum around 1900. On the continental scale, his forecast was incorrect (Bradley, 1987: Fig. 6), but in a regional sense his forecasts were consistent with actual developments: The Great Salt Lake exhibited maximum lake levels from 1910 to 1930. Brückner did not spell out another prediction based on the same reasoning, namely that in the middle of the 1930s the United States would again suffer from dry conditions. Indeed the Great Salt Lake exhibited a sharp water level drop in the early 1930s. Also the "Dust Bowl" dry episode that led to persistent disastrous harvest failures in Central North America took place in the mid 1930s.

series to establish the degree of similarity between them. The sheer work of just collecting the data, checking their consistency and calculating their statistics must have been enormous, and hardly imaginable for a modern scientist. His methodical approach is similar to what is done today when, for instance, compiling records of the global mean temperature. The difference is, of course, that the work is no longer done by human computers but electronic hardware supervised by humans.

> *The number of hypotheses and theories about climate change are numerous. Quite naturally they have caught the public attention, as any proof of past climatic change points to the possibility of future climate change, which inevitably will have significant implications for global economics.*
>
> Brückner (1890:2)

4 CLIMATE CHANGE, CLIMATE POLICIES AND SOCIETY

Today, the concepts of "climate variability", "climate change", and "climate impact" attract an enormous interest not only in the climatological, meteorological, and oceanographic community (von Storch and Hasselmann, 1996) but also in sciences concerned with climate-sensitive systems, such as biometeorology, ecology, coastal defense, or the social sciences. The discussion of "the climate problem"[7] is by no means limited to the scientific community. It has drawn a great deal of interest from a general public (Lacey and Longmann, 1993) perhaps haunted by anticipations of catastrophic developments as a consequence of future anthropogenic climate change (Stehr and von Storch, 1995). Evidence of a public and scientific preoccupation with "the climate problem" is given by such institutions as the "Intergovernmental Panel on Climate Change" (IPCC) and international conferences aimed at the establishment of International Climate Conventions.

[7] We place the expression "climate problem" in quotation marks since it is not well-defined. Natural scientists associate with this expression the understanding, prediction, and, possibly, control of climate variability. Social scientists, on the other hand, consider the perception of climate, and its social and political implications as the "climate problem".

The majority of the scientific and general public interprets the climate problem as a *new* challenge. Yet, although for much of past two centuries most "climatologists" and meteorologists have been convinced, and have considered it to be almost an axiom, that global climate is a constant during historical times,[8] some 19th century climatologists, geographers and meteorologists maintained that climate is *not* a steady phenomenon (e.g., Brückner, 1890; Hann, [1883] 1893:362), recognizing that climate varies not only on geological time scales (thousands of years and longer) but also on decadal and century time scales due to natural and anthropogenic processes.

The processes that were discussed at the turn of the last century as the source of climate variability and change were different. The "*natural* variability", unrelated to man's activities, was speculatively attributed to astronomical factors, such as the solar activity, and to processes in the interior of the earth. In addition, the idea of deterministic periodic processes attracted much attention among climatologists. *Anthropogenic* "climate change" was thought to be the result of human activities, such as de- and reforestation or new cultivation of land in North America. The possibility that anthropogenic emissions of carbon dioxide might alter the global climate was first discussed by the chemist Svante Arrhenius (1896; 1903), but dismissed by him as a realistic perspective for the next few hundred years.

The intensive debate among climatologists at the turn of the century receded into the background when a new disciplinary consensus emerged that remained predominant until recently, namely that the global climate system contained overriding equilibrating processes providing resilience against secular climate fluctuations; fluctuations that did occur were seen as distributed around a fairly stable mean climatic condition. Any anomaly extending for a few years would be canceled by an opposite anomaly at another time. On average, nothing would change. One reason why the perception of climate variations on historical time scales became unpopular may be the rejection of "catastrophism" and the eventual acceptance of "uniformitarism" in geology, as proposed by Lyell in the 1830s. Some of the social

[8] Brückner (1889:2) notes that during the 19th century, a distinct disciplinary division with respect to the issue of climate change could be observed: Geographers and geologists were more inclined to consider a persistent climate change to be a reality while meteorologists defended the thesis that climate is a constant. Brückner (1890:2) offers an explanation why most professional meteorologists and many geographers at the time were rather silent on the issue of climate change; as a matter of fact, he observes that they were embarrassed to engage in research and discussion about climate change. The reason for the reluctance is the wealth of competing hypotheses about climate change formulated earlier in the century. But previous efforts only resulted in many contradictory voices about the nature of climate change, so that climatologists then became reluctant to add to the cacophony of mere opinions. Even in 1959, the prominent climatologist H. Lamb complains that many of his contemporaries consider climate as something static. (Lamb, 1959)

scientific theories about the impact of climate on civilizations, for example, by Sombart ([1911] 1951:324; 1938), Ploetz (1911), or Hellpach (1938), are actually based on the explicit premise of constant climatic conditions (cf. Stehr, 1996). In the public arena at the time, other urgent issues and concerns displaced reflections about climate change and its impact on society.

In the following, we attempt to recover spirited discussions among geographers, meteorologists, and climatologists that occurred toward the end of the last and at the beginning of this century. We make an effort to analyze the dynamics of the discussion, and the degree to which it was introduced to the general public, with the explicit intention of comparing the situation at the time with the present discussions of climate variability and change and of climate policies designed to avoid or mitigate the risk of climate change or to allow for a smooth adaptation.

We concentrate on two of the main contributors to this early discussion of climate variability and change on time scales of decades, namely the already presented Eduard Brückner and Julius Hann, both professors in Vienna for a significant part of their lives. We will discuss their different social roles, their attitudes towards the role of the public, and their understanding of their own work as part of multiple contexts in which they attempted to play different functions. We will show that the two protagonists, Brückner and Hann, represent roles and self-conceptions that resemble present-day roles of climatologists in discussions within and outside the scientific community about the scientific significance and the social impact of climate variability and change. We suggest that the "climate problem", as perceived by scientists and the public at the turn of the century, constitutes a valuable historical analog for present debates on the "climate problem".

4.1 Julius Hann and His View of Climate Variability

Another leading and most influential professional climatologist at the turn of the century was Brückner's Viennese colleague Julius Hann, who was born 1839 in Wartberg, Austria. He studied mathematics, physics, geology, and geography at the University of Vienna. After a career in teaching, he became professor of physics at the University of Vienna and, in 1897, professor of meteorology at the University of Graz. Between 1900 and 1910, he occupied the newly created chair for cosmological physics at the University of Vienna and served as director of the Institute for Meteorology and Geodynamics. He died in 1921 at the age of 83 in Wien.

As Brückner (1923:152) points out in his obituary, Hann may well have been the most important meteorologist of his day and can be considered to be one of the founders of modern meteorology as the science of the physics of the atmosphere (see also Steinhauser, 1951; Kahlig, 1993). He was descriptively oriented, that is, keen to establish the empirical or observational basis for various meteorological phenomena. In meteorology, Hann discovered, independently of Helmholtz, the thermodynamic theory of the Föhn. In climatology and meteorology, he recognized early the importance of quantitative methods and the significance of three-dimensional observation systems, and he initiated the establishment of several mountain observatories. In addition, Hann was editor of the internationally recognized *Meteorologische Zeitschrift* for more than fifty years. Hann was an enemy of speculative thinking; his main goal was to establish the facts (Brückner, 1923:155).

Julius Hann compiled the first textbook on climatology. He first published his *Handbuch der Klimatologie* in 1883. The *Handbuch* appeared in a number of subsequent editions and translations and quickly became a classic in meteorology and climatology (cf. Brückner, 1923; Köppen, 1923:vi; Knoch, 1932:viii). An English edition based on the second edition of the German version of the *Handbook* was published in 1903 (Hann, 1903).

In contrast to later editions of the *Handbuch*, its first edition summarizing the state of knowledge in climatology then still defined as an auxiliary science (*Hilfswissenschaft*) of geography (Hann, 1883:5; also Köppen, 1923:1) did not explicitly deal with the issue of climate variability. Reflecting the preoccupation of the day with the issue of periodicity of climate, Hann distinguishes between two types of climate fluctuations, namely "progressive" (that is, persistent transformations, or, in modern terms "climate change"; e.g., von Storch and Hasselmann, 1996) and "cyclical" changes (that is, fluctuations or oscillations around a constant mean with certain characteristic times or periods; in modern terms "climate variability"). The period for cyclical climate changes could be determined either by deductive reason (by postulating a certain forcing mechanism, such as the sun's activity) or inductive reasoning (by screening the observational record). It should be possible, according to Hann, to trace progressive climate change to either long-term trends of the temperature of the core of the earth or of the output of the sun.

As far as relevant empirical material is concerned, Hann ([1883] 1897:390) referred to both non-instrumental and instrumental observations of temperature and precipitation as well as general accounts or conclusions about climate changes of a wide variety of observers found in disparate historical records. He placed considerable emphasis on the critical examination of the observational climatic record. Obviously such data can be used only if the procedure of observing, archiving and, possibly, correcting the raw data is kept constant (cf. Jones, 1995). The historical data available to

Hann did in general not satisfy this homogeneity condition. He found on close examination that the data recorded in the previous 150 years were almost always contaminated by time-variable biases due to changing observational practices; the oldest instrumental records invariably were started in rapidly expanding cities, and therefore reflect "urbanization", while rain gauges were first placed on higher elevations (e.g., roofs) causing severe biases in measurements (cf. Karl et al., 1993).

On the basis of such methodical pitfalls concerning the quality of the data, Hann was in general rather skeptical of scientific claims identifying climate variability and change in the observational record. In particular, he inferred that the evidence for systematic trends ("progressive changes") of the climate during the historical period based on the available data from different centuries, continents and countries is not substantial (e.g., Hann, [1883] 1897:390). It had been hypothesized that the continental United States of America of the 18th century was subject to an anthropogenic climate change due to the progressive anthropogenic transformation of nature in the course the colonialization. Hann concluded with Whitney (1894), that there is no hard evidence for a resulting climate change on the North American continent (Hann, [1883] 1897:392).

In the case of climate *variability* Hann was less reluctant. He was skeptical about strictly periodic climate fluctuations, especially in regard to any hypothesized connection between variations in sunspot activities and meteorological elements such as temperature, precipitation or changes in the formation of glaciers. On the contrary, he concluded that the influence of sun spot activity on climate patterns is insignificant. Moreover, he rejected the possibility of any predetermination or causal linkage between climatic variations and sun spot activities (Hann, [1883] 1897: 394).

Hann considered Brückner's quasi-oscillatory 35-year cycle much more favorably since it was based on rich data from very different sources. Brückner's discovery seemed valid for many regions and periods, and was supported by many independent observations Hann ([1883] 1897:400) made no serious independent attempts to clarify the dynamics of Brückner's observational evidence. Instead he limited himself to efforts to establish the existence of the patterns of climatic fluctuations. Hann highlighted the fact that Brückner's observations manage to shed light on contradictory accounts of climate variations in specific localities since they "obviously" must have been advanced during different phases of the 35-year period.

Indeed, the second edition of the *Handbuch*, published in 1897, contains a forty-page separate section on climate variability that centers on Brückner's research. In the fourth edition of the *Handbuch*, published in 1932, Karl Knoch had succeeded Hann as author of the *Handbuch*, (Hann and Knoch, [1883] 1932). This fourth edition deals even more systematically with climate variability, even if the summary is rather skeptical. Much prominence is given to contributions that attempt to demonstrate the stability of the climate in historical times and point to the absence of evidence for secular change (see also Berg, 1914).

4.2 Climate Variability and Societal Importance

It was and is common sense that climate variability and climate change have a direct and powerful effect on many aspects of society, including the economy, human health, or even the balance of power among nations.[9] Based on these views, the perspective of *Climate Determinism* emerged suggesting that climatic conditions *determine* virtually all aspects of social life, especially the chances of a society to attain a "high level of civilization".[10] This approach was widely accepted in geography and other disciplines at the time of Brückner and Hann.[11] It is therefore of interest to inquire how Brückner and Hann responded to the challenge of offering their findings to the scientific and general public as warnings of impending climate change but also as instruments to design strategies to deal with climate variations. Interestingly, the two scientists reacted very differently.

Hann disregarded societal impact entirely. He did not even mention possible social consequences of climatic fluctuations. Consistent with the then-prevailing self-conception of climatology as a largely descriptive (e.g., Hann and Knoch, [1883] 1932:3) "young" science (e.g., Köppen, 1923:v), Hann examined the existing evidence on climate variability and change and

[9] The impact of climate on the course of history has been of considerable interest. A more recent account is Lamb's monograph *Climate, History and the Modern World* from 1982 (second edition 1995). Classical views have been put forward by ancient Greek authors such as Hippocrates and philosophers of the enlightenment such as Montesquieu and Herder. Also Friedrich Engels theorized about the influence of climate on society.

[10] The perhaps most prominent representative of modern climate determinism is Ellsworth Huntington (1915, 1945). For a discussion of the climatic determinism, see Stehr and von Storch (1998).

[11] One year after the publication of Brückner's main work on climate variations, in 1891, a certain Professor Umlauff published a scholarly textbook on the "Foundations of Meteorology and Climatology based on most recent research". In his introduction he claims that "*literature of different people is linked in a mysterious manner to the climate of their homeland*" ("*So steht selbst die Literatur eines Volkes in geheimnisvollen Zusammenhängen mit den meteorologischen Elementen des von ihm bewohnten Theils des Erdballes*") and "*Northern Europe has attained his superior level of civilization and moral because of its rain throughout the year, whereas China's success in the past was related to its summer precipitation.*" ("*Nordeuropa habe es seinem Regen zu allen Jahreszeiten zu verdanken, daß es der Sitz der höchsten Gesittung wurde, so wie China seinem Sommerregen die hohe Civilisation in früher Zeit ...*")

attempted to establish whether the data supported arguments for changes in climate phenomena.

Brückner, on the other hand, not only discussed the nature and extent of climatic fluctuations but emphasized their possible consequences for society. In his 1890 monograph, he devoted an entire chapter to these matters: "The importance of climate variability for theory and practice" (see selection 4 in this book). In the terminology of present-day social science, Brückner transformed his academic findings into a form of "practical knowledge" (Stehr, 1991) that was meant to enable strategic responses in the economy, in the field of transportation, health care and agriculture.

He argued that the area covered by ice fields varies, the size, the water level, the appearance or even presence of lakes and rivers, the extent of floods is sensitive to climatic variations. Such disturbances would have a major impact on shipping and commercial patterns. Changing water levels and the duration of ice covers on rivers and streams, in particular, would affect the ability to navigate these waters and therefore the ease with which goods may be moved. Another most important consequence would concern agriculture more directly (see also Brückner, 1894, 1895 and our selection 6) since climatic fluctuations would have a significant influence even if the effects depend to a considerable extent on the harvested product. Brückner concluded that more than two thirds of above-average agricultural outputs in Europe with maritime climate coincided with the warm and dry periods and an equal proportion of poor agricultural yields with the wet and cold climatic periods. In more maritime climates enhanced summer rain would cause harvests to be reduced whereas in continental climates, such as in Central North America or Russia the summer rain would be favorable for agriculture (Brückner, 1894:2, 1915: 137–138).

Thus, the two phases of the 35-year cycle would both have a beneficial effect in certain regions while disadvantages in other regions. Brückner (1915) concluded that as this specific pattern of agricultural productivity change it would leave its marks on the temporal variations of emigration from Western and central Europe to the United States. When conditions were favorable in Europe, namely dry and warm, *fewer* people would emigrate to the United States where a similar dryness and warmth reduced the harvests. On the other hand, when cool and wet conditions prevailed on both sides of the Atlantic, *more* people would travel across the Atlantic, because agriculture in Europe suffered from the climate while productivity in the US was increasing.

Brückner (1890:279–282) also proposed a connection between climatic fluctuations and health. He dealt with one case in some detail, namely the relationship between the appearance of typhus and the level of the ground water, which is controlled by slowly varying precipitation amounts. Having examined records of typhus-mortalities in Central Europe, Brückner attributed at least part of the observed improvement in the mortality—in addition to benefits derived from improvements in the sanitation system—to recovering ground water levels as the result of a shift from dryer to wetter climates.

On the basis of his 35-year "mode of natural variability" and his analysis of the climate sensitivity of civilization, Brückner (1890:279, 287; 1915:132) predicted a number of impending detrimental social consequences of climatic variability, in particular serious economic crises for regions that had benefited from a favorable climate in recent decades, especially areas located within the continental climate regions, such as the United States, Russia and Australia. These regions, Brückner argued, must expect an inevitable shift to dryer weather resulting in significant crop failure.

4.3 The Analogy to the Present State of Affairs

The discussion among scientists at the turn of the century resulted in a series of findings that present-day scientists would consider be sound and perhaps of more recent origin:

1. Climate is not constant but varies on geological as well as historical time scales.
2. Climate variability has to be differentiated between systematic, or, in Hann's words, progressive changes and temporary variations, in Hann's words, cyclical fluctuations.
3. The progressive changes were often related to human action (mainly through land-use changes, often deforestation) while temporary fluctuations were thought to be related to natural processes such a cosmic forcings and processes in the interior of earth (including volcanic activity).

The main difference from today's discussion exists with respect to the last point. At the turn of the century it was acknowledged that the dynamical link between climate and extra-terrestrial variations was not firmly determined. Today, most scientists are convinced that a significant part of the climate fluctuations has its origin in internal climatic processes related to the non-linearity and stochasticity of the climate dynamics.

Scientists, then as today, were confronted with a number of scientific and ethical questions. On the technical side, the problem arose of how to discriminate between human effects and internal processes? Should scientists continue with the conventional curiosity driven research, or were the practical implications of observed patterns so serious that a purely academic orientation should be given up in favor of a more applied research orientation? Because of the perceived importance of climate for political, economic and social institutions, scientists were, and are today again, confronted with the problem whether they should merely inform, or even warn society about impending climate fluctuations and demand active intervention.

The protagonists of anthropogenic climate change, or, in modern terms, environmentally conscious scientists did have an impact on the governmental-administrative level in different societies. Their message was that modifications of the environment were an agent of climate change (Grove, 1975). For instance, the American Association for the Advancement of Sciences (AAAS) in 1878 (quoted after Brückner, 1890:15) demanded reforestation programs to avoid further desiccation that was perceived to take place in the North America. Demands aimed at the abatement of anthropogenic climate change were often met favorably by governments. In the 18th and 19th century, governmental or parliamentary committees were instituted some countries, for instance in Prussia, Russia, France, and Italy (cf. Brückner, 1890: 14–19).

A further noteworthy fact that resonates the present predicament, is the virtual silence of the "soft sciences" in the scientific and public debate. The intellectual boundaries among scientific fields hindered the incorporation of theoretical perspectives and empirical findings about climate that had been advanced in other disciplines such as the evolving social sciences. That is, the domains of the physical and the social milieu and their strict separation had already become part of the social and intellectual structure of the scientific community.

Similarly, considerable energy was then, and is now being spent by politicians, the public and others on the issue of climate change; in each case, scientists played a major role in putting the issue of climate change onto the public agenda. And politics endorsed the issue. In the past, however, the political response was mostly regional and not as it is now demanded, global.

Brückner belonged to a small group of environmentally conscious and socially responsible scientists. He and his colleagues felt obliged to inform the public about the implications of his research. He was convinced that climate varies for natural reasons. He considered the potential for predicting these variations as most benevolent activity since it would allow governments and social institutions to anticipate and prepare for temporary obstacles to social, political and economic developments. The task of the scientist would be to first detect the regularities, and then to formulate and convey the policy options arising from this predictive capability to governments and the public at large. Aside from informing the scientific community about his results, Brückner does appear as far as we know to have not addressed decisions makers directly. Instead he relied as indicated on publishing newspaper articles and presenting public lectures. However, in spite of his formidable insight into the climatological aspects, he, like some contemporaries today, overlooked that he did not have the expertise to anticipate the societal response to pronouncements about pending adverse climatic conditions, for instance by improving hygienic standards (the typhoid forecast), by perfecting the railway system (the forecast concerning the ice on the rivers) or by allowing for artificial watering of agricultural land (the forecast concerning harvests).

Hann, on the other hand, remained an academic and restricted himself to the immediate scientific problems at hand, that is, to the process of monitoring climate and understanding meteorological processes. Why he refrained from communicating more directly with the public and representatives of different social institutions, we do not know. It could be that he did not consider the results sufficiently firmly established, or he understood, unlike Brückner, that leaving one's field of expertise creates a hazardous mixture of scientific and political discourse that in the end may not be of any immediate benefit to society.

In any event, the discussion and concern about climate change quickly faded from the agenda in science and among the public in the first decades of the 20th century. We an only speculate about the reasons. Certainly, some of the practical promises associated with the new findings were found not to be fulfilled, so that the whole story would be a case of "overselling", a process later observed in the 1960s and 1970s with "rainmaking" and "cloud seeding" (Cotton and Pielke, 1992). Also, the attention was diverted to more pressing problems such as the big wars, the deep going social repercussions and the economic disasters. Independently of the reasons, in the end a consensus emerged among climatologists (e.g., Berg, 1914:67; Lamb, 1982) that in "historical times" the global climate has been constant; that neither a warming trend nor a trend toward less precipitation takes place. Moreover, in climate science fascination with the results of new instrumental readings in the 1920s and later shifted research attention away from the issue of climate variability.

5 CONCLUSIONS

Our discussion of climate variability and climate change at the end of the 19th century leads to a number of conclusions that we consider relevant on methodical, theoretical and practical grounds for present-day debates:

1. The discussion about natural climate variability and anthropogenic climate change is not new. A similar debate, almost forgotten today, was going on a century ago. The protagonists found themselves in social roles and situations similar to that of contemporary scientists.

2. The early debate on the nature and consequences of climate change among climatologists, geographers and meteorologists lacked the interaction with the philosophers and in the emerging social sciences who had lively and vigorously written on the impact of climatic conditions on psychological and social processes for decades and centuries.
3. The attention in the academia and in the public concerning the concept of climate change and its societal implications was of limited duration. In the end, the topic lost out in the competition with other economic, political and everyday problems, and eventually also disappeared also from the research agenda of the sciences.

The specific episode we have recounted reminds us that the burgeoning genre of popularized science that surrounds present-day discussions of climate change is by no means new. Nor is it novel to acknowledge the uncertainties that surround scientific data on climate variability.[12] It appears that the issue of climate change lends itself well to such popularization. Perhaps it does so because the issue goes to the heart of our modern common sense understanding of the natural climate as benevolent and trustworthy (cf. Stehr, 1997).

6 BIBLIOGRAPHY

Arrhenius, S. A., 1896: "On the influence of carbonic acid in the air upon the temperature of the ground." *Philosophical Magazine and Journal of Science* 41:237-276.

Arrhenius, S. A., 1903: *Lehrbuch der kosmischen Physik.* Volume 2. Leipzig: S. Hirzel.

Berg, L., 1914: "Das Problem der Klimaveränderung in geschichtlicher Zeit." *Geographische Abhandlungen* 10 (2): 1-70.

Bradley, R. S., H. F. Diaz, J. K. Eischeid, P. D. Jones, P. M. Kelly, and C. M. Goodess, 1987: "Precipitation fluctuations over Northern Hemisphere land areas since the mid-19th century." *Science* 237: 171-175.

Brückner, E., 1888: "Ändert sich unser Klima?" Vortrag, Universität Dorpat (cf. *Neuen Dörptschen Zeitung* No. 68)

Brückner, E., 1889: "In wie weit ist das heutige Klima constant?", pp. 1-13 in *Verhandlungen des VIII. Deutschen Geographentages in Berlin.* Leipzig: Teubner.

Brückner, E., 1890: *Klimaschwankungen seit 1700.* Nebst Bemerkungen über die Klimaschwankungen der Diluvialzeit. Wien and Olmütz: Hölzel.

Brückner, E., 1894: "Rußlands Zukunft als Getreidelieferant." p. 1-3 in Supplement to *Münchener Allgemeine Zeitung* (November 19, 1894).

Brückner, E., 1895: "Der Einfluß der Klimaschwankungen auf die Ernteerträge und Getreidepreise in Europa." *Geographische Zeitschrift* 1: 39-51.

Brückner, E., 1902: "Zur Frage der 35jährigen Klimaschwankungen." *Dr. A. Petermann's Mittheilungen* aus Justus Perthes' Geographischer Anstalt 48:173-178

Brückner, E., 1909: "Über Klimaschwankungen." *Mitteilungen der Deutschen Landwirtschafts-Gesellschaft* 24: 556-561

Brückner, E., 1912: *Klimaschwankungen und Völkerwanderungen.* Vortrag gehalten in der feierlichen Sitzung der kaiserlichen Akademie der Wissenschaften am 13. Mai 1912. Wien: K. K. Hof- und Staatsdruckerei.

Brückner, E., [1912] 1915: "The settlement of the United States as controlled by climate and climatic oscillations." Pp. 125-139 in *Memorial Volume of the Translantic Excursion of 1912 of the American Geographical Society.*

Brückner, E., 1923: "Julius Hann." Pp. 151-160 in Akademie der Wissenschaften in Wien, *Almanach für das Jahr 1922.* Wien: Hölder-Pichler-Tempsky.

Büdel, J., 1977: Klima-Geomorphologie. Gebrüder Bornträger Berlin-Stuttgart, 304 pp. ISBN 3-443-01017-2

Cotton, W. R., and R. A. Pielke, 1992: Human Impacts on Weather and Climate. Geophysical Science Series Volume 2. Ft. Collins, Colorado: ASTER Press.

DeCourny Ward, R., [1908] 1918: *Climate Considered Especially in Relation to Man.* New York and London: G.P. Putnam's Sons.

Grosjean, G., 1991: 100 Jahre Georaphisches Institut der Universität Bern, 1886-1986. *Jahrbuch der geographischen Gesellschaft von Bern,* 56, 1986-90, 175pp.

Grove, R. H, 1975: *Green Imperialism. Expansion, Tropical Islands Edens and the Origins of Environmentalism 1600-1860.* Cambridge University Press

Hann, J., 1883: *Handbuch der Klimatologie.* Volume 1: Allgemeine Klimatologie. Stuttgart: J. Engelhorn.

Hann, J., 1903: *Handbook of Climatology.* Part I: General Climatology. New York: Macmillan.

Hann, J. von and K. Knoch, [1883] 1932: *Handbuch der Klimatologie.* Fourth Edition. Volume 1: Allgemeine Klimalehre. Stuttgart: J. Engelhorn.

Hellpach, W. H., 1938: "Kultur und Klima." p. 417-438 in Heinz Wolterek (ed.), *Klima-Wetter-Mensch*. Leipzig: Quelle & Meyer.

Huntington, E., 1915: "A neglected factor in race development." *The Journal of Race Development* 6: 167-184

Huntington, E., [1915] 1924: *Civilization and Climate*. Third Edition, Revised and Rewritten with Many New Chapters. New Haven: Yale University Press.

Huntington, E., 1916: "Climatic variations and economic cycles." *The Geographical Review* 1:192-202.

Huntington, E., 1945: *Mainsprings of Civilization*. New York: John Wiley and Sons.

Jones, P. D., 1995: "The Instrumental Data Record: Its Accuracy and Use in Attempts to Identify the CO_2 Signal." p. 53-76, in H. von Storch and A. Navarra (eds), *Analysis of Climate Variability: Applications of Statistical Techniques*. Berlin: Springer Verlag

Kahlig, P., 1993: "Some aspects of Julius von Hann's contribution to modern climatology." *Interactions between Global Climate Subsystems*. The Legacy of Julius Hann. Geophysical Monograph 75.

Karl, T. R., R. G. Quayle and P.Y. Groisman, 1993: "Detecting climate variations and change: New challenges for observing and data management systems." *J. Climate* 6: 1481-1494.

Knoch, K., 1932: "Vorwort zur vierten Auflage." p. VIII-X in Julius von Hann and Karl Knoch, *Handbuch der Klimatologie*. Fourth Edition. Volume 1: Allgemeine Klimalehre. Stuttgart: J. Engelhorn.

Lacey, C. and D. Longmann, 1993: "The press and public access to the environment and development debate." *The Sociological Review* 41, 207-243.

Lamb, H. H., 1959: "Our changing climate, past and present." *Weather* 14: 299-318

Lamb, H. H., 1982: *Climate, history and the modern world*. Methuen & Co. London

Le Roy Ladurie, E., [1971] 1988: *Times of Feast, Times of Famine. A History of Climate Since the Year 1000*. New York: Farrar, Straus and Giroux.

Oberhummer, E., 1927: "Eduard Brückner", pp. 195-199 in Akademie der Wissenschaften in Wien, *Almanach für das Jahr 1927*. Wien: Hölder-Pichler-Tempsky

Ploetz, Al., 1911: "Die Begriffe Rasse und Gesellschaft und einige damit zusammenhängende Probleme", pp. 111-136 in *Verhandlungen des Ersten Deutschen Soziologentages vom 13.-22. Oktober 1910 in Frankfurt am Main*. Tübingen: J. C. B. Mohr (Paul Siebeck).

Sombart, W., [1911] 1951: *The Jews and Modern Capitalism*. Translated by M. Epstein. Glencoe, Ill.: Free Press.

Sombart, W., 1938: *Vom Menschen*. Versuch einer geisteswissenschaftlichen Anthropologie. Berlin: Buchholz & Weisswange.

Sorokin, P., 1928: *Contemporary Sociological Theories*. New York: Harper & Brothers.

Stehr, N., 1997: "Trust and climate." *Climate Research* 8: 163-169, 1997

Stehr, N., 1996: "The ubiquity of nature: Climate and culture", *Journal for the History of the Behavioral Sciences* 32: 151-159.

Stehr, N., 1991: *Practical Knowledge*. London: Sage.

Stehr, N. and H. von Storch., 1995: "The social construct of climate and climate change." *Climate Research* 5:99-105.

Stehr, N. and H. von Storch, 1999: An anatomy of climate determinism. H. Kaupen-Haas (Ed.): *Wissenschaftlicher Rassismus - Analysen einer Kontinuität in den Human- und Naturwissenschaften*.. Campus- Verlag Frankfurt a.M. - New York (1999), 451 pp ISBN 3-593-36228-7.

Steinhauser, F., 1951: "Julius Hann." In Österreichische Akademie der Wissenschaften (Hrsg.), *Österreichische Naturforscher und Techniker*. Wien: Gesellschaft für Natur und Technik.

von Storch, H. and K. Hasselmann, 1996: "Climate Variability and Change." p. 33-58 , in: G. Hempel (ed.): *The Ocean and the Poles. Grand Challenges for European Cooperation*. Jena, Stuttgart, New York: Gustav Fischer Verlag

Whitney, J.D., 1894: "Brief Discussion of the question whether changes of climate can be brought about by the agency of man, etc." *United States* Supplement I, Boston. Appendix B: 290-317.

Chapter 1

Groundwater and Typhus*

The description of a disease according to its symptoms, its course and effect, and the measures available to combat it, is the duty of a medical practitioner and it would be presumptuous if a non-professional would venture to participate in such matters. However, it is quite often the case that the occurrence of certain diseases is caused by processes in nature and not in the human body; in this regard those events are of particular concern, which are the object of meteorological and geographical research. Illnesses are triggered by certain meteorological events, e.g., the change between hot and cold, dry and wet weather periods, though modern medicine has taught they are not the actual cause, but contributing factors only. In the course of investigating this correlation, while medicine moved into the territory of meteorology, a new scientific area developed bordering on both, meteorology and medicine, and being successfully researched by both. Its borderline character should not be dismissed and certainly the line separating both sciences should not be moved in favor of one or the other. This situation may justify that the author, though being inexperienced in the area of medicine, during a lecture at the Hamburg--Altona branch of the German Meteorological Society on February 24th, 1888, turned to the subject of the relation between groundwater and typhus, and in the following pages hands his results over to the public, and at the same time would like to thank *Stadtphysikus* Dr. Reincke for the unpublished statistical data of Hamburg's typhus epidemic, offered so obligingly by him.

The incentive for the present study was the major typhus epidemic, which invaded Hamburg between 1884 and 1887. In looking at the results of the Hamburg medical statistics, one recognizes that beginning with the year 1838, the first year when related material became available, the number of the annual deaths due to typhus has decreased more or less continuously, provided we exclude the very first years. Of 10,000 inhabitants, some 14 to

* *Grundwasser und Typhus*, Mitteilungen der Geographischen Gesellschaft in Hamburg, Vol. III, 1887–1888.

19 died each year in the first decade over this time span; by the end of the Seventies and at the beginning of the Eighties, however, only 2 to 3 died. Since 1885 those numbers worsened substantially. In 1885, of every 10,000 inhabitants in the State of Hamburg, 4 succumbed to typhus, in 1886, 7, and in 1887, even 8 to 9, so that in the last year a total of 410 deaths were registered as a result of typhus. How wide-spread the epidemic was is even more evident if one not only looks at the number of deaths, but also at the number of typhus cases: Of all inhabitants in 1879–84, an average of 700 fell ill each year; however, in the 12 months from July of 1884 until June of 1885, this number totalled 1,334, in 1885/86 already, 3,015, and during 1886/87, even 5,330. The last figure is enormous because it indicates that 1% of the overall population was infected with typhus at least once in these 12 months. Fortunately, the outbreak turned out to be fairly mild so that only about 8–9% of those infected died in comparison with 15–16% in the years prior to 1880. One question comes to mind immediately: Where is the cause of this horrible calamity to be found?

His studies of the manner in which these epidemic or endemic diseases spread led *von Pettenkofer* to the conclusion that certain factors in the environment must be responsible for their occurrence, in that they enhance or hinder the multiplication and spreading of the germs involved. All of these external influences combined have been described by von Pettenkofer as a local and temporal disposition. According to him, an epidemic outbreak usually occurs in a particular location and only there if a number of contributing factors is present, among which the meteorological ones play a decisive role. In this way, he tried to explain the frequent and so strangely localized outbreaks of cholera on the one hand, and pointed on the other hand to a connection between the outbreak of typhus and the fluctuations of the groundwater level. Most recently, I. Soyka in Praha has picked up on the latter question, collected much new data and arrived at the same conclusion.[13] It is to Soyka's particular merit to have taken an experimental approach in solving the questions about the role and influence of soil in the development of germs.[14]

Namely, the fluctuations of groundwater appear to be only contributing factors, not the actual cause of the germs. The latter are tiny organisms, fission-fungi, bacteria, bacilli, which when massively present in the human body generate the dysfunctional and destructive symptoms, which add up to

[13] *Zur Epidemiologie und Klimatologie von Frankfurt a. M.* [*Epidemiology and climatology of Frankfurt Main*] Deutsche Vierteljahresschrift für öffentliche Gesundheitspflege XIX, 2. Issue, Braunschweig 1887; *Zur Aetiologie des Abdominaltyphus*, Archiv für Hygiene Vol. VI, p. 257 ff.

[14] Prager Medizinische Wochenschrift, 1885. Nos. 28–31, Zeitschrift für Hygiene Vol. II, 1887, p. 96 ff.

what we describe as typhus. Such germs are omnipresent and everywhere, they continuously penetrate the human body. Our body wages a constant fight against them and tries to destroy these foreign invaders: if the body does not succeed, the individual turns ill. However, there are situations and certain conditions under which these germs multiply particularly fast not within the human body but in the soil and for this reason occur in great numbers, or some conditions under which they very easily come within close proximity of humans. These conditions not only change from place to place, but also over time and the number of germs invading the human body changes accordingly. The immediate consequence is that at those times, where the situation is favorable and fertile for those germs, the number of illnesses grows and the disease takes on an epidemic proportion. A change in these abetting or contributing factors or in their related causes results at the same time in an immediate alteration in the outbreak of the disease. In this sense, according to von Pettenkofer and Soyka, groundwater levels play the role of secondary causes for the outbreak of typhus, the frequency of which changes as fluctuations change.

Wherever solid rock is replaced by loose sediments, such as sand and grit formations in the ground surface, we in our climate zone will find water at a certain depth, which fills all hollow areas of the ground. It is infiltrated rainwater, which collects on strata, such as compact rock or clay preventing any further penetration, and which then accumulates as groundwater. Its level rises or falls in time in the same way as the level of our rivers rises and falls because on the one hand water is continuously added through rainfall and on the other hand withdrawn through evaporation. That rainwater is the source of groundwater has, however, been disputed recently by Volger.[15] It can nevertheless be regarded as a proven fact. Soyka, after all, on the basis of a large amount of data has only recently shown that the fluctuations of the groundwater level do indeed follow the weather patterns on the surface. If more rainwater reaches the ground level than is taken away through evaporation, the groundwater will rise—in the reverse case, it will fall. This is why the groundwater level fluctuates from month to month as well as from year to year. A most striking fact is that the frequency of typhoid occurrences follows exactly the same pattern as these changes of the groundwater: whenever the groundwater drops or is at low levels, typhus occurs particularly often; it is less frequent at increasing or high groundwater levels.

Yet, at present, this opinion has still not found general acceptance in medical circles, and it is safe to assume that more than half of all practitioners remain convinced that not the fluctuations of the groundwater, but impurities in the drinking water are the only cause of epidemic

[15] Meteorologische Zeitschrift, 1887, p. 388.

outbreaks. Indeed, in some cases, as in the most recent typhus epidemic in Zürich, the fact that the illness is limited to houses where drinking water is supplied by one and the same water pipe seems to support this assumption. Improvement of the drinking water is the key word whenever it becomes necessary to fight back the typhus.

Finally, a third theory is added to these which focuses on soil movements as an essential cause for an epidemic outbreak of typhus. If soil that contains germs is turned over or dug up, these germs are brought into direct contact with human beings initiating the outbreak of an epidemic.

Which of these perceptions may be considered the right one has not yet been decided to this date, and it is quite possible that all three are correct because nature is almost always in a position to achieve the same effect in different ways. In my opinion, the typhus statistics of Hamburg in fact seem to confirm this.

It was unavoidable that, during the previous typhus epidemic in Hamburg, its cause was fiercely discussed. Almost unanimously, the drinking water was found responsible for the disease, which admittedly is bad enough in Hamburg: it consists of non-filtered *Elbe River* water with a rich fauna thriving in the water pipe system. V. Pettenkofer's groundwater theory was presented as well, by Dr. Reincke in particular, and finally many voiced the opinion that the epidemic had been caused by the massive soil movements undertaken to facilitate port and customs access. I will abstain from giving a definitive answer to the question: however, I wish to present some of the facts that could apply and may shed some light on a future solution.

Let us first concentrate on the annual pattern of the groundwater levels and the typhus frequency.

The annual fluctuation of the groundwater level depends on the interconnection of the annual rain season and the intensity of evaporation during the annual dry season. While the seasonal summer rains of central Europe force the groundwater to reach its maximum level, evaporation being highest during the summer strives to reduce the groundwater level to a minimum. Depending on the location, wet or dry climatic conditions may be more pronounced as they vary during the year. Yet the groundwater level adjusts to either one or the other factor. Soyka discovered that in this regard there is a contrast between Northern and Central Germany on the one hand and Southern Germany on the other,[16] the influence of rainfall prevailing in one part of the country and evaporation in the other. Consequently, typhus cases vary from month to month.

[16] Soyka, *Schwankungen des Grundwassers mit besonderer Berücksichtigung der mitteleuropäischen Verhältnisse* [*Fluctuations of groundwater with special emphasis on central European conditions*], Penck's Geographische Abhandlungen, Vol. II, Issue 3, p. 49.

Data for München are used to represent Southern Germany as shown in the following chart. The monthly variation in evaporation is expressed as a saturation deficit. The latter is the difference between the actually observed absolute humidity of the air and the degree of moisture usually attributed to the corresponding temperature. It is an indication for the ability of air to absorb water and therefore an excellent measurement of evaporation. In this case, it is arrived at by using the simpler though less exact method of comparing the average temperature of a particular month and its corresponding maximum level of humidity to that month's actual average humidity reading. Figures for the groundwater level indicate its height above sea level. The number of typhus deaths for each month is expressed in percentages of the annual total. Throughout, the maximum is boldfaced, the minimum is marked by an asterisk.

München (1850–85)

		Jan.	Feb.	Mar.	Apr.	May	June
Rain / Precipitation	mm	35	29*	48	56	78	**112**
Saturation Deficit	mm	0.18*	0.42	0.86	1.84	2.43	3.11
Groundwater	51500 cm +	40	42	48	50	52	58
Typhus-deaths[17]	%	11.5	**11.9**	11.2	9.0	7.5	6.9
		July	Aug.	Sep.	Oct.	Nov.	Dec.
Rain / Precipitation	mm	**112**	102	72	54	50	46
Saturation Deficit	mm	**3.54**	3.23	2.06	0.94	0.41	0.22
Groundwater	51500 cm +	**59**	57	45	37	32*	35
Typhus-deaths[17]	%	6.4	6.5	6.3	5.8*	6.9	9.6

It is evident that on the one hand, the maxima of the groundwater level, the rainfall, and the saturation deficit coincide and on the other hand, deaths due to typhus decline to a minimum soon after the groundwater level has peaked. In this case, it seems, the influence of evaporation is almost nil because the effect of the rainfall is so much greater. In reality, however, evaporation manifests itself in the fact that the groundwater level does not increase as much in the summer as the annual period of heavy summer rains might suggest if rain was the only influence. Due to the counteractive influence of evaporation and rain, any fluctuation is quite low: the amplitude amounts to only 27 cm in München, while in other places where rain and evaporation do not counteract to such an extent, the difference is much greater, such as 58 cm in Berlin with a more even rain dispersal, 60 cm in Bremen, 58 cm in Brünn.

The groundwater conditions in the State of Michigan[18] are similar to those in München. The average measurements of two locations showed the

[17] Typhus 1856/85.

following relation between rainfall, groundwater and typhus. The height of the groundwater level is measured upward from a randomly determined fixed point. I could not determine the saturation deficit for these locations.

Michigan (1885–86)

		Jan.	Feb.	March	Apr.	May	June
Rainfall	mm	73	31*	42	62	62	105
Groundwater	cm	25	25	46	**91**	84	71
Typhus Mortality	%	9	5	5	4*	4*	5
		July	Aug.	Sep.	Oct.	Nov.	Dec.
Rainfall	mm	62	**128**	116	51	67	54
Groundwater	cm	51	20	33	20	8	0*
Typhus Mortality	%	5	10	14	**15**	**15**	9

Here again the months of April to July show high levels of groundwater: however, due to the increased evaporation during the summer, the absolute maximum moves to April, while from July to August the water level drops considerably and remains at a low level from August to February. In complete accordance with this development, the incidence of typhus is highest in October and lowest in April. The correlation between groundwater levels [and precipitation] as well as typhus is shown in Figure 1.

Areas with less pronounced periods of rainfall show a very different picture of the annual variation of groundwater levels. Berlin may serve as an example for such an area.

Berlin (1870–85)

		Jan.	Feb.	Mar.	Apr.	May	June
Rainfall	mm	40	35	47	32*	40	62
Saturation Deficit	mm	0.71	0.95	1.55	2.73	3.95	5.13
Groundwater	3200 cm +	72	79	88	**96**	88	69
Typhus Mortality	%	5.7	5.5	5.6	6.0	5.6	5.2*
		July	Aug.	Sep.	Oct.	Nov.	Dec.
Rainfall	mm	**66**	60	41	58	44	46
Saturation Deficit	mm	**5.64**	4.83	3.77	1.72	1.01	.59*
Groundwater	3200 cm +	56	45	40	38*	47	50
Typhus Mortality	%	8.0	11.6	13.9	**14.3**	9.9	8.5

In this case, evaporation is indeed the predominant influence. The groundwater level is highest in the winter months reaching a maximum in April, a month with minimum precipitation; the level then drops with the increase in evaporation although at the same time rainfall is increasing as well. It starts to rise only after falling temperatures have reduced evaporation. The typhus-scale again corresponds with v. Pettenkofer's

[18] American Meteorological Journal, Vol. IV, p. 349 ff.

findings: The months with high water levels show few cases of typhus, the ones with low water levels show many cases.

Let us summarize our results in another way by calculating the annual averages of the groundwater level as well as of the number of typhus cases, as shown in the following chart.

München

	Spring	Summer	Fall	Winter
Groundwater in cm	50	**58**	38*	39*
Typhus Mortality %	27.7	19.8*	19.0*	**33.0**

Michigan

	Spring	Summer	Fall	Winter
Groundwater in cm	**74**	47	20*	17*
Typhus Mortality %	13*	20	**44**	23

Berlin

	Spring	Summer	Fall	Winter
Groundwater in cm	**91**	57	42*	67
Typhus Mortality %	17.2*	24.8	**38.1**	19.7

Figures for Hamburg show the same pattern, as do those of all Northern German cities, at least in the years before 1884. Admittedly, in the case of Hamburg, the observations of the fluctuations of groundwater can only be traced back to the year of 1880: they were carried out with great diligence by the health official C. C. H. Müller in the suburb of Eimsbüttel not far from the *Isebeck* canal, and who most obligingly submitted them to me. It does little credit to the city of Hamburg, the second largest in Germany, that these privately collected and funded data are the only ones available, while many smaller cities have their wells checked continuously over a number of years, such as in Bremen, 10 wells since 1869, in Frankfurt [a. M.], 15 since 1869, of which 6 continue to be monitored, in Berlin, 31 since 1870, in München, 1 well since 1856 with several others recently added, in Salzburg, 1 well since 1860, and a larger number since 1867, etc.

Unfortunately, the location of the observation-point in Eimsbüttel is not very favorable. The drill hole is not far from the local church, in the garden of C. C. H. Müller, within a distance of about 100 m from the Isebeck canal. The latter was dug in 1883 and 1884 replacing the Isebeck, a tiny brook whose water level used to be some meters above the water level of the present canal wide enough for the passage of larger boats. Inevitably, the construction of the canal had some influence on the groundwater level of the surrounding area. Therefore, when we observe that the groundwater level fell by about 420 m from 1880 until 1887, we are certainly not mistaken, if we allot most of the 273 m measured during the years 1884 to 1887 to the predominant effect of the new canal and only to a lesser degree to an overall reduction of rainfall. Due to these factors, the observations cannot be used for the time being as far as the annual averages are concerned. Meanwhile the recently detected strong increase of groundwater levels in the course of 1888 suggests that those interventions by man did not turn out to be quite as permanent or far reaching as it appeared. In any case, these observations may serve well to demonstrate the annual fluctuations. With the exception of course that the gradual drop of the groundwater level must have slightly distorted the annual figures for the years 1884–87 in such a way that the monthly averages at the end of the year appear lower compared with those at the beginning of the year. This is indeed the case, if we compare the distribution of the figures for the annual averages of 1880–83 with the one of 1884–87. Both are quite similar, except the minimum [level] has shifted from July, August, and September to August, September, and October. The December level in 1880–83 was just as high as the January level, whereas in the 1884–87 time frame it was considerably lower. Nonetheless, the overall picture of the scale did not change and nothing keeps us from comparing the annual distributions of typhus and groundwater.

Even a cursory glance at Hamburg's data on typhus and its annual occurrence appeared to advise against expanding upon a common denominator, but to suggest instead to consider the annual cases of 1880–83 and those of the epidemic cycle of 1884–87 separately and to compare the result with the annual fluctuations of the groundwater during the same time frames. The resulting chart shows the outcome. The typhus mortality per month is expressed in percentages of the total annual number of cases.

Hamburg (1880–83)

	Jan.	Feb.	Mar.	Apr.	May	June
Groundwater in cm	17	**19**	19	15	11	5
Typhus Cases %	7	6	5*	5	7	9
	July	Aug.	Sep.	Oct.	Nov.	Dec.
Groundwater in cm	0*	2	3	4	9	17
Typhus Cases %	10	**15**	13	9	8	6

Hamburg (1884–87)

	Jan.	Feb.	Mar.	Apr.	May	June
Groundwater in cm	22	**25**	24	25	20	12
Typhus Illness %	12	8	6	4	3	3*
	July	Aug.	Sep.	Oct.	Nov.	Dec.
Groundwater in cm	4	1	0*	1	4	11
Typhus Illness %	3	5	9	13	15	**19**

For the purpose of clarification, the numbers for Hamburg and also those for Michigan are graphically depicted. In the scale system chosen, an increase by one division equals an overall increase of 2% of the typhus cases as well as an increase of 3 cm of the groundwater in Hamburg, or 15 cm in Michigan.

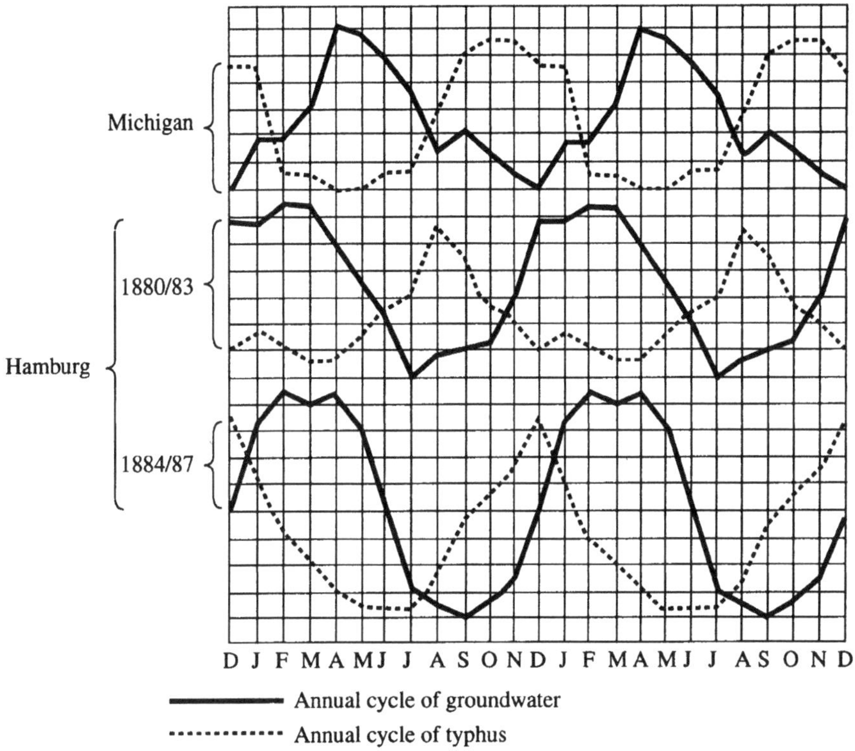

Figure 1.1. Annual Twin Curve Groundwater (solid) and Annual Twin-Curve Typhus (dashed)

At first glance it is evident that for the years 1880–83, the Hamburg figures comply indeed with v. Pettenkofer's and Soyka's findings. The number of typhus cases is highest at low groundwater levels and lowest at high levels. With rising groundwater levels, this number shows a steady decline. For the years 1884–87, the Hamburg figures show an entirely different picture. Thus, contrary to what one would expect according to v. Pettenkofer's theory and the observed tendency of the years 1880–83, the opposite occurs. This contradiction is even more pronounced if we consider that our annual curve is somewhat distorted by generally low levels of groundwater. If we discount this distortion, we notice that the minimum groundwater level moves even closer to the minimum number of typhus cases, and the maximum of typhus occurrences moves closer to the time when groundwater levels are high.

These fluctuations of the groundwater level show the same pattern as those that occurred in 1880–83, except for a minor delay in their onsets. In Hamburg, prior to 1884, typhus used to be a summer epidemic peaking in the months of July, August, and September. Since 1884 it has become a winter illness with a maximum number of cases in November, December and January. In order to visualize this more clearly we average the number of typhus cases per season.

Hamburg: Frequency of Typhus Illness in %

	Spring	Summer	Fall	Winter
1880–83	17*	**34**	30	19
1884–87	13	11*	37	**39**

The outbreak of the epidemic changed the seasonal typhus occurrences completely and disrupted the previously clearly defined correlation between groundwater level and typhus. This may well indicate that this unusual epidemic had nothing whatsoever to do with the fluctuations of the groundwater level, but rather with another unknown cause.

Let us now turn to the question whether, in a year-by-year comparison, the change in the number of typhus cases may be connected with fluctuations of Hamburg's groundwater. Soyka has investigated this for a number of cities, such as Berlin, Frankfurt a. M., Bremen, München, and Salzburg. He came to the conclusion that those years with major typhoid epidemics also showed low groundwater levels and on the other hand years with high groundwater levels appeared relatively typhus-free. To illustrate his findings, we repeat the data for München. The number of typhus deaths relates to 10,000 residents and the groundwater level is measured in centimeters, beginning at 514 m above sea level. The data for the groundwater refer to the Karlstrasse well, however, since 1875 to the one at the [München] Physiological Institute. These data were adjusted by +6 mm to the sea level of the Karlstrasse well based on the measurements for both wells of the years 1869–74. The maximum is always boldfaced and the minimum marked by an asterisk (*).

The years 1858, 1864, 1872, and 1879 stand out both for a high number of typhus deaths and low groundwater levels: the years of 1861, 1867, 1876, and 1878 for high groundwater levels and an extraordinary decline of typhus cases. Even under more detailed examination a certain inter-correlation is maintained; if groundwater rises from one year to the next, the typhus cases decline at the same time. Of a total of 25 cases prior to 1881, 19 follow this norm and only 6 are exceptions. This confirms the pattern that evolves on the basis of the annual averages of both factors. Conformity is lacking from 1882 on due to reasons discussed later.

München: Typhus Mortality and Groundwater

	1856	57	58	59	60	61	62	63	64	65
Typhus	29.1	28.2	**33.0**	17.0	10.7*	11.9	**20.1**	16.1*	**25.3**	20.5
Ground-water	103	91	88*	123	140	**152**	131	120	125	105*
	1866	67	68	69	70	71	72	73	74	75
Typhus	20.8	5.2*	7.4	11.6	**15.0**	13.3*	**23.3**	12.7*	**16.0**	11.7
Ground-water	113	**180**	161	135	125*	**137**	127*	**132**	115*	120
	1876	77	78	79	80	81	82	83	84	85
Typhus	6.0*	**8.0**	5.1*	**10.4**	6.4	1.8	1.7	1.9	1.5	1.7
Ground-water	**186**	180	181	158*	175	**179**	135	141	112	112

Unfortunately, as mentioned before, only since 1880 do we have groundwater levels recorded for Hamburg and even these are inconsistent. Consequently, a direct comparison between the variation of typhus occurrences and of groundwater is not possible. However, we are helped in this predicament by an idea that can be backed up with numbers. The curve of the groundwater level measurements can be replaced with the curve of the river levels without falsifying the outcome. At first I believed this goal was reachable in a different way. Namely, there are two factors that determine the level of the groundwater: in addition to rainfall, which increases the groundwater, we need to consider evaporation, which continuously absorbs the groundwater—as previously mentioned. Only the interaction of both factors determines the groundwater level and its fluctuations. That is why, seen annually, depending on the predominant influence of the one or the other factor, the groundwater level was at times reacting more to changes in rainfall and at times to those in evaporation. For a year-long prognosis, it is therefore impossible to determine a priori whether in a specific location the groundwater level will adjust to the one or the other. The development over several years is a somewhat different matter. Rainy years tend to be both moist and cool, with little evaporation; whereas in very dry years evaporation is high. The following five-year averages for Salzburg seem to prove this correlation.

	Groundwater Level Above Sea Level	Quantity of Rain	Saturation Deficit.
	410 m+	mm	mm
1861/65	2.97*	1058*	**2.38**
1866/77	3.03	1207	1.95
1871/75	2.99	1283	1.70
1876/80	**3.13**	**1431**	1.63*
1881/85	3.04	1285	1.79

The precipitation minimum coincides with the evaporation maximum and vice versa. What applies to five-year averages does, to a certain extent, also apply to each individual year. If for Salzburg individual years are ranked according to the amount of precipitation and then bundled in a group of five, and if, finally, the saturation deficit is determined, it is evident that the following figures correlate with each other:

Average Rainfall Quantity	Average Saturation Deficit
in mm	in mm
1546	1 544
1361	1 752
1278	1 866
1111	1 994
866	2 280

Finally, if one looks at the changes in precipitation and in the saturation deficit on a year-to-year basis, an increase in precipitation is most often accompanied by a decrease in evaporation and vice versa. The figures for München, for instance, show this reverse relationship for the years 1860 to 1885 in 17 cases, while in only 7 cases the two factors are changing in the same direction; for Salzburg the figures are 15 and 7 respectively. Thus, rain and evaporation have an equally strong effect on groundwater levels.

Indeed, it is Soyka's opinion as well that, on a larger scale, the variations in precipitation and groundwater level are the same from year to year. Lang even goes so far as to attempt to replace groundwater fluctuation with total amount of precipitation.[19] Nevertheless, in compiling groundwater and precipitation figures for Berlin, Frankfurt a. M., and Bremen, I became convinced that such an approximation can be flawed as far as individual cases are concerned, and, therefore, does not serve our purpose. I had to look for some other entity and the water levels of the Elbe River seemed what I was looking for.

Soyka pointed out[20] that river levels and groundwater levels of neighboring areas show a similar distribution pattern; he proves this similarity with data for Berlin (Spree River), Frankfurt (Main River), Bremen (Weser River), and München (Würm river).[21] How strikingly parallel the distribution lines for the two entities are is demonstrated in the

[19] *Schwankungen der Niederschlagsmengen und Grundwasserstände in München 1857-1886* [*Fluctuations of the Amount of Rainfall and the Groundwater Levels in Munich 1857–1886*], Beobachtungen der meteorologischen Stationen im Königreich Bayern, Vol. IX, 1887, p. XIII.

[20] *Schwankungen des Grundwassers*, etc., [*Fluctuations of Groundwater*, etc.], p. 80.

[21] The Isar river cannot be used here because the depth of its river-bed changes rapidly.

graphic shown above, which was drafted on the basis of Soyka's data. In this graphic, one grid length corresponds to 0.1 m groundwater level change for Berlin, 0.17 m for Bremen, 0.2 m for München, and to 0.1 m river level change for the Spree, 0.2 m for the Weser and 0.05 m for the Würm.

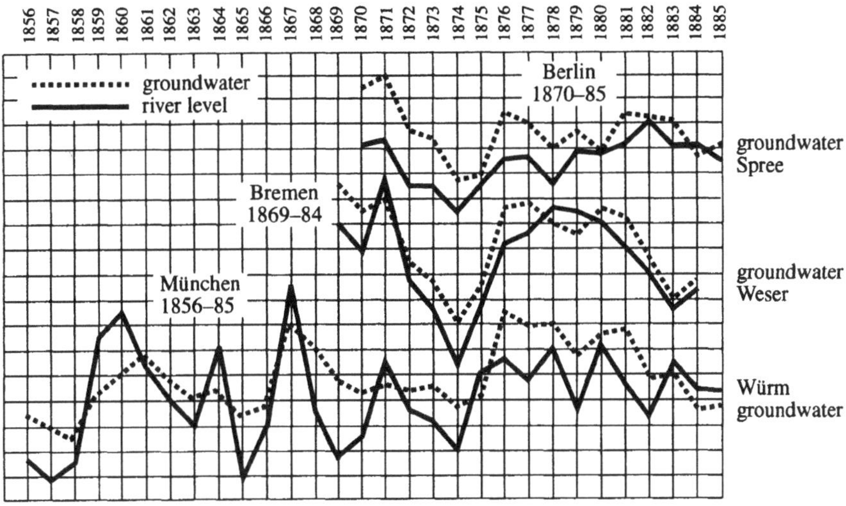

Figure 1.2. Curve of Groundwater Levels (dashed) and Curve of River Water Levels (solid)

This chart gives strong evidence that fluctuations of the groundwater parallel those of recorded river levels and that the one can be used instead of the other without introducing any major distortion. This holds true for long-term changes in water levels as well as for short-term. Groundwater and river levels fall and rise in tandem. The figures for Berlin show changes in the same direction in 11 cases and only 4 in the opposite direction, for Bremen, 13 and 2, and for München, 24 and 4, i.e., of these a total of 48 cases show similarity, and only 11 the opposite. This conformity is even more convincing, when one considers that in only 1 of those 11 cases (near München in 1860–61) the fluctuation is of a larger magnitude. We should not be too surprised though because the amount and distribution of seasonal precipitation and evaporation affect the water level in the ground and in the river in exactly the same way. In addition, in places like Berlin and Bremen, the groundwater interacts with the river water up to a certain degree and consequently follows its movements,[22] except, of course, in the case of München, where the groundwater obviously does not fluctuate or interconnect with the distant Würm River or even with the Isar River, a fact that does not invalidate the observed congruity entirely. In this case it must

[22] Soyka op.cit. p. 58, 80.

be the equally strong influence of rainfall and evaporation alone that causes conformity.

Let us examine the relationship of groundwater and rivers in Hamburg. The Elbe River penetrates the central and southeastern parts of the city by way of numerous canals; the entire northern part of the city, however, is under the influence of the Alster River and its tributaries, such as the Isebeck, etc. The ground in Hamburg consists of either clay or sand occurring in a fairly irregular pattern. In many places, clay boulders have formed ridges up to above the water level with basins in between or detached knolls. Quite often the ground is soft sand down to below the riverbed level. It is more than likely that under such conditions in Hamburg, as in Berlin, Bremen and Frankfurt, groundwater and river water levels are closely interrelated. Regrettably though, I cannot prove this relationship since measurements of the Elbe River levels near Hamburg are available to me up to the year 1869 only, whereas groundwater level measurements did not start before 1880 and were restricted to the Alster area. The nature of the seasonal changes, however, may be alluded to: groundwater and river water levels reach their peaks in February, March, and April and their lowest levels in September. But in view of the conditions found in Bremen, München, Berlin, and Frankfurt a. M., I have no hesitation in using the curve of the fluctuations of the Elbe River levels in place of the curve of the groundwater levels in Hamburg.

The following chart shows typhus deaths per 10,000 residents of Hamburg for the years 1838 to 1887: furthermore, data on typhus cases for the years 1872–87 are included. For this information, as for all the other information on typhus statistics in Hamburg, I am indebted to Dr. Reincke, the city's health official. Unfortunately, the annual averages do not refer to the calendar year but relate to the period from July 1 to June 30 of the following year under which they are listed respectively. On the other hand the table includes the average water levels of the Elbe River for three different locations: Hamburg, Artlenburg about 45 km further upstream, and Magdeburg about 280 km upstream. The numbers for Magdeburg and Artlenburg were arrived at by using a table published by *Bureau der Baudeputation Sektion für Strom- und Hafenbau.* The locations Artlenburg and Magdeburg were added because the water levels of the Elbe at Hamburg are not merely the result of climatic conditions upstream. It is the tide that influences the water level in the city of Hamburg to a considerable degree. High tide, for example, averages 1.8 m annually which is four times the amount between the highest and lowest annual mean water level. In addition, the submitted annual averages are not the result of actual measurements, but are arrived at by using the data on the averages of high and low tide. Therefore, these data carry a flaw that was introduced by those tidal ranges.

The charted water levels are not only altered by the tide but also by the number of months selected as time span. Thus, the water level at Artlenburg, where tidal movement is almost negligible, gives a more accurate picture of those spontaneous changes of the Elbe River levels which can be attributed to climatic events and which can be the only explanation for the changes in the groundwater level of those by far larger areas of Hamburg where the Alster River exerts some influence, whereas the Elbe River does not.

Typhus in Hamburg and the Elbe River Water Level

	Typhus Mortality	Elbe Water Level		
		Hamburg	Artlenburg	Magdeburg
Year	per mil	m	m	m
1838	15.3	—	—	—
1839	16.1	—	—	—
1840	14.7	—	—	—
1841	17.7	—	—	2.1
1842	18.7	—	—	1.3
1843	14.1	1.34	1.25	2.3
1844	14.6	1.38	1.87	2.4
1845	14.3	1.39	1.68	1.9
1846	15.5	1.34	1.42	2.0
1847	14.8	1.27	1.41	2.1
1848	11.4	1.17	0.92	1.6
1849	10.7	1.27	1.18	1.8
1850	11.5	1.47	1.62	2.3
1851	11.1	1.43	1.73	2.3
1852	12.1	1.38	1.54	2.0
1853	9.9	1.28	1.82	2.1
1854	10.2	1.47	1.73	2.2
1855	9.6	1.42	2.34	2.3
1856	5.1	1.29	1.44	1.8
1857	16.2	1.10	0.83	1.4
1858	5.2	1.13	0.88	1.5
1859	10.5	1.26	0.97	1.6
1860	5.2	1.34	1.72	2.0
1861	6.6	1.34	1.45	1.8
1862	7.3	1.27	1.28	1.6
1863	6.9	1.34	0.97	1.5
1864	6.3	1.15	0.94	1.4
1865	11.5	1.14	0.85	1.3
1866	9.4	1.27	0.66	1.2
1867	5.0	1.54	1.87	2.1
1868	6.6	1.45	1.45	1.8
1869	6.8	1.33	1.21	1.7

[Continuation on next page]

[Continuation from previous page]

Typhus in Hamburg and the Elbe River Water Level

	Typhus Mortality	Elbe Water Level		
		Hamburg	Artlenburg	Magdeburg
Year	per mil	m	m	m
1870	8.4	Typhus		1.8
1871	5.3	Cases		2.0
1872	6.7	per mil		1.3
1873	5.4	38.1[23]		1.3
1874	5.4	23.7		1.0
1875	5.5	32.4		1.3
1876	3.5	15.7	Ground-	1.7
1877	3.0	18.7	water	1.6
1878	3.9	24.7	Hamburg	1.4
1879	2.5	19.4	m	1.8
1880	2.6	15.9	4.90	2.0
1881	3.0	13.9	4.45	2.1
1882	2.7	18.9	3.69	1.9
1883	2.5	13.5	3.48	1.8
1884	2.6	18.3	3.43	—
1885	4.2	26.8	2.48	—
1886	7.0	58.1	1.38	—
1887	8.6	100.7	0.70	—

The water levels of the Elbe near Magdeburg are not at all influenced by the tidal range and were mainly included in the study because they are available up to the year 1883, while the Hamburg and Artlenburg measurements go no further than 1869. Also included are the groundwater levels near Hamburg–Eimsbüttel as recorded by C. C. H. Müller since 1880, although these do not measure up to scrutiny because the surrounding ground was disturbed.

As it is more difficult to gain an overall picture from tabulations the results are depicted graphically in the following chart. A variation by 1 grid length is equivalent to 1‰ [per thousand] of typhus deaths and 5‰ of typhus cases, to 0.1 m (one decimeter) of the Elbe water level and 0.5 m of groundwater variance.

[23] For typhus mortality cases the statistical year runs from July 1 to June 30; the annual average is associated with that year whose first half falls into that statistical period, e.g., 38.1 in 1873 refers to the period of July 1872 to June 1873, etc.

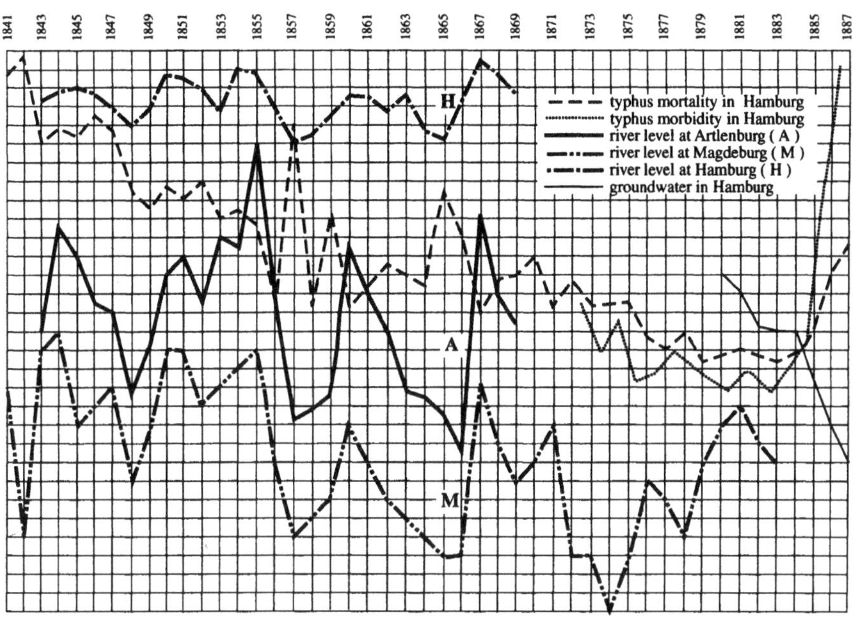

Figure 1.3

As we look at the table, or rather the chart, the most striking feature is the tremendous decline of typhus deaths occurring steadily with minor interruptions over the period from 1842 to 1883. No doubt, this is on the one hand due to improvements in medical treatments leading to lower mortality rates and on the other hand due to improvements to the city's sanitation system leading to an overall decrease of typhus. However, from 1884 onward figures show a rapid and steady increase.

If we compare the three curves representing the water levels, it is obvious that they follow an almost parallel course. This course is completely parallel for Magdeburg (M) and Artlenburg (A), despite the fact that they are 230 km apart. The similarity is much less pronounced between Artlenburg and Hamburg with only 45 km between them. Though in Hamburg—on the whole—the Elbe levels show the same fluctuations as upstream they are far less pronounced. If we take a closer look at the direction these variations follow over several years, they can be categorized as being parallel, opposite or indeterminable in the following number of cases:

- at Magdeburg and Artlenburg—parallel 21, opposite 3, indeterminable 2.
- at Artlenburg and Hamburg—parallel 18, opposite 8, indeterminable 0.
- at Magdeburg and Hamburg—parallel 17, opposite 7, indeterminable 2.

The influence of the tidal range is clearly evident and confirms our assumption that Elbe River levels at Hamburg can be considered in place of the groundwater levels for those parts of the city only where the groundwater is directly affected by the river, whereas the variations of the groundwater in the Alster region being exclusively the result of climatic events can much more likely be represented by the Elbe fluctuations at Artlenburg.

In comparing the typhus curve with those of the water levels, we discover that all of the typhoid epidemics of the years 1842, 1852, 1857, 1859, 1865, 1870, and 1878 occurred when water levels were low, while the years 1843–45, 1860, 1867, 1871, 1876–77 with no typhoid epidemics show high groundwater levels. Exceptions to this correlation are the minor epidemic outbreaks of 1846 and 1872, as well as those years 1856 and 1858 with no typhoid epidemics which show very low levels of groundwater.

If the correlation proves correct, a rise in the groundwater level should correspond with a decrease in the number of typhoid cases and vice versa. If one compares the typhus curve with the water levels at Artlenburg, 62% of all cases seem to confirm the correlation and 38% do not; in the case of Magdeburg this ratio is 55% and 45%, and for Hamburg 54% and 46%. According to Soyka for München, Berlin, Frankfurt a. M., Bremen, and Salzburg this correlation between groundwater levels and the occurrence of typhus is on the average indicated in 61% of all cases. The results are much more favorable if we concentrate on the more significant variations of the curve only. A fall or rise of the Elbe water level near Magdeburg of 0.2 m or more from one year to the next has the opposite effect on the number of typhus cases in Hamburg in 71% of all cases; and a rise or fall in the typhoid curve of 2‰ [per mil] or more corresponds to a counter change of the Elbe River levels near Magdeburg in 66% of all cases.

In Hamburg, therefore, assuming that fluctuations of the Elbe River level result in corresponding changes in the groundwater, everything seems to point to a correlation between groundwater level and typhus in a year-to-year comparison. At the same time, this seems to confirm our earlier findings concerning the year-long observations from 1880 to 1883. However, what about those years 1884 to 1887 when the data did not match the general pattern at all?

Soyka pointed out that in a long-term view the variations in the groundwater level from one year to another show a remarkably parallel development over longer stretches of time. The explanation for this may be seen in the fact that vast expansive areas experience similar climatic conditions. I have previously claimed that the rainfall over continental areas shows a certain general pattern of fluctuations.[24] As a consequence the groundwater fluctuates in the same way. During the drought of early 1860 which occurred in all of Europe, Asia, America, and Australia, in short all over the globe, groundwater levels in Europe were low. Since then, precipitation increased and groundwater levels rose; we may speak of a

[24] Annalen der Hydrographie, 1888, February Issue.

change in climate. Some countries on the globe experience spectacular changes in climate conditions, rainy seasons alternate with fairly dry ones. Already, the present century recorded three series of wet years around 1815, 1850, and 1880 and some years in between had very little rain.

Groundwater levels changed accordingly and, as we can conclude and as Penck[25] has put it first, so did the conditions of typhoid outbreaks. The following small table may demonstrate this correlation.

	Salzburg			München			Elbe River Level		Hamburg
	Rain	Ground-water	Typhus	Rain	Ground-water	Typhus	Artlen-burg	Magde-burg	Typhus
	mm	m	‰	mm	m	‰	m	m	‰
1841/45	—	—	—	—	—		—	2.00	15.9
1846/50	—	—	—	—	—		1.31	1.96	12.8
1851/55	—	—	—	—	—		**1.83**	**2.18**	10.6
1856/60	—	—	—	781	0.09*	23.6	1.17	1.66	8.4
1861/65	1058*	2.97*	—	755*	0.27	18.7	1.10*	1.52*	7.7
1866/70	1207	3.03	9.3	790	0.42	12.0	1.0	1.76	7.2
1871/75	1283	2.99	**14.9**	766	0.27	15.4	—	1.38	5.6
1876/80	**1431**	**3.13**	5.9	874	**0.75**	7.2	—	1.70	3.1
1881/85	1285	3.04	2.6	**919**	0.35	1.7	—	**1.93**	3.0

It is highly regrettable that concerning the occurrence of typhus the morbidity rate is not available to us over a more extended number of years, but that we have to rely on the mortality rate instead. These numbers are misleading whenever long time-spans are involved since they are lowered considerably through progress in medical treatment. Only the actual number of cases can serve as indicator for the frequency in which illnesses occur. That is why in Hamburg, during those years of 1856 to 1865 when river and groundwater levels were low, the typhus death rate did not rise nevertheless. It seems likely that the typhus infection was even higher than during previous lustra: the fact that the general decline of the mortality rate slowed down significantly during those very lustra makes me believe that in fact the morbidity rate was higher. From one lustrum to the next mortality rates declined as follows:

1846/50	51/55	56/60	61/65	66/70	71/75	76/80	81/85	86/87
3.1	2.2	2.2	0.7*	0.5*	1.6	2.5	0.1	4.7

In conclusion, there may have been an overall increase in typhus cases in the Sixties indicated by a stagnation in the decline of deaths, while the more likely time-span with high water levels around 1850 and at the end of the

[25] Münchener Allgemeine Zeitung, 1887.

Seventies shows a rapid decline of mortality. Similar conditions were found in München.

As of 1881 precipitation in Hamburg as well as in all the rest of central Europe declines. Hamburg's rainfall measured in millimeters was:

1876/80	1881	1882	1883	1884	1885	1886	1887
867	695	673	586	758	620	598	466

This resulted in a general drop of the groundwater levels, as for instance Soyka reported for Bremen, Berlin, Frankfurt a. M., München, and Salzburg.[26] Hamburg, too, experienced an enormous drop by a full 4.2 meters as C. C. H. Müller's data verify. However, we also know that the groundwater level of Eimsbüttel sank considerably due to the construction of the Isebeck canal in 1883 and 1884, which might explain most of the unprecedented reduction by 4.2 m in the course of seven years. Still, in part this reduction can most definitely be attributed to the long periods of little precipitation over these years, which, consequently, should be the case for all of Hamburg. It appears only logical to make a connection between this drop in groundwater level and the major epidemic in Hamburg during the years 1884 to 87. If this would be the case, one would expect epidemic outbreaks of typhus in other cities as well where the groundwater level sank significantly. That, however, is not the case, except for Salzburg and Frankfurt a. M. which show a very small increase of typhus cases during a period of dropping groundwater levels in the years from 1883 to 1895. Berlin and München had no such increases, neither did Hamburg's neighbouring city of Bremen in particular, although its water level dropped by 49 cm during 1881–84.[27] Soyka emphasizes this lack of consistency in the correlation between groundwater and typhus during the Eighties and sees its cause in the steady progress of improving the underground sanitation system, which ultimately will eliminate typhus altogether.[28] Could it be assumed that such progress did not take place in Hamburg so that the drop of the groundwater level could cause an epidemic of proportions which by the sheer number of cases might be called unparalleled in the city annals? This is quite unlikely, given that Hamburg is one of the first large cities in Germany that with its system of sluices created a magnificent drainage system. Therefore, it seems probable that the cause of Hamburg's last typhus epidemic is to be found chiefly in some local conditions peculiar to

[26] *Schwankungen des Grundwassers* [*Fluctuations of Groundwater*], pp. 83.

[27] Not by 2.96 meters as Soyka has stated, in *Schwankungen des Grundwassers*, p. 78; he neglected to divide the sum of the annual averages of the different wells in Bremen by the total number of wells in order to obtain the overall average. This mistake has been corrected in Figure 2.

[28] Archive für Hygiene, Vol. VI, p. 282 a. 285.

Hamburg and is not linked to groundwater levels in general. This view is confirmed by the fact that, as previously mentioned, the seasonal outbreak of typhus does not conform at all with seasonal fluctuations of groundwater. These past years of epidemics, therefore, show an abnormal relation between groundwater and typhus.

But what is the possible local cause of the epidemic? "Could it be chance", as I stated less than a year ago,[29] "that the beginning of the epidemic coincides with the beginning of the harbour construction work for the purpose of joining the customs-union? If this massive excavation of soil and dirt was indeed the cause, future will tell. Let us hope so, because in that case we may be about to see the end of the epidemic as harbour construction is about to be completed." Some facts actually support this assumption. The soil movements at the construction sites were completed during the first six months of 1888 and in October of that year the new docks were opened. At the same time, the typhus epidemic was considered retrogressive since the spring of 1888. Of course, the groundwater level rose slightly as well in the wet year of 1888, which weakens this argument somewhat.

Let us review the results of our efforts. In Hamburg, too, the outbreak of typhus is closely related to the fluctuations of groundwater levels. This correlation is quite evident when we compare the annual changes of both factors as observed during 1880–83. Even the changes from one year to the next confirm this relation, as indicated by the fluctuations of the Elbe river levels which were used from 1841 on because proper measurements of the groundwater levels were unavailable.

However, in the epidemic years of 1884–87 this correlation is strikingly absent. The annual pattern is different: what used to be a summer disease has turned into a winter disease. In addition, the outbreak has grown to shocking unprecedented proportions. It cannot be connected to the drop of groundwater level as this was the case in all of central Europe without causing a widespread epidemic of remotely similar proportions. Rather this epidemic seems to originate from massive soil movements in connection with the new harbour construction in Hamburg because it runs its course over exactly the same length of time that soil movements take place and vanishes after they have stopped. During construction, vast masses of moist and drying soil containing numerous bacteria were exposed to air enabling the spread of these carriers of disease to the human body. Whether now, after digging is completed and the epidemic has passed, the occurrence of typhus will again, annually as well as from year to year, show the same pattern of variation as the groundwater level, future research will tell.

[29] Hamburgischer Korrespondent 1888, No. 85, 25 March, Morning Issue, p. 10.

Chapter 2

Fluctuations of Water Levels in the Caspian Sea, the Black Sea, and the Baltic Sea Relative to Weather*[30]

The contemporary opinion, more or less commonly shared by hydrographers and oceanographers, is that the fluctuations of water levels, as observed by water-mark stations at seas which are fairly isolated from open ocean waters, are attributed to the influence of wind. And indeed, if we keep in mind the powerful tidal waves that, under the impact of northerly and northeasterly winds, haunted the lower German coasts repeatedly, this view would seem plausible. As easy as it is to verify and substantiate the direct effect of wind on water levels and their irregular changes from day to day, the more complicated it is, however, to interpret the slow change of the average water levels to be the direct result of the wind. For it was noticed that apart from the short-term irregular variations that occurred differently at different coastal points of the same oceanic waters, long-term general fluctuations take place as well. The Baltic Sea, for example, where average water levels of different months were compared, shows a distinct annual periodicity whose course appears to be more or less parallel at all measuring stations despite their distance. Fluctuations of this kind could no longer be considered a consequence of local winds since they indicated a considerable change in the water volume of the Baltic Sea. General wind conditions and atmospheric wind pressure—which at times pushed the waters of the Baltic Sea towards the South and the West, those of the North Sea towards the

* *Die Schwankungen des Wasserstandes im Kaspischen Meer, dem Schwarzen Meer und der Ostsee in ihrer Beziehung zur Witterung*. Annalen der Hydrographie und Maritimen Meteorologie, Vol. II, 1888.

30 Lecture held at the General Assembly of the German Meteorological Society in Karlsruhe, Easter 1887. The research and its results explained below is part of a larger essay by the author, with the title *Klimaschwankungen seit 1700* [*Climatic variations since 1700*; three chapters of this extensive work are presented in this collection in Chapter 4] which will be published by the end of 1888 in the III. volume of "Geographische Abhandlungen", Wien, Hölzel. Some important addenda to this lecture are the result of additional research by the author and have previously appeared in the commemorative papers.

West, at other times in turn let the North Sea rise eastward and the Baltic Sea surge towards the North and East—could not be the only consideration here. Baensch[31] and H.A. Meyer[32] explained the fluctuations of water levels in as far as they are the result of a volume change this way: northerly and easterly winds cause an increase of flowing water masses into the North Sea, and a decline of water mass in the Baltic Sea; westerly winds on the other hand cause a reduction in the outflow and a rise of the Baltic Sea level. Temperature, atmospheric pressure and precipitation should have no immediate effect on average water levels.

But in contrast to this opinion, Seibt showed in 1885 that the changes of the water-level in the course of one year as well as from year to year do not at all correspond nicely with prevailing wind conditions, as one was inclined to assume:[33] thus, wind could no longer be held responsible as the sole cause of those fluctuations. Seibt thinks the annual seasonal timing could be traced back to a yearly recurring tidal wave. In the summer, movement of water masses of the global oceans occurs from the south to the north, in winter the movement is reversed. Naturally, such a shift would alter water levels including those of seas or lakes interconnected with the oceans. According to Seibt and v. Baeyer[34] this annual tide cycle, which follows the cycle of the sun, can only be observed in the adjoining oceanic waters, while open oceans with a twice daily tide cycle entirely defuse variations. Maydell denotes a similar view in 1884 as being insupportable and useless in defining the annual periods of water level changes in seas with no outlet: he was at a loss to explain the annual variation in the Black Sea.[35] This is noteworthy, because at the same time he was the first to prove that the year-to-year change of the Pontus' [old-fashioned for Black Sea] medium water levels has to be regarded as the direct result of the change in the annual precipitation and, consequently, of the related water volume of the rivers. I may be permitted to demonstrate the overwhelming importance of this last factor, which even Maydell underestimated by far, namely the rivers' capacity, in the water levels of three European seas: the Caspian Sea with no connection

[31] *Studien aus dem Gebiet der Ostsee* [*Studies for the Baltic Sea Region*], 1872.

[32] *Untersuchungen über die physikalischen Verhältnisse des Westlichen Theiles der Ostsee* [*Research of physical conditions of the western region of the Baltic Sea*].

[33] *Das Mittelwasser der Ostsee bei Travemünde* [*The mean water level of the Baltic Sea at Travemünde*], *Publikation des Königlich Preußischen Geodätischen Instituts*, Berlin 1885.

[34] see *Meteorologische Zeitschrift 1885*. November Issue.

[35] Morskoj Sbornik 1884. Vol. 11. November. In an essay of Currents and Water Exchange between the Black Sea and the Mediterranean Sea, published in 1886 in the *Meteorologische Zeitschrift*, p. 532. Not quite correctly it is pointed out that v. Maydell links the seasonal [depending on the months] changing water levels of rivers to the annual fluctuations of the sea's level: this was my own view mentioned in an essay of the magazine "Naturforscher" on February 27th, 1886—opposing v. Maydell's opinion.

to the ocean; the Black Sea only connected by the narrow and shallow Bosporus with the Mediterranean Sea and thereby with the ocean; and the Baltic Sea with a more open connection.

2.1 The Annual Water Level Cycle

In order to substantiate the annual cycle of water levels in the Caspian Sea we have two series of observations which were carried out at *Baku* (over 25 years) and at *Aschur-Ade* in the bay of Astrabad (17 years).[36] I determined the medium values of the two quite congruent series and obtained the following numbers. The + (plus) sign indicates a water level above the annual average, the – (minus) sign a level below the average. We encounter the lowest water level in March and the highest in July and August. The amplitude of variation is more than 0.3 meters and the water volume of the Caspian Sea, determined on the basis of this amplitude and the Sea's entire area, changes by 165 cbkm (cubic kilometres).

		Jan.	Feb.	Mar.	Apr.	May.	Jun.
Caspian Sea	cm	–14	–15	–16*	–8	0	13
Volga (Astrachan)[37]	cm	–50	–60	–70*	–40	120	**220**
		Jul.	**Aug.**	**Sep.**	**Oct.**	**Nov.**	**Dec.**
Caspian Sea	cm	21	**22**	15	2	–9	–11
Volga (Astrachan)	cm	100	–20	–40	–30	–50	–60

That this large variation recurring year after year with near absolute regularity is caused by the annual periodicity of the water supply from several rivers, predominantly the *Volga*, is rather obvious. Indeed, a consensus between the Caspian Sea and the monthly measured water levels of the Volga near Astrachan does exist: at both places winter is the season of low levels and summer of high levels. Noteworthy is a very distinct delay in the peak level, which the Volga reaches in June after the snow melts in Northern Russia, and the Caspian Sea 1 1/2 months later in July/August. Yet it is this delay that confirms the causal relationship between the two phenomena: because the delay of the maximum water level in the basin of the Caspian Sea is a logical consequence if its level changes are the result of changing Volga levels. The water level of the Sea keeps on rising even after the actual floodwaters of the Volga river are fully dispersed, until in late summer the increasing evaporation off the water's surface reduces the level: water accumulates in the Sea.

[36] Published by Filipof, *Variations of Water Levels in the Caspian Sea*, Morskoj Sbornik, 1880 No. 7 and 8 (in Russian).

[37] According to Woeikof, *Klimate der Erde* [*Climates of Earth*] (Jena 1887), Table XXII.

The correlation between the annual fluctuation of the water level in basins connected to the ocean and the annual of their tributaries is not at all as clear. Using the same numbers that v. Maydell was unable to explain,[38] I can prove qualitatively and quantitatively that the annual water level of the Black Sea is also dependent on the rivers' discharge.[39] The following tabulation of figures from the Black Sea, the Donau and the Don will outline this correlation.

	Jan.	Feb.	Mar.	Apr.	May	Jun.
Don at Kälatsch[40]	–0.7	–0.4	0.6	34	37	0.4
Dnjepr at Krementschug[40]	–0.4	–0.4	0.4	23	26	0.8
Donau at Orgsova[40]	–0.5*	–0.3	0.4	1.0	1.0	0.6
Pontus at River Delta	–0.061	–0.056	–0.013	0.075	0.130	0.107
Pontus at coasts without rivers	–0.053	–0.058*	–0.036	0.048	0.109	0.130
	Jul.	**Aug.**	**Sep.**	**Oct.**	**Nov.**	**Dec.**
Don at Kälatsch[40]	–0.4	–2.0	–2.4*	–1.7	–0.7	–0.7
Dnjepr at Krementschug[40]	–0.7	–1.1	–1.3*	–1.2	–0.9	–0.7
Donau at Orgsova[40]	0.0	–0.5	–0.8	–1.0*	–0.5	–0.2
Pontus at River Delta	–0.048	–0.010	–0.058	–0.058	–0.051	–0.051
Pontus at coasts without rivers	0.094	0.020	0.05	–0.084*	–0.079	–0.038

Again, the time delay caused by the accumulated amounts of water collected in the Pontus basin is evident: the highest river-water level occurs in April and May and the highest level of the Black Sea in May/June.

Another secondary, less dominant, but not all that insignificant factor in reaching the summer's maximum level is the thermal expansion of the Sea's water volume by 44 cbkms[†] as a result of the temperature rise between February and May.

The conditions in the Baltic Sea are considerably more complicated due to its freer water exchange with the ocean. Despite the sinuous shape of this basin, the variations throughout the year run parallel at various points along the coast. Generally, levels are at their highest in the summer: after that the sea-level falls and reaches its lowest mark in the spring. This periodicity is not quite as constant at these different measuring stations if we compare monthly averages instead of seasonal ones; it is however constant enough for same time spans, so that one station may serve as representing the entire Sea. I have selected *Swinemünde* for this purpose, since for this station there are carefully researched and absolutely reliable data at hand prepared by Seibt.[41]

[38] Op. cit.

[39] Compare: Meteorologische Zeitschrift, 1886, July Issue, p. 297.

[40] By Woeikof, *Klimate der Erde* [*Climates of Earth*], Jena 1887, Figure XXII.

[†] [cubic kilometres]

[41] *Mittelwasser der Ostsee bei Swinemünde* [*Medium Water Level of the Baltic Sea at Swinemünde*], Berlin 1881.

I am not asserting a general medium value but three decadal means for the decades 1855–64, 1865–74, and 1875–84. It is, as emphasized by Seibt, quite remarkable that these three decades show three very different annual curves: the mean of 1855–64 has no November maximum at all, which in the 1865–74 decade is more pronounced than the August maximum and which to a lesser extent also occurs in 1875–84.

The significant change from one decade to the next is an indication that the annual periodicity of the water level in the Baltic Sea is influenced by a number of factors, each with a different emphasis at different times.

	Jan.	Feb.	Mar.	Apr.	May	Jun.	Jul.	Aug.	Sep.	Oct.	Nov.	Dec.
Baltic Sea at Swinemünde (mm)												
1855–64	13	2	−38	−25	−48	−53	86	**111**	57	3	−45	−74*
1865–74	−38	−56	−67	−72*	−29	25	54	**54**	34	−30	87	35
1875–84	−16	−28	−13	−86*	−71	32	62	**69**	45	−2	17	−12
Oder River at Küstrin[42] (cm)												
1807–35	15	39	**57**	50	10	−22	−23	−35	−32	−40*	−28	−10
Glommen[42]												
1862–76	−99	−114	−123*	−78	78	**226**	119	54	43	12	−65	−94

A comparison between the annual fluctuation of the Baltic Sea level and that of the water level of the major German rivers, in the table above represented by the *Oder River*, indicates that the rivers' run-off can only play a very subordinate role: the Baltic Sea's minimum level coincides with the rivers' maximum levels. Scandinavian rivers, represented here by the *Glommen* river, show a quite different pattern. Their summer maximum may have some impact on the maximum level in the Baltic Sea, which occurs at almost the same time, yet is by no means solely the result of it. The annual amount of precipitation certainly plays an important role increasing the Baltic Sea's water volume to its maximum level during the summer.

Another accountable factor for the high water level in the summer is the thermo-dynamic increase of the Baltic Sea's water volume due to its warmer summer temperatures. This factor alone causes the volume to increase by 30 cubic kilometres or the level to rise by 8 cm. We must remember that evaporation increases during the summer as well, so that water lost through evaporation counteracts the volume expansion by a certain extent.

However, all these factors do not explain the Baltic Sea's minimum level in springtime, which could well be related to the easterly and northeasterly winds prevailing at the time of its occurrence. The high water level in the summer may also partly be attributed to the steady westerly winds at that

[42] According to Wex: *Über die Wasserabnahme in den Quellen und Strömen* [*About water decrease of springs, streams, and rivers*], Zeitschrift des österreichischen Ingenieur- und Architekten-Vereins, 1879.

time of the year. Northeasterly winds force water out of the Baltic Sea while westerly winds have the reverse effect.

We are therefore faced with a multitude of aspects all of which, without any doubt, influence the annual periodicity of the Baltic Sea's level changes; however, at present, we are in no position to weigh their quantitative impact. Results turn out differently when comparing the variations of water levels from one year to the next.

2.2 SECULAR WATER LEVEL VARIATIONS

Rarely does a sea level at the end of a year return to the same level it used to have at the beginning of that year: rarely are water input and water output in equilibrium over the course of one year. Ordinarily, one of the two factors outweighs the other in the annual balance, resulting in a rising or falling water level: the average water level varies from year to year. It is now our intention to observe this change in the three seas selected.

Fluctuations in the medium water levels from one year to the next are quite substantial and this is quite natural in a basin with no outlet such as the Caspian Sea where water is supplied by precipitation and its tributaries and discharged through evaporation. These fluctuations are all the more pronounced since the weather tends to have the opposite effect on water supply and discharge: rainy years adding to the water volume tend to be cool with little evaporation and little water loss.

It becomes evident that these fluctuations of the Caspian Sea do not level off over a period of a few years, but rather that certain major fluctuations of the water level take place over longer periods of time. Let us look at the lustra* averages of water level observations at *Baku* and *Aschur-Ade* published by Filipof;[43] after being adjusted in relation to a fixed point zero as a common denominator, these figures show that from 1851 to 1865, water levels of the Caspian Sea were low, and began to rise steadily from 1866—overall by approx. 3/4 of a meter. This rise is equally prominent at both stations and must be considered an absolutely certain fact. The conformity between both stations contradicts beforehand the assumption that this fluctuation is caused by movements of the ground formation, as one might be inclined to assume of the *Baku* area rich in naphta deposits. The mean of both observation series is shown below as well as in Figure 2.

Caspian Sea

	1851–55	1856–60	1861–65	1866–70	1874–75	1876–78
cm	–21	–26	–19	–19	+16	+56

* [period of five years]

[43] op. cit.

These observations series are not long enough to show similar long-term movements of the Pontus level. The data of the years 1874–1882 reveal however, that the medium water level changes from year to year varying up to 36 cm during this time span. When we take a closer look at the main characteristics of this movement we will again recognize a rise of the water level towards 1880 as was the case with the Caspian Sea.

The lustra means of the Baltic Sea as well reveal long-term water level variations, the amplitude of which is however measured in decimeters and not in centimeters as is the case with the Pontus and the Caspian Sea. These are too pronounced and too universal to be the result of observational errors or changes of the measuring posts.

In order to trace these fluctuations, I determined lustra averages for 10 German water level stations which are, ranging from East to West, *Memel*, *Neufahrwasser* near *Danzig*, *Stolpmünde*, *Wiek* near *Greifswald*, *Stralsund*, *Warnemünde*, *Wismar*, and *Travemünde*, shown non-adjusted in the following table and depicted graphically in Figure 1.

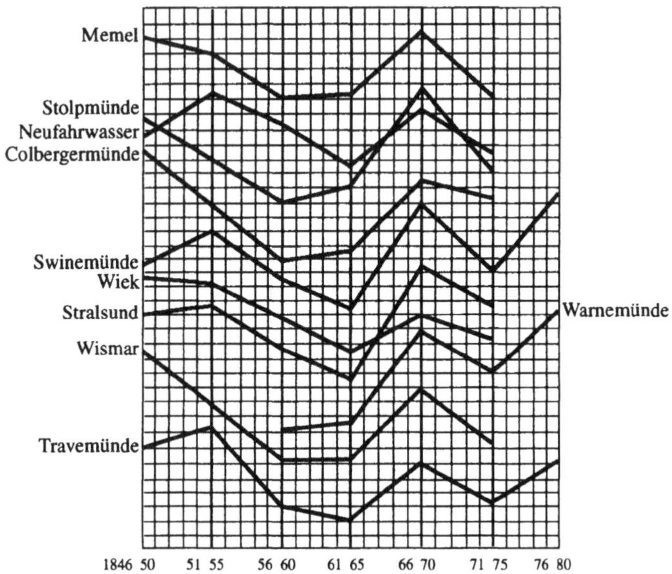

Figure 2.1

It is certainly a drawback that, with the exception of these German stations, no long-term series of water level observations are available at present. However, the number of existing stations in various locations is sufficient and observations at the coastal regions of Sweden and Finland not included here, seem less comparable since those coastlines are supposedly undergoing a secular change in elevation which might distort the independent secular movement of the water level.[44]

A glance at the table and more so at the curves in Figure 1 tells us that the movements of the water level are similar at all stations: a decline starting in 1850 or 1855, reaching a minimum from 1860 to 1865, another increase beginning in 1866, then a sudden jump forward at first followed by a slight decline in 1871–75 and a gradual rise. It is quite remarkable that in the Baltic Sea we again see this rise of the sea level starting in the middle of the Sixties, which we encountered at the Caspian Sea.

Unfortunately, there are not enough precise observation series available to trace secular water level variations far back in time. Only the Swinemünde station provides a series of observations starting at the end of the Twenties, which tell us that the general maximum around mid-century (1845–55) was preceded by a period of low Baltic Sea levels (cf. Figure 2).

Baltic Sea (mm)

Lustrum	Memel[45]	Neufahrwasser[45]	Stolpmünde[45]	Colbergermünde[45]	Swinemünde[46]	Wiek[46]	Stralsund[46]	Warnemünde[47]	Wismar[47]	Travemünde[48]	Number of +	Number of –
1826–30	—	—	—	—	–7	—	—	—	—	—		
1831–35	—	—	—	—	–5	—	—	—	—	—		
1836–40	—	—	—	—	–5	—	—	—	—	—		
1841–45	—	—	—	—	–6	—	—	—	—	—		
1846–50	+19	–4	+25	+9	–0	+26	+12	—	+47	+24	9	1
1851–55	+11	+23	–5	+7	+19	+23	+17	—	+10	+33	9	1
1856–60	–20	+2	–30	–32	–14	–3	–13	–23	–23	–20	1	9
1861–65	–17	–23	–23	–28	–33	–24	–32	–21	–25	–29	—	10
1866–70	+28	+16	+46	+24	+41	+0	+43	+19	+19	+10	10	—
1871–75	–19	–15	–11	+15	–12	–18	–26	–12	–12	–19	1	9
1876–80	—	—	—	—	+47	—	—	+38	—	+11		

As far as the Caspian Sea is concerned matters are much better. Although level observations do not even date back further than the middle of the

[44] Later on I was able to verify similar long-term variations of water levels in these coastal regions as well.

[45] According to Levels and Height Adjustments, etc., made by the *Trigonometrische Abtheilung der Landesaufnahme*. Vol. III. § 11, p. 139. Berlin 1875.

[46] According to Seibt, *Mittelwasser der Ostsee bei Swinemünde* [*Medium Water Level of the Baltic Sea at Swinemünde*]. Berlin 1881.

[47] As in footnote 15 but supplemented by reports of the Kiel Commission.

[48] According to Seibt, *Mittelwasser der Ostsee bei Travemünde [Medium Water Level of the Baltic Sea at Travemünde]* Berlin 1885.

current century, the water level variations, occurring here on a far grander scale, have caused radical changes to the coastlines: there are, for instance, islands which at times vanish and reappear above sea level, at times reconnect with the mainland turning into peninsulas, and then again reappear as proper islands. Consequently, these variations have attracted the inhabitants' attention from very early on. In fact we have numerous reports which include valuable data concerning the changes of the sea level, critically assessed by Lenz and Sokolof. I do not agree with Filipof,[49] who denotes the results of both researchers as being not sufficiently verified. They correspond very well with other publications so that I have to consider them quite reliable and sound.[50] The sea level changed as follows:

Caspian Sea

	Movement	High Level	Low Level
Until 1744	Rise	about 1745	
Until 1766	Fall		about 1765
Until 1809/14	Rise	about 1815	
Until 1842	Fall		about 1845
Since 1847	Rise	about 1850	
Followed By	Fall		about 1860
Since 1866	Rise	(about 1880)	

At this point let us summarize what we have discussed so far. Both, the Caspian Sea and the Baltic Sea show level oscillations; long-term periods of rising levels alternate with falling ones. As far as we can tell from observations available, the rise and fall occurs at both seas at the same time and in a parallel pattern, whereby, though, the amplitude of the Caspian Sea's oscillations is far more pronounced than the one of the Baltic Sea level.

A comparison between these variations and those of glaciers seems obvious since seas with no outlet, such as the Caspian Sea, and glaciers are in size and extent regulated by the same factors: precipitation, which in both cases supplies water, and warm temperatures, which in the one case evaporate the water and in the other melt the ice. As much as we know about glaciers, the variations of the two phenomena do indeed show a parallel pattern. The period of advancement by the glaciers of the Alps at the beginning of the century corresponds with a period of rising and high water levels of the Caspian Sea. The second period of glacial expansion at the end of the Forties is paralleled by rising Caspian Sea levels, and the most recent increase in the Caspian Sea level starting in 1866 finds its counterpart in the

[49] Op. cit.

[50] A more detailed argument is mentioned in another place. Additional references originated only in minor changes of the table. In turn these variations were traced back to 1685. 1685 to 1715 decline; 1715 to 1744 rise, also high level in 1685, low level in 1715.

glacial advancement developing since the late Seventies. Glacial change seems to lag behind the change in the Caspian Sea level by a number of years. What Swarowsky demonstrated for the glaciers of the Eastern Alps and the *Neusiedler See*[51] is repeated, if on a much grander scale at the Caspian Sea. Both phenomena, glacial variations and variations of seas with no outlet, go back to the same cause, to secular weather changes, to climatic changes.

Weather affects the water level of lakes and seas with no outlet by governing the input and output of water. Since long-term observations of river levels, which would show any variation in the river volume directly, are unavailable for the area of the Caspian Sea, I shall have to revert to precipitation measurements.[52] *St. Petersburg* in the Northwest of Russia, *Lugan* in the South, *Bogoslowsk* in the Ural and *Tiflis* in the Caucasus shall represent the different parts of the Caspian Sea's catchment area. Although some of these measuring points seem far away from the Caspian Sea or even beyond its boundaries, we are nevertheless entitled to include them in our inquiry since, by experience, secular weather changes tend to affect large territories.

At all stations we see a decline in rainfall starting around the late Forties and early Fifties and continuing into the Sixties when it starts to increase considerably. These changes in precipitation within the catchment area of the Caspian Sea, become more apparent if we express the lustrum averages of each station in percentages of the mean value of the years 1841–1880 on the basis of which we can then determine lustra averages for the entire area. If we compare these numbers in the table and the diagram of Figure 2 with those of the level variation of the Caspian Sea, there can be no doubt that the latter is caused by variations in precipitation.

It seems of special interest to find the causes for secular level changes of the Baltic Sea because in this case the factors that might have an effect are so numerous. Our findings about the Caspian Sea lead us to an instant comparison between the variations of the Baltic Sea and those of its surrounding tributaries. Long-term water level variations can also be observed in Central European rivers, which allows us to draw direct conclusions about volume fluctuations. It is astonishing that these variations have received little recognition so far, although Hagen[53] and above all Fritz[54] have called attention to it. The following table [following page] shows the lustra averages, based on

[51] XII Bericht des Vereins der Geographen an der Universität Wien, 1886.

[52] Observations of the Volga river level near Astrachan subsequently made available to me, fully confirm these findings.

[53] *Wasserstände preussischer Ströme* [*Water Levels of Prussian Rivers*]. Sitzungsberichte der Berliner Akademie der Wissenschaften, 1880.

[54] Petermann's Geographische Mittheilungen, 1880, p. 245.

data collected by Fritz,[55] of the medium water levels of the *Memel* near *Tilsit*, the *Weichsel* near *Kurzbracke*, the *Oder* near *Küstrin* and *Neuglietzen*, also from the greater vicinity of the Baltic Sea those of the *Elbe* near *Magdeburg*, the *Weser* near *Bremen*, the *Rhine* near *Düsseldorf* and *Emmerich*, and the *Seine* near *Paris*. The [rightmost column] refers to the *Donau* at *Orsova*. These numbers are charted in Figure 2 so that a change by one graduation mark equals a change of the water level by 0.2 meters. The two series for the rivers *Oder* and *Rhine* respectively were integrated into one curve each.

Rainfall (mm)	1841–45	1846–50	1851–55	1856–60	1861–65	1866–70	1871–75	1876–80
St. Petersburg	**482**	436	369	350*	443	**603**	517	518
Lugan	385	287	352	311	280*	379	384	**467**
Tiflis	—	**522**	454	452*	465	**530**	448	520
Bogoslowsk	387	452	412	384	320	282*	**483**	466
Mean	103	89	94	88*	88	105	108	117
Caspian Sea Water Level (cm)	—	—	–21	–26*	–19	19	16	**56**

From the table and even more so from the graphic chart in Figure 2 it is evident that the changes of the water levels of those rivers show in all of Central Europe an almost parallel pattern from lustrum to lustrum. All of these rivers reach their highest levels around 1801–10 which then continue to fall more or less uninterruptedly. Grouped around the year 1830 are the lowest lustra averages. By 1850, or to some extent earlier, we encounter a second maximum, which occurs in the same universal fashion as the first. A very steep decline follows and, almost without exception, a minimum is reached during the years of 1856–65. Since 1866, rivers are once again on the rise, a period which does not seem to have come to a close yet in 1880. Generally, in Central Europe this last variation is more pronounced than the one from the first half of the century.

A comparison between these variations of river levels and the variations of the Baltic Sea level, shown in the Swinemünde measurements of Figure 2, reveals a striking parallel pattern of the two and the surprising finding that the water level of the Baltic Sea rises during years when the water supply from rivers is high, and falls when the situation is reversed. This parallel pattern is so perfect that chance can be excluded entirely. Even the intensity of the fluctuations is alike: a less pronounced variation in the first half of the century (*Swinemünde*), and a more prominent one during the Sixties and Seventies.

[55] *The link between sun spots and magnetic and meteorological phenomena on Earth.* Harlem 1878, p. 135.

It must be determined, whether both events have one and the same cause or whether one is the cause of the other. A common source might perhaps be the wind. To a large extent all precipitation is dependent on wind conditions and the rainfall determines the amount of water the rivers carry; stronger air currents carrying rain must therefore increase the rivers' volume and raise their levels. Also no doubt, wind and storm conditions intensify the drift between Baltic Sea and North Sea either reducing or increasing it. But it is these rain-carrying storms, the westerly winds in particular, which are holding Baltic Sea waters back, while the dry winds from the continental masses support the Sea's outflow through the belts and the Sound. Thus, it would be possible and conceivable to hold wind conditions indirectly accountable for the fluctuations of river levels and directly accountable for fluctuations of the Baltic Sea level. However, a comparison between the figures and charts of individual river and sea measuring stations advises otherwise. The water level of a particular part of the Baltic Sea is more closely related to its neighbouring large waterways than to its more distant parts. This is shown in the following table. With regard to the first maximum and minimum, the *Neufahrwasser* data reveal a delay in comparison with those of its nearest measuring stations *Memel* in the East and *Stolpmünde* in the West, yet correspond in this regard to those measurements taken at the *Weichsel* station *Kurzbracke*.[56]

		1846–50	1851–55	1856–60	1861–65	1866–70	1871–75
Memel	mm	**+19**	–11	–20*	–17	+28	–19
Stolpmünde	mm	**+25**	–5	–30*	–23	+46	–11
Neufahrwasser	mm	–4	**+23**	+2	–23*	+16	–15
Weichsel	mm	187	**231**	140	94*	152	168

The figures demonstrate the direct link between river volume and the Baltic Sea's water level. They do not reflect the impact of the wind conditions which are difficult to measure since they are constantly changing.

Based on the correlation between the amount of rainfall and the level of the Pontus in the years 1874 to 1882[57] it is safe to say that the secular variations of the water levels of the Caspian Sea, the Baltic Sea and the Black Sea are the result of variations in the water volume of the rivers feeding into these seas. This latter fact points to secular variations of the precipitation—to secular climate variations. Similar variations in precipitation and temperature were identified for the Alps by Lang in connection with his most valuable inquiry into glacial changes.[58] It cannot be a mere coincidence

[56] This fact is even more evident when we compare the variation of the Baltic Sea level along the Swedish and German coast.

[57] v. Maydell and other authors.

[58] Zeitschrift für Meteorologie, 1885, p. 113 (433).

Secular variations of water levels of central European rivers (m).

Tilsit	Memel Kurzbracke	Weichsel	Oder Küstrin	Elbe Neu-Glietzen	Weser Magdeburg	Bremen	Rhein Düsseldorf	Seine Emmerich	Donau Paris	Orsova
1766/70	—	—	—	—	2.91	—	—	—	1.24	—
1771/75	—	—	—	—	2.95	—	—	3.52	**1.40**	—
1776/80	—	—	—	—	2.63	—	—	3.24	1.20	—
1781/85	—	—	1.55	—	2.72	—	—	3.09	1.12	—
1786/90	—	—	1.61	—	2.47	—	—	3.18	1.32	—
1791/95	—	—	1.17*	—	2.15*	—	—	3.08	1.22	—
1796/1800	—	—	1.33	—	2.23	—	—	2.74*	1.09*	—
1801/05	—	—	**1.67**	—	**2.54**	—	2.76	**3.10**	1.32	—
1806/10	—	—	1.33	—	2.13	—	**2.98**	3.02	1.37	—
1811/15	—	2.32	1.12	1.78	1.88	—	2.48	2.62	1.11	—
1816/20	—	2.46	1.28	1.88	1.97	1.33	2.93	2.90	**1.46**	—
1821/25	—	2.11	1.00*	1.66*	2.01	1.14*	2.74	2.82	1.14	—
1826/30	—	2.20	1.37	2.06	2.25	1.56	2.65	2.69	1.10	—
1831/35	—	1.67*	1.11	1.71	1.83*	1.22	2.48*	2.41*	1.04*	—
1836/40	—	2.14	—	2.23	2.02	1.38	2.88	—	**1.49**	—
1841/45	—	2.01	—	2.04	2.05	**1.47**	2.85	—	1.29	3.04
1846/50	2.37	1.87	—	2.39	1.99	1.11	2.74	—	1.26	2.97
1851/55	**2.67**	**2.31**	—	**2.69**	**2.21**	1.44	**2.93**	—	**1.43**	**3.15**
1856/60	1.89	1.40	—	1.89	1.71	0.69*	2.29	—	0.93	2.54
1861/65	1.77*	0.94*	—	1.76*	1.52	0.88	2.20*	—	—	2.19*
1866/70	2.06	1.52	—	2.06	1.76	1.02	2.67	—	—	2.65
1871/75	2.00	1.68	—	2.12	1.38*	0.83	2.48	—	—	—
1876/80	2.35	1.65	—	2.35	1.70	—	2.66	—	—	—

that the weather changes we discovered through water level observations at seas and rivers occurred at exactly the same time as those observed by Lang in the Alps.

Let us take a closer look at the entire area that, as was established earlier, is affected by the climatic variation (see Fig. 2): it includes the catchment areas of the Caspian Sea, the Black Sea,[59] the Baltic Sea, the North Sea, also the Seine-basin and the Alps with the Po delta, i.e., nearly all of Europe. Thus, we have to conclude that **almost all of Europe experiences climatic variations at the same time and in the same manner**.

The immediate question arises where the territorial boundaries of these climatic variations may lie, whether they are limited to Europe or extend completely or partly to other parts of the globe. The following table and diagram may lead to some, if sketchy, answers.

We notice the completely identical pattern of the variations of the Lake Constance level (Fig. 2, scale 1:100) and those of the precipitation in the Alps, the water levels of the Baltic Sea, the Caspian Sea etc. Even at Barnaul and Nertschinsk does the amount of rainfall show a significant decline—a fact that was pointed out by Woeikof[60] in 1870 for the years 1845–65, followed by an equally strong increase in 1871 and 1866 respectively. These distant areas of Inner Asia participate in Europe's climate change. The same is the case with tropical Africa as indicated by the levels of the Nile's flood waters changing from lustrum to lustrum with a minimum around 1835, a maximum around 1846/50 and a second less pronounced minimum around 1856/60, followed again by another increase. India, as is evident from rain observations at *Madras*, shows an even closer concurrence with this variation. Here, the amplitude is more than 0.5 meters in rainfall; and the rainfall of the driest lustrums in 1831–35 compares to the three wettest lustra in 1816–20, 1846–50 and 1871–75 at a ratio of 2 to 3.

In the New World, climatic variations reveal themselves in the amount of water discharged in the Mississippi delta as well as the variations in the water level of the five Great Lakes—here represented by Lake Michigan. Despite some discrepancies with its closest neighbours, the Great Salt Lake is also part of this variation. By 1856 the latter reached an, if somewhat low, maximum followed by a drop of the sea level by 1.5 meters in 1860. Starting in the early Sixties the lake increased by a full 3 meters. **The surprising result is that all of the countries of the Northern hemisphere[61] are at the present time experiencing the same secular climate variations: a fairly dry period around 1830, a wet period around 1850 and a second dry period around 1860, followed again by a wet period** (around 1880?).

[59] Danube, Rainfall (South of Russia).

[60] Wild's Repertorium für Meteorologie I 2, p. 199.

[61] and, according to my latest research, I may include the Southern hemisphere.

We are dealing with atmospheric variations or climatic changes of a grander scale than those eleven-year ones which some tried to link to sun spot activity. They are grander because of their larger amplitude as well as their greater length—variations in which all hydrographical phenomena on earth participate: **glaciers, rivers, lakes, and the relatively enclosed inlets—all of which grow at the same time and recede again at the same time.**

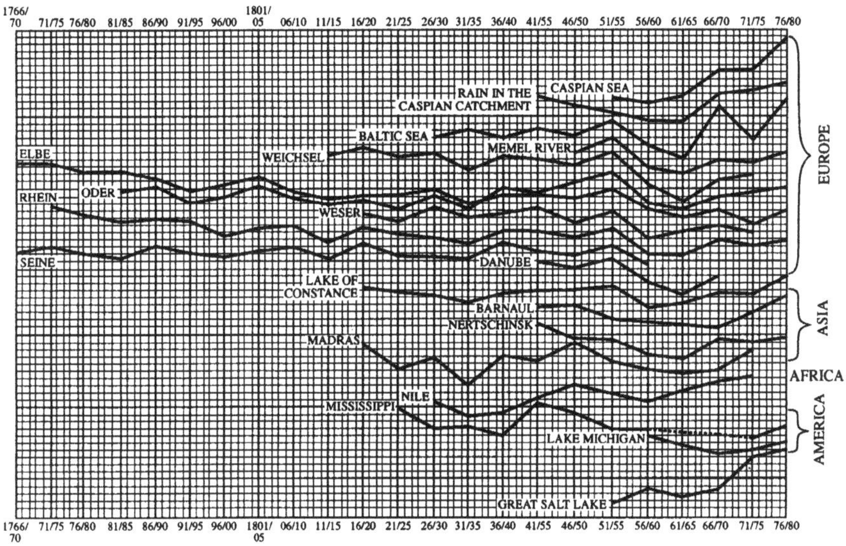

Figure 2.2.

	Europe	Asia			Africa	North America		
	Water level of Lake Constance[1]	Rain Fall at Barnaul[2]	Rain Fall at Nertschinsk[2]	Rain Fall at Madras[3]	Nile Level[4]	Water volume of the Mississippi[5]	Water level of Michigan Lake[6]	Water level of Great Salt Lake[6]
	m	mm	mm	mm	m	cbkm	cm	cm
1816/20	+4	—	—	1458	—	—	—	—
1821/25	+0	—	—	1133	—	619	—	—
1826/30	−2	—	—	1303	6.80	527	—	—
1831/35	−14*	—	—	914*	6.44*	523	—	—
1836/40	−3	—	—	1311	6.55	479*	—	—
1841/45	+1	282	**514**	1262	6.97	**659**	—	—
1846/50	+1	**291**	431	**1486**	**7.30**	608	—	0
1851/55	**+10**	221	421	1262	7.13	503	—	79
1856/60	−17*	203	351	1174	6.92*	523	84	52*
1861/65	−16	194	325*	1112*	7.20	—	61	88
1866/70	+5	173*	432	1130	7.39	—	45*	274
1871/75	+5	258	408	1432	7.52	483	48	311
1876/80	+28	369	430	—	—	534	—	—

[1] Beiträge zur Hydrographie des Grossherzogthums Baden, I. Issue, 1884; Honsell, Der Bodensee, Atlas Table. IV.

[2] Annalen des physikalischen Centralobservatoriums zu St. Petersburg.

[3] Fritz, *Beziehungen der Sonnenflecken und den magnetischen und meteorologischen Erscheinungen der Erde* [Relation of Sun Spots to Magnetic and Meteorological Phenomena on Earth], Haarlem 1878, p. 129.

[4] Fritz in Zeitschrift für Meteorologie, 1878, p. 363.

[5] v. d. Groeben, op. cit., 1884, p. 1 ff.

[6] Woeikof, op. cit., 1881, p. 288.

Chapter 3

How Constant is Today's Climate?*

The question of climatic change in historical times has always been of great interest as it has far reaching practical consequences. Once the evidence is before us that climatic changes took place on our globe within man's sight, we have to conclude that changes will continue to happen in the near future. Such changes, however, will not pass without the profoundest impact on the life activities of people. Thus, to a certain extent, the general answer to this question is a prognosis of man's future fate and that of his accomplishments.

This question has been answered in many different ways, and there are few conceivable cases of climate change whose occurrence has not been claimed for either smaller or larger areas.

There can be no doubt that the climate of the geological past was different from today's. Fossils found in the polar zone give us the same clear evidence of a once tropical vegetation during the earliest periods of Earth's history, such as the Cretaceous Period, as the moraines of the diluvial glaciers in the moderate zones and even in certain tropical mountains. Climate has changed from the Cretaceous Period up to the ice age and since the ice age; but can we prove this latter change in historical times? The answer has been a multiple "yes". In 1882, Whitney, for instance, tried to show that the whole planet experienced a general drought.[62] A similar climate change is claimed by Theobald Fischer when he is talking about the advancing deserts in the Mediterranean.[63]

However, more often than these geological changes of the climate, as one is inclined to call them, during historical times one has claimed to have

* *In wie weit ist das heutige Klima konstant?*, Verhandlungen des VIII. Deutschen Geographentages, Berlin, 1889.

62 Whitney, *Climatic changes in later geological times*, Memoirs of the Museum of Comparative Zoology at Harvard College. Vol. VII. Cambridge, 1882.

63 Th. Fischer, *Beiträge zur physischen Geographie der Mittelmeerländer* [*Essays of Physical Geography of the Mediterranean*], Leipzig 1887, p. 25 ff. and *Studien über das Klima der Mittelmeerländer* [*Studies of the Climate of Mediterranean Countries*]. Additional issue no. 58 to *Petermann's Mitteilungen*. Gotha 1879, p. 41.

observed local changes restricted to a particular area and caused by human actions. Numerous researchers held for instance the opinion that the clear-cutting of forests may diminish the amount of rainfall and the water supply in springs and rivers, while it increases with reforestation. On the other hand Americans living in the dry zones of the western United States firmly believe that extensive cultivation has increased the rainfall of the region.

The many relevant hypotheses and theories which in this or a similar fashion are defending an ongoing unidirectional change of the climate are contradicted by an equal number of findings denying a change of this kind in historical times. It is notable that it is usually the meteorological science that rejects the idea of a continuous change of climate, while geologists, geographers and hydrographers have frequently supported it. For meteorologists, the constancy of the climate is, to a certain extent, axiomatic.

It appears quite inexplicable how such a conflict of opinions could develop.

The question entered into a new phase when a one-way change was no longer investigated and meteorological data material were instead examined with respect to secular weather variations. This was instigated by the mysterious fluctuations of the alpine glaciers whose only cause had to be the meteorological changes.

Forel[64] and Richter[65] were the first to attempt to prove this theory statistically. It gained general acceptance through the research by C. Lang in 1885[66] which, based on more extensive data about the entire region of the Alps, identified alternating long-term periods of fairly cool and moist and relatively warm and dry weather in relation to the glacial changes.

This was the situation two years ago when I discovered analogous weather variations with a completely different approach and for very different areas and presented my first findings at the German Meteorological Society's meeting in Karlsruhe.[67] Hydrographical research made me aware of unusual fluctuations of the water levels of the Baltic Sea, the Caspian Sea and the Black Sea, all of which showed a rhythmic pattern, as Swarowsky

[64] Forel in *Archives des sciences phys. et naturelles*, 1881, No. 5.

[65] Ed. Richter, *Der Obersulzbachgletscher* [*The Obersulzbach Glacier*], Zeitschrift des Deutschen und Österreichischen Alpenvereins, 1883, p. 75 ff.

[66] C. Lang, *Der säkulare Verlauf der Witterung als Ursache der Gletscherschwankungen in den Alpen* [*The Secular Weather Changes as a Source of Glacial Variations in the Alps*], Zeitschrift der Österreichischen Gesellschaft für Meteorologie, 1885, p. 443.

[67] Ed. Brückner. *Die Schwankungen des Wasserstandes im Kaspischen Meer, im Schwarzen Meer und in der Ostsee* [*Fluctuations of Water Levels in the Caspian Sea, the Black Sea and the Baltic Sea Relative to Weather*], Lecture, Annalen der Hydrographie 20. 1888, February issue. A short version of the lecture was published in the Meteorologische Zeitschrift, 1887, June issue, p. 232. [Reprinted as Chapter 2 in this publication.]

had discovered it at the Neusiedler Lake[68] with a distinct similarity to the rhythmic variations of glaciers. At all locations and at the same time, periods of generally high water levels alternated with those of fairly low levels. At the Caspian Sea it was obvious to seek the cause in a change of the amount of water supplied by its rivers and in the lake's evaporation. An assessment of water level observations at the Volga as well as observations of precipitation carried out by a number of meteorological stations in the Russian Empire changed assumption into fact. It turned out that the same variations of rainfall Lang had identified for the Alps were also occurring in the enormous drainage area of the Caspian Sea. Furthermore, the areas of the Baltic Sea and the Black Sea were subject to the same rainfall variations as well, and the peculiar long-term level variations occurring at these seas are partly a result of the water supplied by their tributaries which fluctuates along with the rainfall. The average water levels of the Weichsel, Oder, Elbe, Weser, Rhine, Danube, and even the Seine, all of them reflect clearly the variations of rainfall with the same long-term periodicity. In short, we find these secular variations of weather conditions all over Europe. In testing some of the meteorological and hydrographical data available we learnt that more or less all countries of the northern hemisphere participate in these variations: their universal occurrence and length enable us to identify them as climate variations. I was very pleased when in the fall of 1888, based on his research on the level variations of a number of lakes, Sieger[69] confirmed the majority of my findings.

Meanwhile I was able to gather additional data relevant to the subject and to extend the research to the southern hemisphere as well.

The initial results were primarily achieved on the basis of hydrographical phenomena and consequently of a qualitative nature. In examining the meteorological data, it was now necessary to also give a quantitative account of the climate variations.

Today we have standardised observations from nearly 600 meteorological and hydrographical stations adding up to a total of 30,000 years of observations, and with the help of these data we will already be in a position to gain a clear picture of climate fluctuations experienced on our globe. I

[68] A. Swarowsky, *Die Schwankungen des Neusiedler Sees* [*Fluctuations of the Neusiedler Lake*]. Bericht über das XII. Vereinsjahr des Vereins der Geographen der Universität Wien, 1886, p. 15.

[69] R. Sieger, *Die Schwankungen der hocharmenischen Seen seit 1800 In Vergleichung mit einigen verwandten Erscheinungen* [*Variations of the Upper Armenian Lakes since 1800 in comparison with some related phenomena*], Mitteilungen der K.K. Geographischen Gesellschaft, Wien, 1888.

may be permitted to briefly submit some of the main results from my nearly completed research.[70]

Already, the characteristic variations of hydrographical phenomena such as measurements taken at glaciers, lakes and rivers, made it seem most likely that climatic changes would be particularly evident during rainfall. In fact, this proved to be the case.

The following diagrams (Figs. 1 and 2) show the graphic curves of the rainfall variations. The first series of curves reflects the entire northern hemisphere, the second series spans from the northern hemisphere through the tropics to the southern hemisphere. Each curve represents the variation of a large territory based on the average measurements from many stations. The consulted stations are:

1. Scotland (Arbroath, Laurick Castle, Loch Leven Sluice, Northesk Reservoir, Glencrose, Swanton, Fernielaw, Edingburgh, Inveresk, Haddington, Culloden, Sandwich, Arrdaroach, Castle Toward, Cameron House, and Bothwell Castle). 16 Stations.
2. England (Chillgrove, Nash Mills, Oxford, Exeter, Orleton, Podehale, Boston, Bolton and Kendal). 9 Stations.
3. Northern France (Rouen, Paris, Vendôme, Pannetière, La Collancelle, Clamecy, Avallon, Laroche, Montbard, Poully, and Dijon). 11 Stations.
4. Northern Germany (Kleve, Trier, Köln, Boppard, Gütersloh, Frankfurt (Main), Gießen, Bremen, Kiel, Heiligenstadt, Torgau, Dresden, Stettin, Berlin, Küstrin, Frankfurt (Oder), Posen, Görlitz, Breslau, Königsberg and Tilsit). 21 Stations.
5. Austria–Hungary (Bodenbach, Praha, Deutschbrod, Lemberg, Kremsmünster, Klagenfurt, Wien, and Hermannstadt). 8 Stations.
6. West Russia (Helsingfors, St. Petersburg, Riga, Warsaw, Moscow, and Kiev). 6 Stations.
7. East Russia (Lugan, Ssimferopol, Astrakhan, Baku, Tiflis, Bogoslows, Jekatherinenburg, and Slatoust). 8 Stations.
8. West Sibiria (Barnaul). 1 Station.
9. East of Sibiria (Nertschinsk (metallurgical plant), Nikolajewsk (Amur), and Peking). 3 Stations.
10. United States, North America, Interior, (Toronto/Ontario, Milwaukee/Wisconsin., Detroit/Michigan, Madison/Iowa, Steubenville/Ohio, Marietta/Ohio, Cincinnati/Ohio, Leavenworth/Kansas, and St. Louis/Mississippi). 9 Stations.
11. Central Italy (Parma, Modena, Bologna, Genoa, Florence, Sienna, and Rome). 7 Stations.

[70] This inquiry will be published by early 1890 in *Penck's Geographischen Abhandlungen* (Wien, Hölzel) under the title *Klimawandel seit 1700* [*Climate Changes since 1700*]. It will include all the research material in detail which served as basis for the curves.

12. India (Madras, Calcutta, Jablapur, and Bombay). 4 Stations.
13. Mauritius (S. Louis, Alfred-Observatory). 1 Station.
14. Australia (Adelaide, Bathers, Bukelong, Deniliquin, Goulburn, Melbourne, and Sydney). 7 Stations.

The graphical charts are based on observations from a total of 111 meteorological stations. After averages for each country were derived from these stations, the lustra averages of rainfall were determined for each station and expressed in percentages of the thirty-year mean of 1851–80; the averages of each country based on data from several stations were then smoothed according to the following formula: $(a + 2b + c)/4$, i.e., the first and last lustrum according to the formula $(2a + b)/3$ and $(a + 2b)/3$.

The curves of the chart give a clearer picture of the data trend.

A rise and fall of the curve by one increment refers to an increase and decrease of rain by 5%. The distance between the top and the bottom of each curve shows the amplitude of the variation, in relative not in absolute terms. The wider the gap, the greater the difference between the maximum and minimum amount of rainfall.

We see a surprising similarity in the shape of the curves with an overall decrease towards the year 1860 and an increase towards 1850 and 1880, from the west coast of the old world to the east coast as well as in the interior of North America, from Germany via Italy and India up to Australia. It is evident that more or less all of the countries around the globe go through a rainy period and a dry period at the same time. In the current century the rainfall maxima are grouped around the years 1815, 1850, and 1880, the minima around 1830 and 1860.

In absolute terms of course, the epochs do not coincide perfectly; in some areas the minimum precipitation occurs in 1856–60, in others around 1861–65, in one case even not until 1866–70, and the maximum shifts accordingly. In none of our cases, however, does a minimum coincide with a maximum. No minimum occurs in the years 1841–55 and 1871–85 and no maximum in the years 1825–40 and 1856–70. Thus, too much rain in one region does not mean too little rain in another; a compensative process does not take place on the global continents concerned. The slight deviations derived from the epochal mean are irregular and mostly confined to smaller areas. However, this cannot be otherwise. Just as despite a clearly defined annual periodicity the maximum of a meteorological element, such as temperature, may occur earlier or later in the year depending on the actual weather conditions, it is the case here.

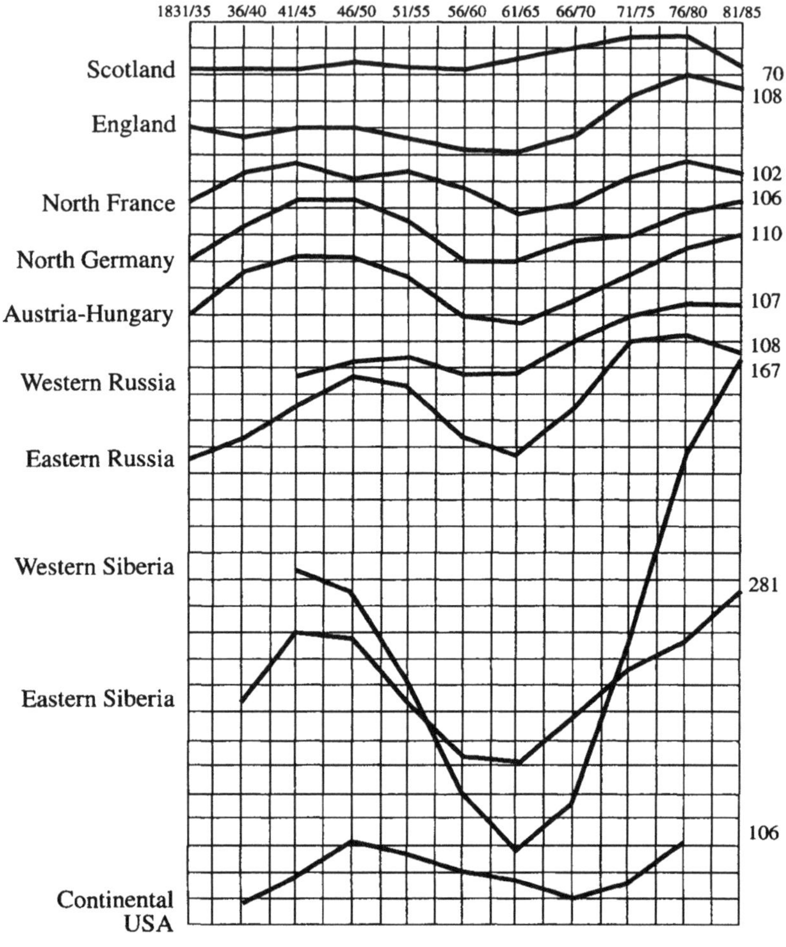

Figure 3.1. Secular Variations of Rainfall:10 Section from West to East

Also the relative intensity of the maxima and minima is not the same everywhere: in Australia the maximum around 1850 is more pronounced than the one in the 70s; in a number of regions the two maxima are equally intense, while in the majority of cases the maximum around 1880 is stronger than the one around 1850.[71]

[71] Some of these instances can most likely be attributed to the improved methods of measuring the precipitation, particularly during the winter, as the general upward trend in the curves for Russia, Siberia, and Central Italy seems to indicate.

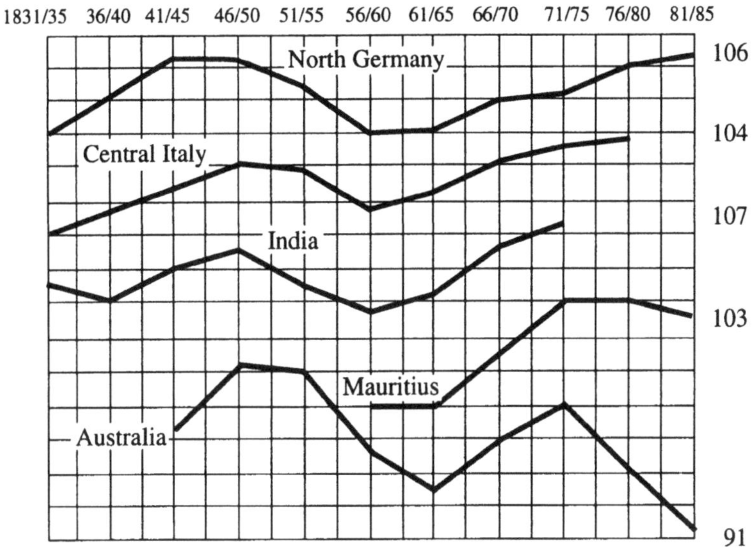

Figure 3.2. Secular Variation of Rainfall: Section from North to South

However, certain areas seem to be the exception to the rule. There is southern Italy and Sicily, as well as southern Spain, there are the lower valleys of the Indus and the Ganges, furthermore the eastern United States, where a maximum rainfall occurs in the 60s, during a time, that is, when other countries have little rain. Iceland seems to be an analogous case. Also Scotland displaying variations that are in part extremely vague is an exception while England adheres to the rule. These exceptional cases are, as far as we are able to tell from the present state of meteorological and hydrological observation methods, quite insignificant in comparison with the bulk of the land masses exposed to the variations.

Another pattern emerges clearly from the curves: the variation is more distinct as it advances further into the interior of the continental land mass. In Scotland it is vague. In Germany it is distinct and the rainfall of the driest lustrum around 1860 compares to the amount of the lustrum with the most rainfall in 1880 at a ratio of 1:1.09; in the eastern parts of European Russia this ratio is 1:1.24, and in West Siberia even 1:2.26, where more than twice the amount of rain fell during the wet 5 years of 1881–85 than during the dry years of 1861–65. In East Siberia the ratio falls again to 1:1.36. In view of this fact, and especially when considering the geographical location of some of the exceptional regions along the coasts of the Atlantic Ocean, it seems logical to conclude that the rhythmic pattern of the rainfall variation has to do with the landmass of continents and that compensation might take place over parts of the open ocean.

Rain is not the only meteorological element displaying this kind of rhythmic variation. Similar ones have been identified for the temperature in a two-fold approach: one based on records about the length of time of the rivers' ice cover during winter and the other based on temperature measurements. The following chart, Figure 3, will illustrate these temperature variations. The resulting curve is based on lustra averages of the mean values reported by Köppen[72] and determined by those stations in Europe and New England which carried out observations even prior to 1820. An increase of the curve by one equals a temperature increase of 0.05 degrees Celsius.

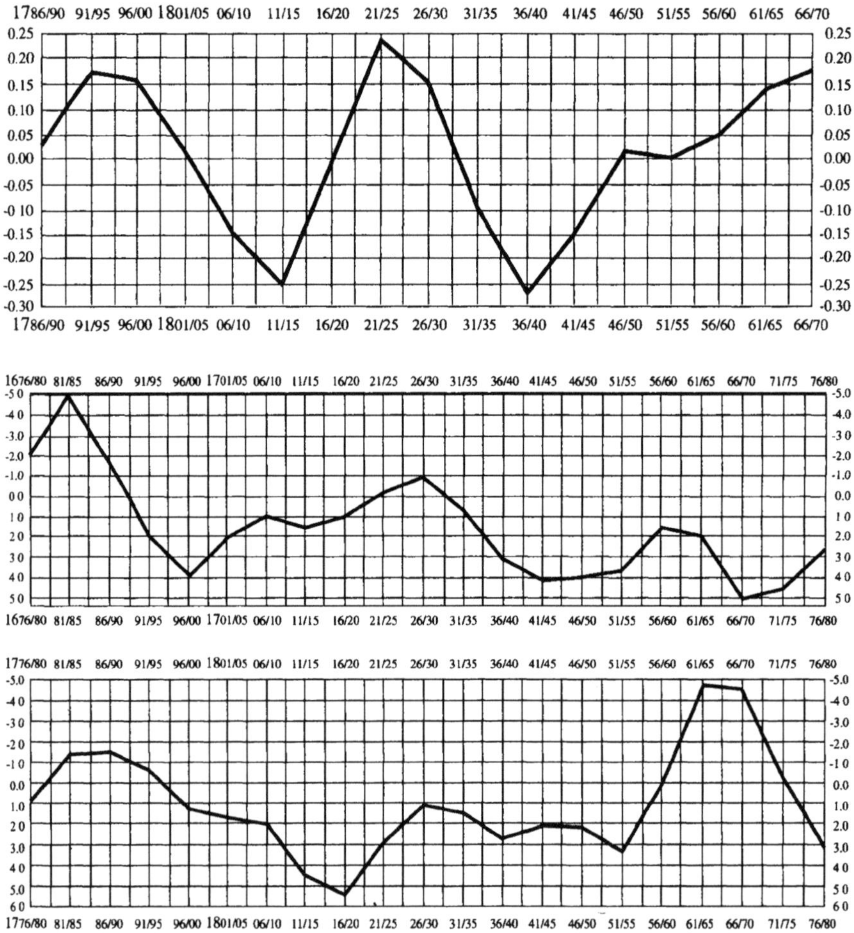

Figure 3.3. Secular Temperature Variations

[72] Köppen in Zeitschrift der Österreichischen Gesellschaft für Meteorologie, 1874, p. 266.

One general finding is obvious: the rhythmic pattern of temperature variations is such that **wet periods are at the same time cool, and dry periods warm**. For the current century we encounter, for instance, a generally cold period around 1815 in Europe and New England, a warm period around 1825, a cool period again around 1840, the years of the 60s are warm again followed by the last cool period at the end of the 70s.

As a next step, the question arises whether these climatic variations can also be identified for earlier centuries. One succeeded by relying on a number of different data. Of course, reliable meteorological observations do not date back very far into previous centuries; yet other valuable data were available, based partly on hydrographical and partly on plant-phenomenological factors. Among them are, for instance, the water level variations of the Caspian Sea recorded since 1685; there also are observations of the freeze-up and break-up of the ice on several of the Russian rivers, which with few interruptions go back to the beginning of the previous century. Foremost however, there are the dates of the beginning of the grape harvests in the wine areas of France, Southwest Germany and Switzerland recorded over many years and starting as early as 1400. Not until the year 1550 however is the number of recording stations large enough to allow cross-checking and to obtain an average.

It is common knowledge that grapes tend to mature late during cool, wet years and early in dry, warm years. As we have evidence that the changes in the harvesting dates until the middle of the previous century coincided with the annual changes in temperature and rainfall, they may be considered the ultimate indicators of climatic variations.

The lustra averages of the harvesting dates from 30 stations in the three countries mentioned were determined (Verdun, Argenteuil, Foug, Loches, Les Riceys, Coupignon, Denainvilliers, Auxerre, Vendôme, Vesoul, Dijon, Beaune, Lois-Le-Soulier, Renne, Pichon-Longueville, the Médoc, Tain, Castres, Stuttgart, Kürnbach, Altstetten, Veytaux, Vevey, Lausanne, Lavaux, Aubonne, Rolle, and Pully near Morges).[73]

For each station, these lustra averages were expressed in deviations from the 30-year mean of 1851–80 and based on the stations' averages. The total average was then determined for each lustrum. These values have been adjusted as before and are shown in the curve above allowing us to trace the climatic variations back to 1670. I could even have traced them back to 1550, though with less accuracy, because as of 1670 the number of recording stations goes down rapidly. But the sample provided here should suffice.[74] An increase

[73] The majority of data series collected by Angot Vendanges en France. Annales du Bureau Central Météorologique de France 1883, I. Paris 1885.

[74] The curve starts in 1550, and, as all other observational material, will be explained in more detail by the author in the extensive publication mentioned in footnote 70.

of the curve by one increment equals an earlier start of the grape harvest by one day, that is an increase in temperature and a decrease in rainfall.

Around the years 1880, 1851–55, 1816–20, 1766–70, 1741–45, 1696–1700, and 1671–75 we register cool and wet periods, around the years 1861–65, 1820–30, 1786–90, 1756–60, 1726–30 and 1681–85, warm and dry periods. The time spans between two maxima are not completely the same: climatic variations do not take place with the precise periodicity of a certain length of time and when we talk about a periodicity of 36–37 years it is not more than an average value.

What causes these characteristic variations of the most basic climatic elements? The ultimate reason is still a complete mystery. Data about rainfall are the only ones that can be explained reasonably, and only for the limited region of Europe. Here, the variations in rainfall go exactly hand in hand with secular variations of the air pressure. These take place over the North Atlantic Ocean and over all of Europe in such a way that during rainy periods the pressure differences are reduced, and increased during dry periods. During most parts of the year and on average the air flows from the continent to the ocean; when the continental system is weakened, there is less resistance to the oceanic air mass that can then move inland causing precipitation. When the system gains strength, the opposite effect is achieved. Fluctuations of air pressure are nothing but the results of fluctuations in temperature. An increase of the latter leads to a sharper thermal contrast between water and land in areas of moderate and upper latitudes causing high-pressure ridges over land with less precipitation. Similarly, the fluctuations of the rainfall should, in general, also be attributable to those fluctuations of temperature, though at present this has only been proven for Europe. He who can explain the temperature variations, has found the cause of the entire phenomenon of climatic variations.

Are climate variations so significant that they are of practical impact? Indeed they are. In dry areas in particular where water is notoriously scarce the hydrographic conditions change dramatically during periods of climatic variations. Lakes disappear during dry periods and reappear during wet ones, as for instance Lake George in New South Whales which in 1820 and again in 1876, and to a lesser extent in 1850, used to be a large lake of 12 to 18 kilometres in length, 10 kilometres in width, and 5 to 8 meters in depth, yet disappeared completely in the dry periods in between; or some central African lakes such as the Tshad, Tanganyika and Nyassa, which, according to Sieger,[75] at times rise so high that their overflowing waters create an outlet lasting for some years and which then lose this outlet again when the dry period begins. Rivers and creeks dry out for a full decade; swamps dry up and reappear in the

[75] R. Sieger, *Schwankungen der innerafrikanischen Seen* [*Fluctuations of Inner Africa Lakes*]. Bericht über das XIII Vereinsjahr des Vereins der Geographen an der Universität Wien, 1888.

next wet period. I could give many more examples in Africa and Australia, if time would permit it.

Consequently climatic variations deeply affect human life. River navigation to a great extent depends on the amount of water in the river bed which determines its depth. In those dry years around 1830 and 1860 shipping problems increased and soon a lot of speculation began about the possible cause of the lower river-water levels. In most cases the increasing practice of deforestation was found to be the source. Now we know better: it is because of climatic changes.

Another way in which temperature variations are affecting traffic is through the length of time of the rivers' freeze-up. For example, during the cold spill from 1806 to 1820, the Newa and the harbour of St. Petersburg remained blocked by ice for more than three weeks longer than they did during the warm period from 1821 to 1835.[76] This means that during cold years harbours in a more westerly location and with shorter closure times handle part of St. Petersburg's shipping traffic which they lose again during warm periods. Of course, the navigational period may also change in length from one year to the next. In that case, however, the following year may make up for the loss of the previous year; not so in the case of climate variations when average values have changed and favourable and unfavourable conditions cannot be reversed so soon. Thus certain changes in shipping traffic go hand in hand with climate changes.

Another area that is affected by climatic variations is agriculture. This is of special significance for the interior of the continents where variations are more extreme, e.g., Siberia which suffered famines as a result of the droughts in the 60s; or Egypt where the Nile's floodwater levels determine the fertile productivity of vast areas of agricultural land.

A future climate change may turn out to be quite disastrous for the arid areas surrounding the Salt Lake in North America. From the early 1860s to the middle 1870s the water level of the Great Salt Lake rose by 3 m—filling its tributaries whose waters were used for the irrigation of recently cultivated fields and grass lands.[77] We may recall the prevailing view that extensive cultivation of previously desert-like areas has increased rainfall considerably.

Contrary to this, I would like to point out that the climatic improvement coincides exactly with a period of increased rainfall experienced all over the globe and in continental areas in particular. On the other hand, similar changes of climatic conditions have occurred in America before, as we were able to demonstrate through observations of rainfall and river levels in the

[76] Based on data collected by Rykatschew (Closure and Reopening of Russian Rivers), see charts.

[77] Compare Gilbert in Powell, *Report upon arid Regions*, Washington 1879, p. 57 ff.

neighbouring Mississippi area for the current century. They are the same variations as the ones which in Europe could be traced back to 1550, and by this large number of verified variations we are convinced to believe that they will most likely continue to occur in future. It seems to me highly probable that the climate improvement in the Salt Lake area that began in the 60s will now be followed by an arid deterioration in climate. Early signs of this development can be seen in the recent arid years which in fact caused the Salt Lake to fall: in 1888 the lake's water level had again returned to its low marks of 1864.[78] If this hypothesis holds true, then this area will be unable to avoid a major economic crisis in the very near future; because agricultural land which was cultivated during the years 1870–80 would soon stop to produce because of aridity. We would see the same development here, as we have seen in Egypt and Siberia, namely that **the size of cultivable land changes along with climatic variations**.

However, time is running out, let us conclude!

We were able to identify general variations of the climate; and I have tried to outline the situation briefly. At first glance it may seem unusual that these variations had escaped scientific scrutiny to this day. However, there has been speculation about them before: every now and then, based on unusual sightings at some source or body of water the view emerged in publications that the climate of certain locations, their rainfall in particular, was probably subject to periodic changes, as a.o. by Hann for the area of the Caspian Sea,[79] Schweinfurth for parts of the Mediterranean countries,[80] and most of all Fritz for many areas on the globe.[81] However, for the Alps the only meteorological evidence was provided by Forel, Richter and Lang. And the universal occurrence of the phenomenon, its global importance and simultaneous course could by strictly meteorological standards not be verified to this date, before a large number of meteorological stations had recorded the dry period of the 60s and the wet period around 1880.

These general climatic variations are the key to the prevalent great confusion about the issue of climate changes, which we attempted to describe at the beginning: they are the explanation for the fact that such

[78] According to a handwritten curve by G. K. Gilbert which I gratefully received from Dr. R. Sieger.

[79] Hann in Zeitschrift der Österreichischen Gesellschaft für Meteorologie Vol. II, 1867.

[80] Schweinfurth in the preamble to Bädeker's *Egypten* I. 1877, p. 79.

[81] Fritz in *Petermann's Mittheilungen* 1880, p. 245 ff.

contradictory opinions could exist side by side: the climate changes over time first in one direction and, then again, in another—the climate fluctuates and with it fluctuate rivers, lakes, and glaciers.

Chapter 4

Climate Change Since 1700*

4.1 THE CURRENT STATUS OF THE INQUIRY INTO CLIMATE CHANGES**

Definition of the Climate Concept. **I. The climate in the geological past.** Lyell's view. Heer's homogenous climate of the pre-Tertiary age. The cooling of global climate during the Tertiary (Heer and Engler). Neumayr questions the homogenous climate of the Jurassic, Cretaceous, and Carboniferous periods. The Carboniferous Ice Age. Nathorst and Neumayr advocate a shift of the earth axis during the Tertiary period. The Climate of the Diluvial Age. — **II. The Question of climate changes in historical times.** *a) Changes of rainfall.* Statements of general changes of precipitation around the globe. 'Desiccation' Theory by Whitney. Th. Fischer. Critique of research material. Partsch contra Fischer. Purported local variations of rainfall in temperate climate zones and in the tropics: Decrease of water resources from deforestation or increase from reforestation. Water level reductions of running water in cultivated areas: Wex. Comparative observations on rainfall unacceptable for deciding whether forest has any influence on same. The experimental research of Blanford and Gannet. Rainfall reductions unverified by observations. There is no continuous decrease of water resources in cultivated regions. Increase of rainfall due to deforestation in Australia. Alleged influence of cultivation of land on rainfall in the interior regions of North America. — *b) Temperature changes.* General cooling of the climate in the Northern Hemisphere, sometimes suggested—sometimes refuted. Constant temperature in historical times: Ideler, L. Dufour. Changes of wind conditions. Summary: A continuous thread is missing in the maze of hypotheses about climate changes. — **III. Meteorological Cycles.** Hypotheses of recurrent seasonal weather periods within a certain time span. Multi-year periods of cold cinters: Krafft, Renou, Köppen. Influence of sunspots on meteorological

* *Klimaschwankungen seit 1700*, Wien; E.D. Hölzel, 1890.

** [Chapter One: Der gegenwärtige Stand der Frage nach den Klimaänderungen.]

factors: temperature, rainfall, air pressure, etc. Glacial variations in connection with secular variations of the weather: v. Sonklar, Forel, Richter, Lang.

Just as wind and weather change daily, as rain follows sunshine and sunshine follows rain, dry and wet, cold and warm years alternate in the same manner. If for a particular location the annual average temperature is determined on the basis of temperature readings taken at certain times daily and compared to averages of the previous or next years, it is not unusual to find differences of 2° to 3°C or even more. When these temperature variations from one year to the next are depicted graphically, the result will be an irregularly shaped 'zigzag' line. However, its shape is not completely arbitrary but shows distinct deflective clusters around an imaginary centre line. The latter represents the average annual temperature of that particular location, whereas the irregular curve represents the weather of consecutive years. "Climate, as we understand it, is an aggregate of meteorological phenomena defining the average atmospheric condition at any particular location on the globe. What we call 'weather' is only one phase, one single event in a series of climatic phenomena, the entire more or less constant (annual) cycle which determines the climate of a particular location. Climate is the summation of weather events over a longer or shorter period of time as they usually tend to occur at that specific time of the year."[82] Hence, to us weather is linked to unstableness or variability, climate to stableness, changing from location to location only, but not from one time to another. People's awareness of the invariability of the climate is deep-rooted and is expressed in the firm conviction that unusual weather conditions in one season or year will have to be compensated for in the next.

Nonetheless, even climate did not remain the same over long periods of time. Numerous hypotheses and theories were formed about the change of climate in the past and, naturally, occupied the more or less keen interest in more than one quarter, because a clear confirmation of climate changes in the past immediately raised the possibility of future change which in turn could not be without decisive effects on the economic situation of future generations. It is this practical aspect, to be sure, that may explain why there are so many different hypotheses that hardly an incident of climate change imaginable remains which has not found an advocate. In this confusion of contradicting opinions, often with little foundation, it seems no wonder if presently it almost violates conventional rules among meteorologists to even deal with this question of climate change, let alone to add a new hypothesis to the existing ones.

It cannot be our intention to compile all these different opinions on climate change; such an even remotely complete collection would easily fill

[82] Hann: *Handbuch der Klimatologie* [*Handbook of Climatology*], Stuttgart, 1883, p. 1.

many volumes. But it may be acceptable to roughly sketch the difference of opinion prevailing to this day by categorising these legions of hypotheses and theories according to the most common viewpoints.

4.1.1 Climate of the Geological Past

Our knowledge that the climate of the geological past was different from our climate today is as old as the science of geology. However, there are very few areas where speculation ran as far ahead of establishing facts as concerns this issue. Even before a mere fraction of data on this pertinent subject became known, many were prepared to offer theories on climate variations in geological prehistoric times, which were believed to have been recognised as such. As a result, from the beginning of this century to the present day one hypothesis followed the next, and while most were of no value, others prepared the future for a more realistic basis of theories and speculative discussions of different explanations for climate changes. Of all these, we can discard the purely speculative theories since not one of them comes even close to satisfying scientific research standards. We will revert to a brief description of those opinions, which were uttered about the climate of different geological formations based on facts and which remain valid even to this day.

It had to be discovered quite early that the prehistoric environment of a particular place, remnants of which were found in the form of fossils in geological formations, was exposed to a different climate than it is now. The idea of a changing climate seemed therefore plausible. Two different routes could be taken to find out more about the prehistoric climate: Either one compared organic forms of different geological formations with those of the present and tried to identify climate changes over time by looking at the characteristic transformations of those forms; or one focused on a comparison of different flora and fauna from one and the same prehistoric period and strived to establish the presence or absence of climatic patterns for that time period, as they can be observed today from pole to equator. Both methods have been used.

The first approach is the older of the two and has been chosen quite frequently since the beginning of this century. It first led to the theory of a gradual progressive cooling of the global climate. Lyell, however, came to a slightly different conclusion.[83] Although he, too, was led to believe that past climates were generally warmer; but this warmer climate had temporary periods of lower temperatures not only during the Diluvium but also in the

[83] Lyell: *Principles of Geology*, 10th ed., Chap. X–XIII.

Miocene, in the Eocene and in the Permian. And just as his predecessors, he hardly touched the issue of climate zones.

Heer's approach was quite different: he carefully compared plant remnants of the same time period originating from different places of the globe. To him, his studies of the fossil flora seemed to reveal the fact that over long periods of time the global climate had been homogeneous and tropical, which is inconsistent with the solar climate of today and its climate zones between equator and pole. Modern arctic research and the journeys of Nordenskjöld in particular, who travelled to the polar regions with the intention to examine prehistoric climate conditions, confirmed the assumption that findings from the hot and temperate zones could be applied to the far North as well.

This homogeneity of the climate could, of course, only be claimed for some eras of our geological past, mainly for the Carboniferous Period. Nares' polar expedition[84] discovered the same abundant flora in the lower carbon strata below the 74° and 76° latitude, which grew in temperate climate zones in the Old and New World at that period, while they discovered tropical corals and cephalopods in the dolomite at the northern coast of Grinnell Island (83° latitude).[85] Oswald Heer describes carbonised plant species and types discovered on Bear Island and Spitzbergen which are identical with those found in Europe.[86] Some parts of the Mesozoic era seemed to show the same results. Nathorst, for example, introduced us to the same Jurassic flora in Spitzbergen and in India,[87] and Nathorst and Heer[88] showed us how deciduous trees appeared at the same time in Northern Greenland, Europe and North America during the Jurassic Period.

According to Heer and Engler,[89] not until the Tertiary Period is the homogenous global climate interrupted and do today's climate zones evolve starting from the pole. Both, flora and fauna clearly changed with more advancing cooler temperatures. Heer's standard groundbreaking work has shown that in Switzerland plant life grew in the same consecutive

[84] G.S. Nares: *A Narrative of a Voyage to the Polar Sea during 1875–76 in H. M. Ships Alert and Discovery*, London, 1878. Volume II, p. 331 f. cit. by Whitney.

[85] O. Heer: *Flora Fossilis Arctica*. Vol V. Part 1, p. 17.

[86] Heer: op. cit. above, Volume II Chapter 1, Volume III. Chapter 1 and Vol IV. Chapter 1.

[87] Nathorst: *Polarforskningens bidrag till forntidens växtgeografi* [*The contribution of polar research to prehistoric plant geography*] in A. E. Nordenskjöld: *Studier och forskningar föranledda af mina resor I höga norden* [*Studies and research related to my travel in hte nigh north*]. Stockholm 1883.

[88] Heer: *Flora Fossilis Arctica*, Volume I, p. 60 and in other places of this huge publication. In Volume I, p. 53 op. cit., and Volume VII, p. 226, a short summary of all of Heer's climatological findings.

[89] Engler: *Versuch einer Entwicklungsgeschitchte der Pflanzenweit* [*A Tentative History of Plant Life*], 2 Volumes, Leipzig 1879 and 1882.

pattern and hence fossilised in the same vertical layers during the tertiary period as those that are imbedded in layers anywhere between pole and equator.[90] The tropical Eocene flora is followed by the subtropical Miocene flora which in turn, by the end of the Tertiary Period, changed to a *flora borealis* similar to what we have today. Lesquereux confirmed these results by examining the fossil flora of North America's Rocky Mountains.[91]

The advancement of cooler climates from polar regions to lower latitudes, as indicated by these successive layers of fossilised flora, was fully confirmed by the discovery of the remains of tertiary plant life within the polar circle which Nordenskjöld in particular used for his studies. Once again the collected material was put exclusively into Heer's expert hands, who described the findings in an imposing number of volumes, the "Flora fossilis arctica". A flora identical in characteristics to the subtropical flora from the Swiss Miocene was discovered in East and West Greenland, Grinnell-Land, the Lena River delta, and on Spitzbergen.

The question arose whether the arctic tertiary flora belonged indeed to the same time period as the Swiss Miocene flora or was perhaps older and had developed at an age when the Swiss Alps still had a tropical climate and a tropical vegetation. A reliable answer to this question could only be given by studying the Miocene layers in continuity from the South to the North, which cannot be done as the continent is cut apart by the ocean. Heer[92] and Engler[93] maintained that both, the Miocene flora of the Swiss Alps and the closely related plant life of the polar region were of the same age, and assumed, therefore, that the temperature difference between the polar region and central Europe had been relatively small. Other researchers however, such as J. H. Gardner and Saporta, saw the arctic tertiary flora with its subtropical characteristics as an equivalent to the tropical Eocene and Oligocene flora rather than the subtropical Miocene flora of central Europe.[94] In this case, the age difference between the subtropical tertiary flora of the arctic polar region and the closely related subtropical Miocene flora of the Swiss and West German regions would perhaps be comparable in likeness and age to the fossil Miocene flora of central Europe and today's closely related subtropical vegetation of Japan and of the Southern United States; and the temperature difference between central Europe and the polar region would have been greater than Heer had assumed.

[90] Heer: *Flora Fossilis Helvetiae*.

[91] Lesquereux: *A Review of the Fossil Flora of North America*, Bull. of the Geol. and Geog. Survey of the Territories, II. Ser. no. V.

[92] Heer: *Flora Fossilis Arctica*, Volume I p. 73, Volume VII, p. 22.

[93] Engler, op. cit. see also Volume I, p. 2.

[94] Penck in: *Verhandlungen des V. Deutschen Geographentages*, Berlin, 1885, p. 33; Neumayr: *Erdgeschichte* [*Earth's History*], Volume II, Leipzig, 1887, p. 510.

In either constellation, the arctic Tertiary can be separated into two climate zones, the outer one of which, reaching up to the 75° Northern latitude, contains the Tertiary flora, which shares a number of subtropical species with the Swiss Miocene, while the inner region below the 80° latitude exhibits vivid similarities to today's boreal flora.[95]

According to Heer and Engler we can detect the creation of our present day solar climate developing in the course of the Tertiary. Cooling can first be perceived at the pole, which as early as the Oligocene Age may have taken on such dimensions that the tropical vegetation was forced back to temperate zones and a subtropical flora had spread in the southern parts of the polar region, while the cooler climate of the region north of the 75_ latitude only permitted the growth of a boreal flora. This process of cooling continued during the Miocene and Pliocene periods moving the vegetation further south in ever widening circles with the regions of tropical vegetation becoming smaller and smaller. By the end of the Tertiary Period, in the Pliocene, climate conditions as we know them today had more or less evolved and a flora very similar to today's grew in central Europe.[96]

It has been attempted to determine the extent of the cooling of the climate; it was found to differ greatly from one latitude to another. While today's tropical regions maintained a tropical vegetation from early on and no noticeable cooling process took place, the southern part of Central Europe with its tropical Eocene climate being replaced by the subtropical Miocene climate and finally in the Pliocene Period by the boreal climate was found to have suffered a temperature loss of 14°–15°C; at the polar region, where by the end of the Cretaceous Period plants were still growing whose descendants are nowadays restricted to the tropics, and where by the end of the Tertiary the snow and ice cover expanded with some scanty arctic vegetation still present, temperatures sank by almost 30°C.[97]

Heer's view, as we have just recounted above, can be summarised as follows: It is a homogenous climate up to the Carboniferous Age and possibly in some epochs of the Mesozoic era as well—a cooling process in the Tertiary Period started in the polar region and progressed from there.[98]

The cause of this peculiar cooling of the climate during the Tertiary Period is still unknown. In any case, the old hypothesis that the cooling of

[95] Heer: *Flora Fossilis Arctica*, Volume I, p. 73, Volume VII, p. 22.

[96] Geyler und Kinkelin: *Oberpliocänflora aus den Baugruben des Klärbeckens bei Niederrad und der Schleuse bei Höchst a. M.*, Separate print of the Abhandlungen der Senckenbergischen naturf. Gessellschaft Frankfurt a. M., 1887, p. 43 ff.

[97] Also compare Penck: *Deutsches Reich* [*German Empire*], Wien, Prag, Leipzig, 1887, p. 107.

[98] Penck gives an excellent description of this idea: *Die erdgeschichtliche Bedeutung der Südpolarforschung* [*The Geological Significance of the South Polar Research*], Verhandlungen des V. deutschen Geographentages, Berlin, 1885, p. 25, cont'd.

the earth's inner core had something to do with it, has no merit anymore today. All serious doubts about the physical methods applied set aside, we learn from Heer's history of the earth as he deciphered it from the fossil remains, that a gradual cooling of the climate occurred during the Tertiary Period and only then, neither before nor after that period. It is highly unlikely that the cooling process of the terrestrial globe due to a temperature loss to the universe should have been slow and hardly noticeable during the long geological periods prior to the Tertiary—only to accelerate enormously during the relatively short period of the Tertiary.

Although the evidentiary material submitted by said researcher is substantial and his conclusions may seem logical, doubts have been raised most recently against his views, and, above all, the theory of a homogenous climate during the earliest prehistoric periods is nowadays again seriously being questioned.

There were three categories of arguments favouring a homogenous, predominantly warm climate on earth during earlier periods. One category concerns arguments based on the abundant prehistoric vegetation which had created the huge coal deposits; the second refers to the fact that the organisms of pre-tertiary periods show a closer analogy and similarity with today's life forms in the tropics rather than with organisms of colder regions; the third conclusion assumes that the prehistoric fauna and flora from very different geographical regions are identical.

As Melchior Neumayr points out, these arguments are partly inconclusive, incorrect, and erroneous.[99]

The first argument is completely wrong, as, by the way, has been known for a long time. Today it is in the cold regions with rich vegetation that deposits of carbon substances are formed while in warm regions plant material is destroyed quickly in the process of decomposition.

The second type of argument is disputed on the ground that the adaptability of organisms to different climate conditions has been underestimated; as Neumayr attempts to show, this change could have happened over time when competing species evolved with better protection against the harsh climate than the older species, which were then forced to retreat to tropical regions. Even today wild growing plants of the tropics can be cultivated in severer climates as long as they are protected in their fight for

[99] M. Neumayr: *Über klimatische Zonen während der Jura- und Kreidezeit* [*About the climatic zones during the Jurassic and Cretaceous*], Denkschrift Wiener Akademie Mathematisch-naturwissenschaftliche Classe, Volume XLVII. Also: *Die klimatischen Verhältnisse der Vorzeit* [*Climatic conditions of the prehistoric age*], Schriften des Verein zur Verbreitung naturwissenschaftlicher Kenntnisse in Wien 1889. Several comments in his *Erdgeschichte* [*Earth's History*] Volume II (Leipzig 1887).

survival from better equipped competitors, that is if they are grown in fields and beds free of "weeds".

On closer look, the often-emphasised identity between fauna and flora from different latitudes is not entirely correct either. In contrast, Neumayr showed that during the Jurassic and Cretaceous Periods the spreading of certain sea organisms occurred in a zonal structure, which leads him to conclude that climate zones were present and corresponding in their distance from the poles to today's climate zones.

According to Neumayr, the flora of the Carboniferous P eriod too is not at all absolutely homogenous. On the contrary, some findings can be linked to different climates, such as the rare occurrence or total lack of *Sigillaria* [a lycopodium plant indigenous to the carbon age] in carbon layers from northern latitudes and the lack of a characteristic carboniferous flora in the tropics. No doubt the polar regions' annual temperature was higher then and, above all, winters were milder than today.

The theory of a homogenous climate in the Carboniferous Period may well suffer the strongest blow, however, if identifying certain carbon deposits as glacial formations should prove correct.[100] For, in the southern part of Africa, in Australia, and in India[101] peculiar conglomerates or composites were observed in the top layers of carbon, which, embedded in slate clay and fine meshed sandstone, contain scratched blocks of glacial appearance; in India and in South Africa[102] the layers of the deposits even show [glacial] scratches in certain regions. Therefore everything seems to join in the effort to insure the presence of extensive glaciers by the end of the Carboniferous Period. Although in light of the many deceptions, which even skilled geologists have fallen victim to by mistaking pseudo-glacial formations for real ones,[103] the issue of a carboniferous ice age remains unsolved at present. The seemingly simultaneous appearance of glacial formations found in three different locations of some distance apart and in layers supposedly of the same age might be called striking, if the

[100] A detailed explanation to the question is given by W. Waagen: *Die carbone Eiszeit* [*The carbon ice age*] (Annual Report of the k.k. geolog. Reichsanstalt, 1887, pp. 143–192) and by Feistmantel: *Über die Pflanzen und Kohlen führenden Schichten in Indien (bezw. Asien), Afrika und Australien und die darin vorkommenden glacialen Erscheinungen* [*About the Flora- and Carbon-Containing Layers in India, (resp. Asia), Africa and Australia and their Glacial Indications*], Sitzungsbericht der königlichen böhmischen Gesellschaft der Wissenschaften, Praha 1887, pp. 1–109.

[101] Waagen, op. cit., p. 147.

[102] A. Schenck: *Die geologische Entwicklung Südafrikas* [*The geological development of South Africa*]. Petermann's Mittheilungen, 1888, p. 229 f.; also: *Über Galcialerscheinungen in Südafrika* [*About Glacial Indications in South Africa*], Verhandlungen des Deutschen Geographentages, Berlin, 1889.

[103] Compare Penck: *Pseudo-glaciale Erscheinungen* [*Pseudo-Glacial Indications*], Ausland, 1884, p. 641 ff.

conclusion that they are all of the same age had not been based in part on those very same alleged glacial formations. As the case may be, the theory of a homogenous climate during the Carboniferous Period and the pre-tertiary periods definitely stands on very shaky ground; according to Neumayr,[104] "almost everything points against the assumption that the global temperature was uniformly hot from the equator to the poles at one time."

The latest research by Nathorst and Neumayr also sheds new light on the cooling process of the global climate during the Tertiary Period which according to Heer and Engler, was to have been a continuous, homogeneous process originating at the poles; it seems, the course of events was not as simple as assumed.

Neumayr has disputed the continuity of the cooling process. He tries to show the likelihood that Europe's climate was slightly cooler during the lower Eocene than during the middle and upper periods, with the highest temperature in the latter.[105] The beginning of a progressive cooling trend, which lasted up to the Diluvial Ages, can be dated from that time on. Nathorst suggests that in Spitzbergen for example the climate was severer during the upper Jurassic period than during the entire Tertiary Period and that in Japan no cooling has been observed since the Miocene.[106] Also on both the Sachalin and Camtschatka peninsulas the tertiary (Miocene) flora indicates a climate condition only marginally warmer than today's. Consequently, the cooling during the Tertiary Period was not a global event, but manifests itself in those parts of Europe, which were first explored for their fossil flora, and is strongest in Greenland where together with Grinnell-Land the temperature dropped by almost 30_ C, as mentioned above. Nathorst points out that especially Japan, a region with no cooling, and Greenland, a region with a strong cooling, are both located along the same meridian. The geographical disposition of the Miocene flora from different climates could be explained by assuming that the pole's location was below 70° northern latitude and 120° eastern longitude of Greenwich, i.e., was shifted by 20° toward Japan, while its present location is not in line with those tropical flora zones. In this scenario the flora of Camtschatka, of the Amur region, and the Sachalin peninsula would be located within the polar circle, those of Spitzbergen and Grinnell-Land would even be north of the 60° latitude. The subtropical flora of Japan, Alaska, the Mackenzie River, of Greenland and Iceland would be located

[104] Neumayr: *Die klimatischen Verhältnisse der Vorzeit* [*Climate Conditions of the Prehistoric Age*], op. cit., p. 27 of the separate print.

[105] Neumayr, op. cit., p. 30.

[106] Nathorst: *Zur fossilen Flora Japans* [*The fossil flora of Japan*]. Paläontologische Abhandlungen, publ. by Dames and Kayser, Volume IV., no. 3, 1888, p. 53, 51 ff.

between the 50° and 60° northern latitude, those of Switzerland, even more tropical in nature, below the 36° northern latitude. Nathorst states that, based on what we know today, we are led to assume that a shift in the poles' location during the Tertiary Age seems most likely. Today, he is not the only one to foster this view.

Independently from Nathorst, Neumayr came to similar results[107] and provides paleontological facts in support of this theory not only for the northern but also for the southern hemisphere.[108] Still, this does not solve all problems, for in whatever position we put the pole within the circle of tertiary fossil flora, the findings of tertiary forest trees will always be closer to the pole than the northern tree line of today. Neumayr concludes that the climate of the Tertiary was on the whole slightly warmer than today's but to a much lesser degree than those countries benefiting from a shift of the poles, such as Grinell-Land, Greenland, Spitzbergen, West- and Central Europe, might suggest.[109]

If we look back, we must admit freely that we know very little or almost nothing about the climate conditions of the pre-diluvial geological past. Only a few years ago the theory of a gradual cooling of the global climate was the predominant one and it still has its followers; Heer's work on the homogeneity of the climate and a gradual cooling during the Tertiary Period was done in the 1870s and the early 1880s; and the most recent work is that of Nathorst and Neumayr, who reject both those ideas. Even today both views have supporters. Consequently v. Czerny fails to recognise facts when he writes in 1881 that it is necessary now to explain the causes for those changes which the climate has undergone since prehistoric times, because these changes have long been a fact established and proven by geologists.[110]

Our knowledge about the climate during the diluvial age when the glacial cover increased tremendously, is somewhat better; we will devote a full chapter to this subject; therefore we can be brief at this point and anticipate our findings recounted in that chapter.

It was once believed the remarkable expansion of the glaciers during the diluvial period had local causes. Mountains were alleged to have been much higher then and this played a major role. Charpentier[111] mentioned

[107] Neumayr: *Erdgeschichte* [*Earth's History*] *II*, p. 511–514.

[108] Nathorst: op. cit., p. 55. Recently Schiaparelli did not fully exclude the possibility of a shift of the earth axis, a possibility denied by many astrologers. Compare with Neumayr: *Erdgeschichte* [*Earth's History*] *II*, p. 513.

[109] Neumayr: *Klimatische Verhältnisse der Vorzeit* [*Climate Conditions of Prehistoric Times*], p. 38.

[110] v. Czerny: *Die Veränderlichkeit des Klimas und ihre Ursachen* [*The Change of climate and its Cause*], Wien, Pest, Leipzig, 1881, p. 76.

[111] Charpentier in the *Annales des mines*, 1835.

this idea at the beginning of his glacial studies and many others followed his view. Nowadays however, after glacial research has discovered that the ice age was a general phenomenon proportional in intensity to the size of today's glaciers, this view must be considered refuted. There can be no doubt that, during the ice age, global climate conditions were somewhat different from today's.

One is generally inclined to assume that this difference refers mainly to the amount of precipitation that was believed to have been much higher during the ice age. That however is not correct; in our opinion the old view from the year 1841 held by Agassiz and Charpentier, in which glaciers are looked upon as thermometers, is not without warranty. We will show later that even during the ice age the temperature must have been lower although the difference to today's temperature was probably less than 5°C globally. Thus the ice age with its cool and wet climate occurs between the end of the Tertiary Period whose climate conditions closely resembled today's, and the next period. However there was not just one ice age but at least two, and both were separated by a fairly dry inter-glacial age. Recurrent climate variations are typical for the Diluvial Age and in strange contrast to the prolonged gradual cooling of the Tertiary Period.

To this time we do not know what caused these climate variations during the Diluvial Age nor do we know what caused the cooling process during the Tertiary Period. So far, no convincing theory has been presented which would be in complete accordance with facts. We will not revert to speculations at this point as Croll,[112] Blytt,[113] also Schmick[114] have done, though we do acknowledge these as initial efforts.

The climate of the ice age is different from today's: climate has changed since the ice age. The change occurred when man was already present on earth,[115] and the question comes to mind immediately, whether man has passed on reports of at least some of these events. It is the issue of climate change in historical times that we are approaching.

[112] Croll: *Climate and Time* (1875), as well as a number of recent publications.

[113] Blytt in *Biologisches Centralblatt,* Volume IV, p. 33 ff.

[114] Schmick's numerous essays were listed in the Geographisches Jahrbuch, Volume V, Gotha 1874, p. 236.

[115] Penck: *Mensch und Eiszeit* [*Man and the Ice Age*], Archive für Anthropologie, Volume XV, 1884, Issue 3.

4.1.2 Views and Opinions about Climate Change in Historical Times

Many different hypotheses and theories about climate change have been established in historical times, and have been defended more or less successfully against never ending attacks. At one time general causes are invoked, resulting in general climate changes without human interference; at other times it is man's interference with nature and modifications to the flora in particular, which are held responsible. In most cases it is a change in the most important meteorological elements, such as temperature or rainfall, that is claimed to have been observed. All other climate factors play only a minor role in the literature on climate change. If occasionally the attempt is made to emphasise, for instance, wind direction or wind velocity, it is almost always done with the intention to gain better insight into the change of the temperature or the rainfall.

Man's survival depends on rain, i.e., water, almost even more so than on warm temperature conditions. Man can, to a certain degree, protect himself against the cold by wearing appropriate clothing, but succumbs instantly to arid conditions. Consequently, changes in rainfall have always been a much-discussed issue in historical times.

Naturally, we are here mainly interested in those approaches trying to verify the change in rainfall as it occurred since the end of the ice age.

Among the many works on this subject those of J. D. Whitney deserve undoubtedly first rank.[116] According to his theory the ice age is a local phenomenon, occurring for each single mountain formation, and a necessary aspect of the general cooling process on earth. The latter, by reducing the evaporation at the surface of the oceans more and more in the course of time, is to be followed by a gradual drying out of the land masses. This is supposed to be the gradual drying of the climate which, we are led to conclude, is the cause for the observed gradual dissipation of the diluvial glaciers. To establish this drying process from historical data was one of Whitney's main goals. Of course, he is not the first to defend the idea of a reduction in rainfall on earth caused by forces other than man's tampering. Among Whitney's predecessors Theodor Fischer should be mentioned who argued in several publications for the theory that the climate of the Mediterranean countries had become drier since ancient times.[117] But he

[116] Whitney: *Climate changes of Later Geological Times*, Memoirs of the Museum of Comparative Zoology at Harvard College. Cambridge, 1882.

[117] Theobald Fischer: *Über Klimaänderungen an der Äquatorialgrenze der subtropischen Regenzone* [*About climatic changes at the equatorial border of the rainy zone*], Ausland, 1877, p. 891, published without name; Beiträge zur physischen Geographie der Mittelmeerländer, Leipzig, 1877; *Studien über das Klima der Mittelmeerländer* [*Studies

also, to some extent, blames deforestation, which has repeatedly changed the face of the Mediterranean landscape completely during the course of historical times. He feels, however, that the drying process and the advancement of desert-like conditions towards the Mediterranean Sea is a much too general phenomenon to be explained by such localised human interventions, which moreover take place north of the 34° latitude. Rather, he considers it to be an aspect of an, in general, increasingly drier climate of the subtropical zones bordering the equator.

While Fischer limited his research to the surroundings of the Mediterranean Sea, Whitney has gathered data on climate change in historical times for almost all the countries of the globe. These data are to be found in a special chapter together with the prehistoric geological evidence of the climate's gradual drying trend since the ice age. Relevant observations have been collected for the region of the Aralo–Caspian Basin, for Persia, Central Asia, the Mediterranean Sea, the Sahara, Inner Africa, and South America. He draws the conclusion: the climate has become dryer all over the globe in historical times. This confirms the views expressed for certain areas by many others, apart from Fischer, namely by Humboldt,[118] Schmick,[119] W. T. Blanford,[120] von Richthofen,[121] O. Fraas, Chavanne,[122] lately by Jadrinzew,[123] Venukof,[124] W. Götz,[125] all of them assume a development towards a drier climate unrelated to human interference.

Varied are the facts from which conclusions for this drying process are drawn. Only to a small extent is the evidence geological in nature insofar as it is based on hydrographical changes which in all respects are beyond any human interference, such as reduced water levels of lakes with no outlet; therefore, the much larger diameter of the Lob-nor in the Tarym Basin

---

on the climate of the mediterranean countries], Petermann's Mittheilungen Ergänzungsheft No. 48. Gotha, 1879, p. 41 ff.; *Zur Frage der Klimaänderung im südlichen Mittelmeergebiet* [*About climatic changes in the Mediterranean area*], etc., Petermann's Mittheilungen, 1883. p. 1 ff.

[118] Humboldt for the Aralo-Caspian Basin in: *L'Asie centrale* [*Central Asia*], Paris, 1843, Volume II, p. 142.

[119] Schmick: *Die Aralo-Kaspische Niederung und ihre Behandung* [*The Aralo-Caspian Basin and its Cultivation*], Leipzig, 1874.

[120] Blanford for Persia in *Quarterly Journal of the Geological Society*, Volume XXIX, London 1873, p. 493.

[121] v. Richthofen for the territory of Lob-Nor in: *China*, I, p. 110.

[122] Chavanne: *Die Sahara* [*The Sahara*], Wien, 1879, p. 627.

[123] Jadrinzew about the disappearance of the West Siberian lakes, in Iswestija of the Imperial Russian Geographical Society, XXII, No. 2.

[124] Venukof in *Revue de géographie*, X, Paris 1886, p. 81 ff.

[125] W. Götz: *Die Verkehrswege im Dienste des Welthandels* [*The Traffic Routes Serving World Trade*], Stuttgart, 1888. p. 418, 506, 610, 669.

about 4,000 years ago leads von Richthoven to the conclusion that aridicity increased since that time. The disappearance of the Belt of Aibugir at the Aral Sea as well as the allegedly continuous reduction of the Caspian Sea's water level were once considered to be further examples of this process. To a much larger part, however, the evidence relies on changes to plant life on earth, the disappearance of some oases due to water-shortage for instance, in short on appearances which are not entirely excluded from man's influence and consequently are not complete or absolute proof of the theory. As we will see below, they are to a large extent the same facts that are used to prove climate change as a result of localised deforestation.

In fact, this theory of the climate's gradual drying has not escaped opposition. The legitimacy of the assumption has been contested for almost all of the regions that were claimed to be under the influence of a dryer climate. It was pointed out that the fact that the shore lines of a lake with no outlet are forced back, may well be attributed to a local siltation, and the decline in the agricultural land of the Mediterranean regions to the increasing indolence of their inhabitants. Just as determined as Fischer has been in defending a climate change for the African regions of the Mediterranean, Zittel[126] rejects this postulation, and Partsch sided with him just recently, based on the extremely critical assessment of his research material: The level of the *Schott el Djerid* in Tunisia has not changed since pre-historic times, etc.[127] Tietze[128] voiced a similar opinion.

Very old and wide-spread is the opinion that forests have an important impact on rainfall. And indeed, *a priori*, this seems quite likely. First of all forests are natural barriers to wind-driven air masses, which are then, as when encountering hills and mountains, forced to rise. No matter how light this upward drift may be, in theory it will have to lead to more condensation at its windward side. But the forest's influence is also felt in the fact that the air above stays relatively humid. Forests slow down the swift runoff of the rainwater and store the water in the ground, which is then evaporated back into the air through the treetops. This process again must bring about an increase in precipitation above the forest, the more so as because of the strong friction between wind and forest surface and the resulting delay in air flow the moist air tends to stagnate above the forest. If

[126] Zittel: *Beträge zur Geologie und Paläontologie der Libyschen Wüste* [*Contributions to the Geology and Paleontology of the Lybian Desert*], etc., Paläontographica, Volume XXX, p. 42.

[127] Partsch: *Über den Nachweis einer Klimaänderung der Mittelmeerländer* [*Evidence for a Climate Change in Mediterranean Countries*], Verhandlungen des VIII Deutschen Geographentages [Discussions of the VIII German Geographical Convention]. Berlin, 1889, p. 123.

[128] E. Tietze: *Ueber Steppen und Wüsten* [*About savannahs and deserts*], Schriften des Vereins zur Verbreitung naturwissenschaftlicher Kenntnisse, Wien, 1885, p. 160.

forests enhance the amount and frequency of precipitation simply by being there, deforestation as part of agricultural expansion everywhere, must necessarily result in less rainfall and more frequent droughts. This view is most poignantly expressed by the saying: Man walks the earth and desert follows his steps![129]

Perhaps no other area on earth has been mentioned more often in connection with the effects of deforestation on rainfall than the regions of the Mediterranean Sea, and the increasingly drier climate since prehistoric times that Fischer and Whitney interpreted as a general phenomenon, has been attributed more frequently to man's localised destructive efforts of turning woodland into arable land. And indeed, if we compare the previously prosperous cultural life along the eastern shores of the Mediterranean Sea with the Orientals' bare existence in those areas today, we are struck by the tremendous cultural decline and are only too readily inclined to see its causes in what saps our own energy upon arrival in the orient: the scorching dry heat. To today's cultural leaders who live in the cool, wet northern hemisphere it seems inconceivable that the blossoming antique culture could have prospered under the present climatic conditions of the orient: the climate must have grown warmer and dryer since antiquity. Consequently, the clear-cutting of forests carried out in these regions since ancient times offers a convenient explanation: man has destroyed his own culture by destroying the forests and has devastated the land, on which he is now left with a meagre existence. A comparison between descriptions of the landscape as it used to be and as it is today appears to confirm this situation.[130] Other authors voice their opinion in a similar manner: Herschel, Arago, Kämtz, Lecoq, Clavé, David Milne Home, Mathieu, Wilson Flagg, G. vom Rath, Fautrat, Marsh, Simony,[131] Denza,[132] etc. Also Theobald Fischer believes that the drying of the Mediterranean countries north of the 34th latitude is at least partially due to deforestation. But it was often overlooked that in antique times the

[129] Quoted by Simony: *Schutz dem Walde* [*Protect the Forests!*], Schriften des Vereins zur Verbreitung naturwissenschaftlicher Kenntnisse, Wien. Volume LXX 1876/77, Wien, 1877, p. 425.

[130] A summary of the current literature on the influence of deforestation on the climate was written by D. Milne Home in the Journal of the Scotish Meteorigcal Society, New Series IV, 1870, p. 35 ff.; also Löffelholz-Colberg: *Die Bedeutung und Wichtigkeit des Waldes* [*The Impact and Importance of Forests*]. Leipzig 1872; finally Whitney, op. cit. We only refer to authors not named in any of the mentioned publications.

[131] Simony: *Schutz dem Walde!* [*Protect the Forest!*], Schriften des Vereins zur Verbreitung naturwissenschaftlicher Kentnisse, Wien, 1877, p. 451.

[132] Denza: *La meteorologia e la fisica terrestre al III, Congresso geografico internationale diVenezia* [*Meteorology and geophysics at the III International Congress on Geography*]. Roma, 1882, p. 16 ff. (Cited by Günther.)

descriptions of the subtropical environment were done by its inhabitants while today's scientific efforts take mainly place in the temperate climate zones of Europe. The inhabitant of a southern country had to see the same phenomena with different eyes and painted them in different colours than the inhabitant north of the Alps.

This increasing aridicity still continues today; Trottier[133] for example substantiates this trend from 1838 on and especially since 1855 for Algeria based on his observations of rainfall at Port d'Alger; and like Niel he expects an improvement only through reforestation on a large scale. According to Marmont,[134] whose observations are repeated by N. Gräger[135] without quoting him, in Upper Egypt where rains used to be frequent just 100 years ago, it apparently does not rain anymore since the Arabs have felled the trees in the mountains bordering the Nile valley to the east and the west. In Gräger's and also Anderlind's[136] (1888) opinion, the opposite result was achieved by extensive tree planting in the surroundings of Cairo during this century, a view mentioned much earlier (1835) by Marmont. Here, rainfall supposedly became more frequent whilst it was lacking almost entirely before. Likewise, according to Murphy,[137] the area of the Kidron creek near Jerusalem enjoys heavier rainfalls after a mulberry grove was planted, etc.

It is not surprising that under such circumstances the issue of a link between forests and climate has now and then been addressed by governments. Lately, the Italian government has been paying special attention to reforestation in Italy and its expected improvement of the climate.[138] Father Denza emphasises the goal of such replanting efforts with these few words: It must be prevented that periods of heavy rainfall alternate with droughts.[139]

Indeed, observations about other areas seem to support these conclusions regarding the forest's influence and the expectations tied to reforestation.

It was often believed that the climate in Germany had improved since the early ages, resulting in less cloud cover and precipitation from increasing deforestation. And, in fact, a comparison of the rather gloomy description of 'Germania', as for instance given by Tacitus, with the

[133] Trottier cited by O. Niel: *Géographie de l'Algérie* [*Geography of Algeria*] 23 éd. T. I., 1876, p. 178.

[134] Marmont referred to by Berghaus: *Allgemeine Länder- und Völkerkunde* [*General Geo- and Ethnology*], Volume II, Stuttgart 1837, p. 309.

[135] N. Gräger: *Sonnenschein und Regen* [*Sunshine and Rain*]. Weimar, 1870, p. 153.

[136] Anderlind in *Meteorologische Zeitschrift*, 1888, p. 154.

[137] J. J. Murphy: *Are we drying up?*, Nature X, 1876, p. 6.

[138] David Milne Home, op. cit.; Günther: *Geophysik*, Volume II, p. 290.

[139] Denza, op. cit., p. 16 ff. Cited by Günther.

Germany of today seemed to point to a climate change; it was not taken into consideration though that this Roman's portrayal, naturally, had to be subjectively distorted. But even in recent times multiple efforts have been made to prove a link between changes in the climate and deforestation for parts of Central Europe. Van Bebber expresses views along this line in his publication on rainfall in Germany,[140] and so does Studni_ka for Bohemia.[141] According to Wessely, in Hungary the climate of the steppe has been gradually advancing since the lifetime of Maria Theresia. In his opinion, resolute reforestation measures alone promise help in preventing the impending drought.[142]

In 1836 Rivière[143] advocated the theory of deforestation for parts of southern France, namely the Vendée, the Provence, and particularly the Département du Var, at the Academy in Paris; frost damage followed by the clearing of olive tree plantations has presumably caused a considerable reduction in rainfall and dried-up springs in the years from 1821 to 1822. A similar situation exists in the former Poitou and the Department of the lower Charente according to Fleuriau de Bellevue.[144] Actually, the question of climate change due to destruction of forests has been raised in France many times, i.e., in 1858 by Ladoucette, who pointed out before the French Chamber of Deputies that the climate of the Départements Pyrénées Orientales and the Hérault had turned dryer and warmer after the destruction of forests.[145] Because of these reports the French legislature took a serious look at the subject of reforestation.[146]

As early as the 1820s Kasthofer attributed the increasingly drought-like climate conditions to the continued clear-cutting of forests in the alpine areas of Switzerland and recommended reforestation.[147]

In the Unites States deforestation plays an important role as well and is seen as the cause for a reduction in rainfall, which is believed to have been observed in the New England States and also in the Pacific States;[148] F. B.

140 van Bebber: *Regenverhältnisse Deutschlands* [*Rainfall in Germany*], München, 1877, p. 119.

141 Studni_ka: *Hyetographie von Böhmen* [*Hyetography of Behomeia*], Archiv für Landesdurchforschung von Böhmen. Volume VI. No. 3. Prag, 1887.

142 Wessely in Simony: *Schutz dem Walde!* [*Protect the Forest!*], op. cit., p. 497.

143 Rivière: *Effets des défrichements*, Comptes Rendus II, 1836, 2, p. 358.

144 Fleuriau de Bellevue cited by Berghaus: *Länder und Völkerkunde* [*Geo- and Ethnology*], Stuttgart, 1837, Volume II, p. 30.

145 Hough: *Report on Forestry*, Washington, 1878, p. 395.

146 Marsh: *The Earth as modified by human action*, New York, 1877, p. 395.

147 Kasthofer, *Bemerkungen auf einer Alpenreise* [*Notes from a Journey through the Alps*]. Aarau, 1822. Annex: *Klima des Alpengebirges* [*Climate of the Alps*], p. 329.

148 Compare with Whitney's compilation: *Climate Changes in later Geological Time*, Cambridge, 1882, p. 162 f.

Hough[149] in his capacity as committee chairman of the American Association for Advancement of Science demands decisive steps to extend woodland in order to counteract the increasing drought.

Major droughts followed by famine characterised the sixties of this century in Siberia. In general, one did not hesitate to link the drier climate directly to the increased practice of deforestation. Van den Brinken even blames the drought in the southern Russian steppe on the destruction of woodland by nomads! These may be enough examples for the situation in the temperate northern hemisphere.

Advancing deforestation was also thought to be responsible for the purported dryer climate in some areas of the tropics.

Blanqui[150] considers the dry climate of the Isles of Cap Verdes to be the result of deforestation, while claiming that St. Helena's rainfall increased since the time of Napoleon I due to minor additions to the woodland. Likewise, rainfalls reportedly have occurred more frequently on Ascension after the British implemented reforestation on parts of the island.[151] The island of Madeira lost all its forests through fire in the early 15th century and as early as 1450 a reduction in rainfall was claimed to be noticeable.[152]

Meldrum mentions in the late sixties that the inhabitants of the wetlands on Mauritius cleared all forests with the purpose to gain dryer agricultural land. However, clear-cutting went too far and resulted in terrible droughts in the early sixties. Meldrum recommends reforestation to avoid their reoccurrence.[153]

In 1846 Gibson reported to the government[154] that according to the native population India's climate in the areas surrounding Bombay and the Nilgiri Hills has become drier since deforestation practices had increased. According to Bidin the deforestation of the Coorg landscape in the western parts of Ghats had a similar effect.[155] Just recently Blanford[156] established the reverse effect for a region in central India's southern provinces: an increase in rainfall coincides with an increase in woodland area

Reporting about his second expedition to New Mexico,Wheeler writes that from year to year the climate becomes more desert-like because of

[149] Hough, op. cit.

[150] Blanqui is mentioned by Marsh (op. cit.) on page 184.

[151] J. J. Murphy in Nature, Volume XV, p. 6.

[152] Peschel: *Neue Probleme der vergleichenden Erdkunde* [*New Issues in Comparative Geography*], Leipzig, 1870, p. 163.

[153] Meldrum in Quarterly Journal of the Royal Meteorological Society, Volume IV, London, 1868, p. 187.

[154] D. Milne Home, op. cit.

[155] Bidin reports in Geographisches Jahrbuch, IV, p. 30.

[156] Blandford: *The Rainfall of India*. Part II, p. 135 ff. and in Meteorologische Zeitschrift, 1888, p. 35.

deforestation. In St. Cruz in the West Indies precipitation decreased to such a degree due to deforestation that the island's wastelands and depopulation figures are taking on sad proportions.[157] Sachs reports of similar situations in parts of Venezuela's coastal region[158] and Hartt[159] does the same for Brazil, where deforestation has already had an obvious impact on the climate of Bahia.

Of the many examples of the situation in the tropics one may be emphasised in particular since it seems to give compelling evidence of the influence of forested land on rainfall and has been used as such many times. We are talking about the observations made in the surroundings of Lake Tacarigua or Valencia in Venezuela. Alexander von Humboldt visited its shores in 1800 and verified that the lake's dimensions had decreased since the founding of the city of Valencia, and in particular over the last 30 years of the previous century. Humboldt did not hesitate to attribute the lake's descending water levels to the vast deforestation during the second half of the 18th century. Twenty-two years later Boussingault visited the lake and learnt that water levels had been rising again considerably for a number of years; islands that had emerged in 1796 had disappeared again, and large areas of acreage that used to be completely dry were in danger of being flooded. Once again the forests are the answer. During the early years of this century, bloody battles took place in Venezuela, particularly within the immediate vicinity of the lake. Industry and agriculture were deteriorating considerably and the lush growth of the tropical virgin forest reclaimed the territory that man had taken away. As soon as the forest area grew, the rainfall increased again and Lake Tacarigua began to rise. A most compelling proof, indeed, of the forest's impact, one could not wish for a more splendid one![160]

Analogous observations are available for the temperate climate of the southern hemisphere. Milne Home[161] observed an area in Australia which only twenty years ago had a precipitation of 37 inches; this amount is now—Homes wrote in 1870—reduced to only 17 inches due to deforestation. As a result, one of the Australian colonies took the initiative and founded a special department for the preservation and planting of forests. In fact, large-scale reforestation has been practised in certain areas and Landsborough points out that Australia's climate is beginning to show signs of an increase in precipitation and soon the continent will not only be

[157] Simony, op. cit., p. 465 f.

[158] Sachs is mentioned by Fritz in Petermann's Mittheilungen, 1880, p. 251 f.

[159] Hartt: *Geology and Physical Geography of Brasil*, Boston, 1870, p. 321 (quoted by Whitney).

[160] Boussingault: *Influence des défrichements sur la diminution des cours d'eau*, Annales de Chimie et de Physique. T LXIV, 1837, pp. 118–122.

[161] D. Milne Home, op. cit.

suitable for cattle farming but also for agriculture.[162] The climate in Tasmania, on the other hand, as reported by Strzelecki, has become much dryer thanks to the destruction of woodland.[163] According to S. Fritsch, Wilson and Livingstone South Africa's Cape Colony is experiencing drier climate conditions. Fritsch, who travelled in that area in the years 1863–66, found traces of aridicity everywhere;[164] he sees its cause in increased deforestation. Böhm and Bernsmann made the same observations in the Herero Plains.[165]

In looking at the views recounted above, whether they apply to parts of the old or the new world, to temperate or subtropical zones, and in consideration of the names of the scientists involved, there seems to be no doubt: deforestation has played a major role in reducing rainfall in many regions of the globe while reforestation has enhanced it.

This statement was substantially supported by research directed to verify a decrease of the surface water in all countries with large agricultures. The fact that spring waters are reduced or even completely destroyed by deforestation has been known for some time. Becquerel emphasised it,[166] as did Boussingault[167] before him, the Swiss scientist Marchand,[168] and many others who have already been mentioned as supporters of this theory about a link between deforestation and rainfall.

However, twice in this century the problem turned into a pressing topic of the day: in the thirties and the early seventies. The former decade produced an ample amount of literature about reduced water levels in the rivers and streams of central Europe in the wake of the epochal research on German rivers by Heinrich Berghaus,[169] and particularly on the Rhein by P. Merian.[170] From long-term observations over many, sometimes even 100 years, Berghaus determined that, overall, water levels of the rivers Elbe and Oder had been falling continuously. He thought this reduction had to be attributed to lower water volumes and related those to cultivation and drainage of wetlands which used up more water due to evaporation. He even expresses apprehensions that, if the trend continues, in 24 years, i.e.,

[162] Landsborough: as reported in Nature, Volume XVI, 1877, p. 217.

[163] Strzelecki: *Van Diemens-Land*, p. 192 f. (quoted by Milne Home).

[164] More about Fritsch, Wilson and Livingstone in the essay of von Hann published in the Zeitschrift für Meteorologie, 1869, p. 18 ff.

[165] Böhm and Bernsmann in Petermann's Mittheilungen., 1878, p. 307 f.

[166] Becquerel in *Atlas méteorologique de l'Observatoire de Paris*, 1867.

[167] Boussingault: *Rural Economy*, p. 686. Cited by Whitney.

[168] Marchand: *Über die Entwaldung der Gebirge* [*Deforestation of Mountains*], Bern, 1849, p. 29 ff.

[169] Berghaus: *Allgemeine Länder- und Völkerkunde* [*General Geo- and Ethnology*], Volume II, Stuttgart 1837, p. 300 ff. and p. 325.

[170] Merian in *Poggendorff's Annalen*, LVII. 1842, p. 314 ff.

around 1860, the water levels of the Elbe River will be too low for standard boat traffic. That was in the year 1837; later he extended his research to the Weser, Weichsel, and Memel and came to the same conclusions.[171] Both times he identified the cause of the water level reductions in this way, although putting some blame on deforestation in accordance with Pfeil's theory[172] and using a more cautious approach than most of his colleagues who attributed the reductions entirely to the decline in rainfall due to deforestation.

Those concerns were not just restricted to central Europe. A similar water shortage occurred in Russia. The headwater regions of the Oka and Don rivers, which were still rich in forests and waters during the first two decades of this century, have become dry and barren according to a report by the author A... written in 1842.[173] This situation is found almost everywhere in Russia. In 1836 the low water levels of the Wolga raised considerable concerns; they were attributed to massive deforestation and threatened shipping operations. That is why Tsar Nicholas established a commission to investigate the correlation between the reduction of woodlands and the dropping water levels of the upper Wolga, which indeed seems to have been confirmed.[174]

In the early seventies G. Wex[175] published his well-known work about the reduced amounts of water in springs, rivers, and streams. His research material was by far more extensive than that of Berghaus but focused again entirely on observations of water levels. Wex concluded that in cultivated areas decreasing water levels result in a continuous decline in precipitation, which in certain cases he even tried to measure. Based on observations of the levels of the Rhein river near Basel during the years 1857–72 he arrives at an annual drop of 1.97 cm and based on this reduced amount of water at

[171] Berghaus in his *Annalen der Erd-, Völker- und Staatenkunde* [*Annals of Geography, Enthnology and Political Sciences*], Series III, Volume IV 1838, p. 95.

[172] Pfeil: *Rührt der niedrige Wasserstand der Flüsse etc. von der Verminderung der Wälder her?* [*Is the low water level of rivers, etc. the result of deforestation?*], Berghaus' Annalen, Series III. Volume IV, 1837, p. 289 ff.

[173] A...: *About the causes of crop failures in Russia and Preventative Measures*, Journal of the Empirial Russian Ministerium for the Empire's Crown-Land and State Forests, 1842, Part IV, p. 135 f., (in Russian). [Only the first character, "A", of the author's name is given in the German original.]

[174] See P. v. Köppen's report to the commission: in v. Baer and v. Helmersen: *Beiträge zur Kenntnis des russischen Reiches* [*Essays about the Russian Empire*], Volume 4, p. III f., with a preamble by the editors which along with v. Köppen's report vehemently opposes the commission's findings.

[175] Wex: *Über die Wasserabnahme in den Quellen, Flüssen und Strömen* [*Water Level Reductions in Springs, Rivers and Streams*], Zeitschrift des österreichischen Ingenieur- und Architektenvereins, 1874. Also Wex: *II Abhandlung über die Wasserabnahme*, etc. [*II Essay about water reduction*, etc.], ibid., 1879.

an annual decline in rainfall by 6.95 mm in the upper Rhein's catchment area upstream from Basel.[176] His findings led Wex to the general conclusion: a continuing decrease of the water levels of springs, rivers and streams takes place in cultivated areas caused mainly by the increased practice of deforestation and its resulting decrease in rainfall. M. W. Schmidt came to the same results for the river Elbe.[177]

This proof had to raise some serious concerns. In 1873, in Wien, the congress for agriculture and forestry discussed the problem in detail;[178] and when the Prussian house of representatives ordered a special commission to examine a proposed law pertaining to the preservation and implementation of forests for safeguarding, it pointed out that the steady decrease in the water levels of Prussian rivers was one of the most serious consequences of deforestation only to be rectified by reforestation programs.[179] It is worth mentioning that at the same time or only a few years earlier the same concerns were raised in Russia as well and governmental circles reconsidered the issue of deforestation.[180]

As often as the link between deforestation and rainfall is attested to, as numerous are the voices opposing it, and it is difficult to decide who is right.

It has only been a relatively short time ago that the section of meteorology pertaining to forestry received a boost, in particular from Ebermayer's work.[181] Not surprisingly, only a few reliable observations are available regarding the influence of woodland on the climate in general and on rainfall in particular. Those that are available seem hardly suitable to support the theory of a decline in rainfall due to deforestation. Based on his observations Ebermayer himself acknowledges that in plains or flatlands with identical geographical characteristics, woodlands have very little influence on the amount of rain; it increases however with higher elevations above sea level.

Not all scientists were as cautious and prudent in drawing their conclusions as Ebermayer was, and some achieved different results based, in part, on verifiable fallacies. From their observations of rainfall over forests and open country Fautrat and Sartiaux, for instance, attempted to

[176] Wex: *II Abhandlung,* etc. [*II Essay*, etc.].

[177] M. W. Schmidt: *Wasserstandsbeobachtungen an der Elbe im Königreich* [*Water level observations of the river Elbe in the Kingdom of Sachsen*], Civilingenieur, Leipzig, 1878, p. 559.

[178] Hough, op. cit., p. 292.

[179] Schlichtung, in *Deutsche Bauzeitung*, 1875, p. 274.

[180] Compare Journal of the Empirial Russian Ministry of the Empire's Crown Land and State Forests. 1863, April, Report on Forestry, p. 167.

[181] Ebermayer: *Die physikalische Einwirkung des Waldes auf Luft und Boden,* etc. [*The Physical Impacts Forests Have on Air and Soil,* etc.], Berlin, 1873, p. 202.

prove an increase in the amount of rain over the forest areas.[182] One reviewer (Hann?) is however correct[183] in pointing out that the two observation stations used in this comparison were operating under entirely different conditions. The same can be said of analogous results by Mathieu who over a period of six years carried out comparative observations in the middle of a forest and in an area free of trees near Nancy.[184] His measuring showed that the amount of rain that fell on the treed area was 6% higher than the one that fell on the area without trees. However this result seems quite shaky if one considers that the two stations, though being set up at the same sea level, were positioned differently and 17 km apart. In general, all attempts to solve the question under discussion through comparative observations in a forested and a non-forested area can be contested on the ground that they are prone to uncontrollable factors that may cause local differences in rainfall, but have nothing to do with forests. This also applies to A. Woeikof's research—who has vigorously supported the forest theory.[185]

How numerous such local factors are, some of which can be attributed to the placement of the measuring devices, and how much influence they have on the difference in rainfall between two locations even within close vicinity in a flat area with no elevations, has been proven through observations initiated by Hellmann[186] and carried out at the test fields for rain-gauging near Berlin. In 1886 and 1887 observations were made at 10 stations placed within a perimeter of just 40 square kilometres on a treeless area with buildings and fields. Nonetheless the two-year averages of precipitation differed by up to 16 percent. This shows that by using the approach of comparative observations taken at the same time at different stations, it is very difficult to prove the link between forest and rainfall.

Recently Blanford took steps to establish the effect of forested areas in an approach we consider the only possible one.[187] We have already briefly referred to his findings. In the southern part of India's central provinces there is a large area which had been massively clear-cut in the past, but where lately trees began to grow again. Deforestation ended and reforestation began in 1875, and today all of this area is woodland again. Blanford measured the average rainfall for the area in question at 14

[182] L. Fautrat in *Comptes Rendus*, Volume 83, Paris, 1876, p. 514.

[183] Zeitschrift für Meteorologie, 1874, p. 384.

[184] Compare A. Woeikoff: *Einfluss der Wälder auf das Klima* [*Influence of Forests on the Climate*], Petermann's Mittheilungen, 1885, p. 81.

[185] Woeikoff: *Die Klimate der Erde* [*Climate on Earth*], Volume I, Jena, 1887, p. 290 ff., also several times in numerous publications of this highly reputable scientist.

[186] Hellmann in Meteorologische Zeitschrift, 1887, p. [62] and in Wetter, 1888, p. 165 ff.

[187] Blanford: *The Rainfall of India*, Part II, p. 135 ff., and in Meteorologische Zeitschrift, 1888, p. 35.

stations from 1866–75, as well as from 1876–85, and determined a substantial increase in precipitation. This increase was not noticed at seven other stations close by yet outside the area of reforestation. He does not hesitate to attribute this increase to the forest's impact, even more so since it turned out to be a continuous process. Therefore, according to Blanford, at least as far as tropical areas are concerned reforestation enhances rainfall and deforestation, consequently, has to decrease it.

Gannet achieved entirely different results for the temperate North American climate by using a similar method intended, again, to determine quantitative changes in rainfall in areas that were undergoing extensive gradual transformations of their vegetation.[188] First there are the prairies, including Iowa, northern Missouri, southern Minnesota, most of Illinois, and a small part of Indiana, an area of 5,000 square kilometres altogether, where reforestation was done on the grandest scale. As it occurred on a continuing basis, each series of observations if divided into two equal time segments, should for one and the same location measure rainfall at a time of little forestation in its first half, and at a larger extent of forestation in its second half. Gannet discovered, however, that rainfall was 4% lower in the second time period despite increasing reforestation. The opposite was happening in Ohio and in the New England states where deforestation continued. Still, Ohio showed a reduction in rainfall by only 0.5%, and the New England states an increase of 7% based on the average measurements from 12 stations done prior to 1860; the averages from 14 stations after 1860 showed no change at all. Gannet concludes from these data that deforestation and reforestation has no significant impact on the amount of precipitation.

On the one hand, based on their respective experimental research Ebermayer and Gannet refuse to accept the theory that deforestation brings about a substantial change in the amount of rain, at least not in temperate climate zones, while Blandford seems to have proof of the opposite for the tropics. On the other hand, the discussion of long-term meteorological observations has by no means resulted in a confirmation of the acclaimed reduction in rainfall either. Schott[189] and Draper,[190] for instance, point out that there is no evidence of such a reduction in the eastern part of the United States over the last 60 years. Jamieson[191] shows in 1859 that rainfall has not changed since the middle of the last century in Great Britain.

[188] Gannet's essay is discussed in detail in Das Wetter, 1888, pp. 97–105.

[189] Schott, in Smithsonian Contributions to Knowledge, Volume XXIV, Washington, 1885, p. 288.

[190] Draper, in Meteorologische Zeitschrift, 1874, p. 239.

[191] Jamieson mentions Wex's research in his report to the Wiener Akademie: *Sitzungsberichte der Wiener Akademie, II Abtheilung* 1874, p. 642. A lecture was also published in Kämtz' *Repertorium für Meteorologie* [*Repertory for Meteorology*].

Burton[192] and Schweinfurth[193] shared his view with regard to Egypt—contrary to Anderlind et al. The commissions of most of the academies in Europe to which Wex had submitted his report about the reduction in water levels and rainfall in cultivated regions for appraisal could not from meteorological observations in their own countries support this view—despite increased deforestation. This was the response of the academies in Paris,[194] Petersburg,[195] Wien,[196] and also of the commission of the Austrian Society of Engineers and Architects.[197] However, very few accurate meteorological observations date back further than the middle of the past century so that they are at best predicative for the duration of one century. Efforts have been made to find out more about the changes in rainfall in earlier time periods by comparing meteorological observations that have been carried out without instruments. By comparing meteorological notes from the diary of Tycho de Brahe with today's meteorological data, La Cour infers that cloud and rain patterns have not changed over the past 300 years at the coast of the Sound.[198] And R. Wolf concluded from a lecture held by J. Gessner in 1747 that during the past 140 years the average amount of precipitation in northern Switzerland did neither increase nor decrease.[199] These samples are, however, too small to serve as conclusive evidence. As far as exact observations are available, we are able to draw the following conclusion from the research we introduced above: Despite the fact that deforestation has taken place to this day in ever growing proportions, according to meteorological records rainfall has not decreased in those regions. Any reductions in the water levels of rivers should be linked to other factors. In what way this may be done has for instance been pointed out by A. Marié Davy.[200] He again attributes the falling of water levels in France to the expansion of cultivated land but relates it not to less precipitation but to an increase in evaporation which is highest over cultivated land and higher than over woodland.

[192] Burton: *The Gold Mines of Midian*, London, 1878, p. 26.

[193] Schweinfurth, in Bädecker: *Ägypten* [*Egypt*], Volume I, 1878, p. 79.

[194] The Report of the Paris Academy mentioned by Grebenau in *Deutsche Bau-Zeitung,*, 1876, p. 426.

[195] Bulletin de l'acad. des sc. de Pétersbourg. T. XXI, 1876, pp. 293–302.

[196] Sitzungsbericht der Wiener Akademie, mathematisch-naturwissenschaftliche Classe, Volume LXIX, Part II, 1874, p. 642.

[197] Report in the journal of this society, 1881, p. 86.

[198] La Cour: *Tycho Brahe's Meteorologische Dagbog holdt paa Kransborg for aarene 1582–1597* [*Tycho Brahe's meteorological diary from Kransborg for the years 1582–1597*], Kopenhagen, 1876.

[199] R. Wolf, cited by Günther: *Geophysik* II, p. 294.

[200] Marié Davy, in Meteorologsiche Zeitschrift, 1874, p. 145 ff.

This disproves the meteorological conclusions Wex had drawn from the claimed reduction in water levels in accordance with the deforestation theory. Even this reduction, the basis of Wex's inquiry, was declared without proof. In a very critical essay written in the thirties even Pfeil[201] cautioned against blindly accepting the observations by Berghaus and others about sinking water levels as the symptom of an actual reduction in flowing bodies of waters. K. E. v. Baer and Helmersen,[202] and also Berghaus[203] agreed with him and declared the forests not responsible. Later technical aspects were occasionally added to the discussion initiated by Wex's inquiry. For instance, J. Schlichting,[204] Sasse,[205] and others point out that only then will water levels accurately reflect the quantity of the flowing water if the cross-section of the river and speed of flow have not changed. That, however, is an assumption that will never apply to bedload rivers. Such changes are actually produced by correcting the flow of the body of water, which in this century has affected our rivers enormously. Grebenau[206] for instance, interpreted the general reduction in water levels, on the average by 1 cm annually, as the result of lower river beds, as an act of erosion. Fessel[207] expresses the same opinion on this subject.

Others again claimed that the lower averages could have been caused by a change in the regimen of the rivers due to increased deforestation and drainage of the river catchments. Deforestation has completely changed the way rainwater drains: now rain runs off much faster than before when the

[201] Pfeil: *Rührt der niedrige Wasserstand der Flüsse und insbesondere derjenige der Elbe und der Oder, welchen man in neuerer Zeit bemerkt, von der Verminderung der Wälder her?* [*Are the low water levels of the rivers and the Elbe and Oder in particular, which have lately been noticed, caused by deforestation?*] Pfeil's kritische Blätter für Forst und Jagdwirthschaft, Volume XI, 1837, issue 2, p. 62 ff.; reprinted in Berghaus's *Annalen der Erd-, Völker und Staatenkunde* [*Annals of Geography, Enthnology and Political Sciences*], Series III, Volume IV, 1837, p. 289.

[202] Baer and Helmersen: *Beiträge zur Kenntnis des russischen Reiches* [*Essays on a better knowledge of the Russian Empire*], Volume 4, Petersburg, 1841. Preamble to the Report of P. v. Köppen to the Commission about his inquiry into a possible link between the decrease in woodland and the decrease in the water of the upper Wolga river.

[203] Berghaus: *Allgemeine Länder- und Völkerkunde* [*General Geo- and Ethnology*], Volume II, Stuttgart 1837, p. 310.

[204] Schlichting, in *Deutsche Bauzeitung*, 1875, p. 273. Also compare with the reports of academies and societies mentioned above.

[205] Sasse, in *Deutsche Bauzeitung*, 1873, pp. 259, 268.

[206] Grebenau: *Resultate der Pegelbeobachtungen an den elsass-lothringischen Flüssen Rhein und Model von 1807–1872* [*Results of Water Level Observations at the Alsatian Rivers Rhine and Mosel from 1807–72*], Strassburg, 1874 (III Heft der statistischen Mittheilungen über Elsass-Lothringen) Also: *Flusssenkungen und die amit zusammenhängenden Erscheinungen* [*Lowered river levels and related phenomena*]. Lecture in Deutsche Bauzeitung, 1876, p. 425.

[207] Fessel, *Deutsche Bauzeitung*, 1873, p. 329.

forest's soil absorbed it like a sponge and released it gradually into the streams. As a result, the off-seasonal fluctuations of water levels have become more frequent and stronger, the floodwaters more numerous and higher, the low waters more frequent and lower. Such a process must necessarily result in lower annual water levels. For every river the water level corresponding to the annual mean stream flow is higher than the average water level; for, the stream flow is not linearly proportional to the depth but proportionally to a higher power. Only in rivers without any annual fluctuations are both water levels, the one corresponding to the average rate of flow and the arithmetic average of all water levels per year, the same. However, the greater the difference between the levels of low-water and high-water, the more the average annual water level will fall below that corresponding to the average stream flow. Therefore, as a result of deforestation the annual water levels will be lower while the stream flow is unchanged. The hydro-technical committee of the Austrian Society of Engineers and Architects[208] commented on the subject along these lines, as did von Helmersen and Wild,[209] as well as Whitney,[210] Markham, Hann,[211] v. Wagner,[212] and others. Deforestation changes the regimen of bodies of water and consequently their water levels, but does not affect rainfall or water volume—that is the gist of it.

But even this relatively small role which forests play in this process has been successfully and expertly disputed by rejecting the fundamental fact from which Berghaus and Wex proceeded, namely the continuous decrease in the water levels of streams in cultivated areas. Schlichting pointed out that parts of Wex's own data do not even show a continuous lowering of water levels if one determines the averages in a different way than Wex did.[213] As far as a reduction undeniably did occur, it might partly be explained by the fact that ice blockages have become less frequent and smaller.[214] Hagen mentioned that a reduction of water levels is only found

[208] Report in Zeitschrift des österreichischen Ingenieur- und Architektenvereins, 1875, p. 157 ff.; conclusion.

[209] v. Helmersen and Wild's commentary on Wex's study. Bull. de l'acad. des sc. de St. Péterbourg, 1876, p. 293 ff.

[210] Whitney op. cit. , p. 179 ff.

[211] Hann: *Thatsachen und Bemerkungen über einige schädliche Folgen der Zerstörung des natürlichen Pflanzenkleides der Erdoberfläche.* [*Facts and comments on some damaging results of the destruction of plant life on earth*], Meteorologische Zeitschrift, 1869, p. 18 ff. Also Markham is mentioned here.

[212] v. Wagner: *Hydrologische Untersuchungen an der Weser, Elbe, dem Rhein und mehreren kleineren Flüssen.* [*Hydrological Research of the Weser, the Elbe, the Rhine and a number of smaller rivers*], Braunschweig, 1881, p. 24.

[213] Schlichting, in Franzius and Sonne: *Handbook of Engineering*, Volume IV: "Der Wasserbau" Leipzig, 1883, p. 73 of the second edition.

[214] Schlichting, in Deutsche Bauzeitung, 1875, p. 144.

in some of Prussia's rivers, and then again in others it is not,[215] and by comparing observations done at several gauging stations at the Elbe, Pralle[216] as well as Schlichting[217] showed that the lowering of the water level is a local occurrence while in other spots along the same river levels are rising. No doubt this points to local changes to the riverbed and not to less water in the river. Graeve came to the same conclusions—only earlier. At the same time he emphatically opposes Grebenau's theory that rivers in general are running at lower levels. He invalidates the theory simply by pointing out that, if it was correct, almost all rivers, not long ago, would have been running well above the average level of the older cities and communities along their paths.[218] Recently Honsell expressed his opposition to the idea of a general reduction of water levels in rivers. Thus, neither Wex's nor Grebenau's theory is necessary, since what both wanted to explain, the general decrease of water levels in the rivers of cultivated areas, does not exist and the observed fluctuations of water levels are discontinuous, irregular events related to local conditions.

Let us review the many hypotheses mentioned! Deforestation is the cause of a dryer climate everywhere, it reduces the water levels in our springs, creeks and rivers—this is the opinion of one group; no trace of a drier climate and no trace of lower water levels—argue the others. Two opinions that are mutually exclusive and yet their advocates are first-rate scientists. At present we cannot decide between the two. Only one thing is evident: We are still very much in the dark as to the link between forests and rainfall.

This becomes even more evident when we learn that deforestation not only reduces rainfall and river water, but supposedly also increases both in some blessed spots on this earth. I am not referring to the ancient story by Theophrast according to which, as Seneca reports,[219] the river *Haemus* had plenty of water after deforestation—but to more recent authors who have written on this subject. Strangely enough voices were raised favouring an overall increase in river levels, and again deforestation had to be the culprit. As Schmid writes in 1858, it was widely believed that due to the practices

[215] Hagen: *Über die Verminderung der Wasserstände in den preußischen Strömen* [*Reduced water levels in rivers in Prussia*], Abhandlungen der Königlichen Akademie der Wissenschaften in Berlin, 1880.

[216] Pralle: *Wasserstandsverhältnisse in der Oder* [*The Situation of the Odra's Water Levels*], Zeitschrift für Bauwesen, 1882, p. 188.

[217] Schlichting: *Elbestromschau 1869 und 1873,* [*Inspection of the river Elbe 1869 and 1873*], Deutsche Bauzeitung, 1975, p. 274.

[218] Graeve: *Über den Wasserreichthum und die Senung der Flüsse in Culturländern* [*Water volume and its decline in the rivers of cultivated*], Deutsche Bauzeitung, 1877, pp. 261, 271 ff.

[219] Seneca: *Quaestiones naturales III*, 11.

of deforestation and amelioration in Poland the river Weichsel now (1858) carried more water than before.[220] However, he argues against this theory and relates the acknowledged rise of the water levels to major ice blockages caused by dyking. His argumentation is similar and yet reversed to Schlichting's, who linked the observed decrease in water levels to a smaller amount of ice blockages.

We have mentioned before, that the increasing dryness of Australia's climate had in the past been related to massive deforestation. But according to the latest research carried out in New South Wales and other regions of Australia this assumption is entirely incorrect, as mentioned by R. Abbay,[221] M. E. Abbott[222] and R. von Lendenfeld.[223] The opposite has been observed in many regions where since the late sixties it was deforestation that increased the area's water resources. For example, Lake George and Lake Bathurst, two lakes in New South Wales with no outlet, rose considerably since the fifties. According to Abbay, obviously rain water was able to reach the lakes much faster than in the past without evaporating after forests had been cut down. Many dry or sparsely running creeks have turned into permanently flowing murmuring brooks after the valleys' forests were cut down. Consequently, Abbott and Lendenfeld recommend to further extend the practice of deforestation in order to provide dry regions with more water veins. "This way Australia would be able to draw more and more amounts of rain water and at least in part conserve them". One explanation for this phenomenon, so incompatible with findings from the 'old world', may be the fact that Australian trees absorb larger amounts of water which dries out the forests' soils more quickly.[224] But it nevertheless seems strange that the same deforestation of Australian woodlands which in previous years was generally held responsible for the increasing dryness of the climate, should now attract rain and bring water to the soil. One is inclined to assume that, as so often, the truth is caught somewhere in the middle and that deforestation has been as innocent of causing less rainfall and lower water levels in the past as it is today. In that case, increase as well as decrease of rainfall occurs irrespective of deforestation for completely different reasons. In any case, this unusual link between deforestation and climate in Australia, which at times reverses,

[220] Schmid: *Nachrichten über die Ströme des preußischen Staates* [*News about Prussia's Rivers*], III; Weichselstrom, Zeitschrift für Bauwesen, VIII, 1858, p. 158 f.

[221] Abbay in Nature, XIV, p. 47 f.

[222] Abbott in the Publications of the Royal Geographical Society of N. S. Wales, referred to in American Meteorological Journal, Volume IV, 1887, Oct., p. 247.

[223] R. v. Lendenfeld: *Der Einfluß der Entwaldung auf das Klima Australiens* [*The Impact of Deforestation on Australia's Climate*], Petermann's Mittheilungen, 1888, p. 41 ff.

[224] von Lendenfeld, op. cit., p. 43.

certainly proves the point we made earlier—that we know nothing about the influence of forested areas on rainfall.

The situation seems not to be any different regarding another human interference with the climate conditions on this earth. I am referring to the improvement of the climate by expanding the cultivation of originally dry regions and areas with very little vegetation. This view dates back to ancient times; Theophrast reports that the water sources and lakes around the city of Arcadia on Crete dried up when agriculture was abandoned. But after the city had been destroyed and the land was once again used for agriculture, water, too, came back.[225] Lately, this hypothesis was applied on a grand scale to the dry regions in the Western United States of America.

In this case, the facts from which such far reaching conclusions regarding climate change are drawn, are not meteorological in nature; rather they have to do with the expansion of agriculture into areas which 30 years ago were considered totally unsuited for any kind of cultivation. In 1856 the eastern border of the infertile "Great American Desert" ran between the 96th and 97th meridian through the States of Dakota, Nebraska, Kansas, the Indian Territories and Texas. Gradually settlers moved beyond this line and step by step agriculture was pushed further west. Today, it has exceeded the 100th meridian and partly reached the 102nd. The official 1885 census for Kansas puts the population in the sector between the 97° and 100° latitude of this state at more than half a million, who settled there during the past 20 years. It is the fundamental belief of the settlers that they have changed the climate and increased rainfall by cultivating the dry land and planting wheat and corn. They attribute the transformation of the land and its climate to their own energy.[226] Similar cases are reported from Montana and Dakota.[227]

The same happened around the Great Salt Lake. Here it was observed that from 1860 on, rivers were rising; one was able to drain their water for the irrigation of fields, and agriculture took possession of regions that only a short time ago were considered infertile.[228] The Great Salt Lake rose by more than 3 m and its area increased from 4,532 sq. km to 5,609 sq. km. The rising water level of the lake coincided with the expansion of settlements around the Lake and the increase in rainfall and water volume

[225] Ideler: *Über die angebliche Veränderung des Klimas* [*About the alleged changes of climate*]. Berghaus: *Annalen der Erd-, Völker und Staatenkunde* [*Annals of Geography, Enthnology and Political Sciences*], Volume V, 1832, p. 425.

[226] Compare with the lecture of Heyer about N. R. Hilton: *Report of the Kansas State Board of Agriculture*, for the quarter ending March 31, Topeha, 1888, in Wetter, 1889, p. 223 f. The original was not available to me.

[227] American Meteorological Journal, Volume IV, 1887, Oct., p. 242.

[228] See also Gilbert, in Powell: *Report on the Lands of the Arid Region of the United States*, 2nd Edition, Washington, 1879, p. 57 cont'd.

in streams and lakes is directly attributed to the expansion of cultivated land recovered from the desert. In 1869 Cyrus Thomas commented on the noticeable improvement of the climate over the past 8 years;[229] he is convinced that as the population increases so will the rainfall. In 1878 Hough voices a similar opinion. He states that the diligent Mormons have good reason to expect further increases in rainfall if cultivation continues; with more trees planted, the air will become more humid until finally precipitation will be sufficient.[230] Such statements reflect common opinion. Whether rainfall actually increased or not has evolved into a big controversy.

Gilbert, the unremitting explorer of the Great Basin, comments on the cause of water increase in the Great Salt Lake region in the critical manner peculiarly his own. He leaves open whether it is the result of a climate change or of the settlers' changes to the land's drainage structure.[231] If the former should be the case, he rules out anything but a general climate change that came about without any human effort and is comparable to geological climate changes. He considers the changes in the water resources to be constant over longer periods of time and not just the sign of a short-term oscillation of the climate around a mean.

The view that rainfall has increased on the wide plains of the far West is lately being shared by Morrow, Snow, Ch. F. Adams and Greely, while Dorsey[232] and also H. A. Hazen[233] were unable to see the increase from looking at observations of the rainfall, and S.R. Thompson[234] declares the entire issue unfit for judgement at this point because it has not been fully investigated. Three recently published inquiries try to approach the problem in a more exacting manner by using meteorological data.

It is evident that a decision can be reached immediately by comparing two rain charts composed of data collected from different time spans. Mark W. Harrington has taken this approach by comparing two rain charts of the United States, the one created by Blodget on the basis of older observations prior to 1856 and the other by Ch. Denison based on observations by the *Signal-Service* during the years 1870 to 1883. He discovered that the isohyetes of the prairie areas between the 45° and 30° northern latitudes

[229] Gilbert, op. cit., p. 71.
[230] Hough, op. cit., p. 92.
[231] Gilbert, op. cit., pp. 67–77.
[232] Harrington refers briefly to Adams, Snow, Morrow, Greely, and Dorsey in American Met. Journal, Vol. IV, p. 369; and Curtis ibid., Volume V, p. 69 ff.
[233] H. A. Hazen: *Variations of Rainfall West of the Mississippi River*. Signal Service Notes. N. VII. Washington, 1883.
[234] Thompson, in American Meteorological Journal, Volume I, p. 59.

have undergone a general shift to the west, which corresponds to an increase in rainfall.[235]

Gannet achieved the exact opposite result. Applying the method described earlier (p. p.) he concludes that in the region between Missouri and the Rocky Mountain range precipitation did not increase and consequently the cultivation of the land had no influence on rainfall.[236] The latest research by G. E. Curtis led to the same negative result.[237]

Some time ago Whitney argued strongly against the hypothesis that increasing cultivation of the land enhances rainfall;[238] to him it appears unthinkable that around the Great Salt Lake the cultivation of a mere 1/400 of the total area (equal to 1/12 of the lake's area) should have been able to increase rainfall and raise the lake's level to such an extent as seems to be the case. Moreover neighbouring regions show increased water levels as well, even though they experienced deforestation but no cultivation. Powell, again, rejects this view in his article published at the end of 1888.[239]

We are observing the same conflict of opinions here as in the issue of the link between forests and precipitation. First it is claimed that the climate in the interior of America became more humid thanks to the practice of cultivation; then again it is stated that agriculture is not responsible; and finally an increase in rainfall cannot be established at all.

Let us summarise the results of our brief historical excursion by outlining in a few words the present status of the issue concerning a change of precipitation in historical times.

An increase in rainfall in historical times is nowadays claimed for a few limited regions only and in most cases related to some form of human intervention.

On the other hand, the number of researchers who opted for a decline in precipitation is quite impressive. However, the cause of this decline is a matter of completely different opinions. Some are willing to blame everything on increased deforestation without knowing much about the effect of woodlands on precipitation. Others, led by Whitney, reject the idea that deforestation could have such a major effect on rainfall and assume instead that the global climate is in the process of drying up and the many single occurrences are to be considered as nothing more than symptoms of this development.

235 Harrington, in American Meteorological Journal, Volume IV, p. 309 ff. Compare with my article in Meteorologische Zeitschrift, 1888, p. 43.

236 Gannet, op. cit., p. 103.

237 Curtis: *The Trans-Mississippi Rainfall Problem Restated*, American Meteorological Journal, Volume V (June 1888), p. 66 ff.

238 Whitney, op. cit., p. 179.

239 Powell, in Proceedings of the Royal Geographical Society, London 1888, p. 793.

The strongest opposition to all these researchers comes from those scientists who deny any change in rainfall in historical times. It is noteworthy that the majority of meteorologists, whose observations do not date back very far, belongs to this group of opponents.

A consensus between these fundamentally different views seems impossible and we cannot help but ask ourselves: how could this confusion of opinions come about?

The number of researchers who favour the theory that the temperature has changed in the course of historical times is anything but small. The majority of these do not rely on temperature readings but on symptoms evident in the flora and fauna on earth, or on changes in hydrographical phenomena which have been discovered by comparing the historical data from different time spans.

The fact that today's population of Greenland is rather small compared to the 15th century, and above all that Greenland's east coast is said to have been inhabited as late as the end of the middle ages while today those regions are covered with ice and surrounded by an almost impenetrable belt of pack-ice, has frequently been used as evidence for a trend towards a cooler climate in historical times. It was talked about a shift of the isotherms to the south toward the North Atlantic Ocean. This view is not new and has been repeated many times up to this date. It was strongly reinforced by the term "Greenland" which indeed is not a very suitable name for the Greenland of today. Arago mentions this view in his essay about temperatures on earth,[240] as does Bernhard Studer.[241] Even more recent authors such as Czerny[242] and lately Michelier[243] have tried to draw major conclusions from this supposition that Greenland's climate was so much milder in the past. Whitney[244] again insists that a climate change took place in Greenland, yet uses a more critical approach than his predecessors. Nonetheless, today, all these conclusions have essentially lost their basis since Conrad Maurer established that there never have been any Norman settlements on Greenland's east coast and that those small Norman hamlets on the West Coast built in the 10th century and resembling Eskimo settlements did not succumb to a climate which had become inclement, but

[240] Arago: *Oeuvres complètes*. T. VIII. Paris 1858, p. 243.

[241] B. Studer: *Lehrbuch der physikalischen Geographie II*, [*Textbook of physical geography*]. Bern, Chur, Leipzig 1847, p. 306.

[242] Czerny: *Veränderlichkeit des Klimas* [*Variability of the Climate*], Wien, Pest, Leipzig 1881, p. 5.

[243] Michelier: *Etude sur les variations des glaciers des Pyrénées* [*A study about the vaiations fo the Pyrenean glaciers*]. Annales du Bureau Central Météorologique de France, 1885. Part I, pp. 207–234.

[244] Whitney, op. cit., see p. 239.

to an invasion by Eskimos against which they were—abandoned by the motherland—quite defenceless.[245]

It has often been claimed that the European climate shows some signs of a general cooling trend. This idea seemed to harmonise perfectly with Greenland's climate change. The decline of cultural life in Iceland, for instance, is linked to such a temperature change,[246] where, in addition, the tree line is said to have retreated indicating a deterioration of the climate. Signs for this cooling trend are apparently to be found all over Northern Europe and Asia.[247] Tree lines are retreating southward on the Shetland Islands, in Iceland and Scotland according to Czerny,[248] in Lapland according to Whitney,[249] in Siberia according to v. Middendorf[250] and F. Schmidt.[251] Similar observations were made for the Alps by a number of researchers, such as Kasthofer[252] and Tschudi[253] for Switzerland, whose findings are unconditionally confirmed by Theobald,[254] Muret,[255] Leresche,[256] and Coaz,[257] as well as Whitney for the entire Alps.[258]

A retreating tree vegetation is almost always interpreted as a sign of climate change. Ideler however maintains as early as 1837[259] that a shift of the polar tree line to the south may well be the result of human intervention because the largest and best trees are continuously cut down leaving the smaller trees exposed to the wind and making their survival much harder than it used to be. Coaz[260] draws similar conclusions. He attributes the fact that mountain treelines are descending to lower elevations, to the damage caused by alpine farming and its herds of cattle, but not to changes in temperatures. L. Dufour agrees and rejects the conclusion that this fact was the necessary outcome of a climate change.

[245] Compare with article by Kirchhoff about Czerny, Leopoldina 1881, p. 176.
[246] Whitney, op. cit., pp. 239 ff. and 236.
[247] Ideler, in Berghaus' Annalen, 1832, Volume V, p. 421.
[248] Czerny, op. cit., p. 49.
[249] Whitney, op. cit., p. 236.
[250] v. Middendorf: *Sibirische Reise* [*Siberian Journey*], Volume IV 1, St. Petersburg, 1867, p. 612.
[251] F. Schmidt: *Resultate der zur Aufsuchung eines Mammuthcadavers etc. ausgesandten Exepedition* [*Results of an Expedition sent to Explore the Remains of a Mammoth, etc.*] Bulletin of the St. Petersburg Academy, Volume VIII, 1872, p. 26.
[252] Kasthover: *Bemerkungen auf einer Alpenreise, etc.* [*Notes from a Journey through the Alps, etc.*] Aarau, 1822.
[253] Tschudi: *Die Alpen* [*The Alps*] 1859, p. 305.
[254] Theobald, in the 1868 Jahrbuch des Schweizer Alpenclubs.
[255] Muret, in Dufour, Bulletin Soc. Vaud. des sc. X., p. 373.
[256] Leresche in Dufour as above, p. 378.
[257] Coaz: Letter to Dufour—in Dufour, op. cit., p. 379.
[258] Whitney, op. cit., p. 236 f.
[259] Ideler, op. cit., p. 420 f.
[260] Coaz, op. cit., p. 375.

At the end of the last century Hamilton advocates this change towards a cooler climate for Great Britain and Ireland relying on the judgement of farming experts.[261] Glaisher showed however, that London's temperature has risen considerably over the past 100 years. But it turned out that Glaisher's conclusion applies to London only, the temperature increase being a result of the ongoing expansion of the city, which in time completely enclosed the meteorological observation site.[262] Finally, based on a series of long-term temperature readings Buchan came to the conclusion that the temperature in Scotland had not changed at all since readings began.[263]

One country which has been the focus of especially thorough and varied investigations regarding the issue of climate change, particularly in historical times, is France. The identified changes are in part contradicting. According to Picot France's climate has become considerably warmer since early history as a consequence of the gradual deforestation in this country.[264] The same view was later expressed in Government Chambers by Ladoucette for the southern part of France.[265] Arago,[266] on the other hand, draws the opposite conclusion from the fact that in the past there used to be wine growing much further to the north; he claims that in the course of the past centuries France's summers became cooler and winters warmer. And once again increasing deforestation is blamed! In agreement with this hypothesis and based on phenological observations Bourlot[267] tried to establish in 1870 that the climate of the Alsace had deteriorated considerably since the 13th century. Of course, Arago does not suggest that this climate change took place in all of France but in most of it. He refers to the development in other countries where deforestation had the same effect on temperatures. Apart from the fact that Arago's allegation is inconsistent with any observations made so far about the link between forested areas and temperatures, even the preposition for his findings, that is the shift to the south of once northerly wine-growing regions, is nowadays according to Ideler and Ch. Martins[268] attributed to other than climatic impacts. In France, just as in Germany and England where in previous centuries vineyards were cultivated in areas of higher latitudes than today, their

[261] Hamilton, in Transactions of the Irish Academy, Volume II, 1788; cited by Günther: *Geophysik II*, p. 294 (was not available to me).
[262] See Whitney, op. cit., p. 228.
[263] Buchan: *Climate of Scotland*, Athenaeum, 1876, p. 329.
[264] Picot, cited by Ideler, op. cit., p. 425.
[265] Ladoucette cited by Hough: *Report on Forestry*, Washington, 1878, p. 293.
[266] Arago: *Oeuvres complètes*. T. VIII. Paris 1858, p. 230 ff.
[267] Bourlot in Bull. d l'Ass. sc. France, 1870, 23 January.
[268] Ideler, op. cit., p. 449; Ch. Martins: *Le climat de la France a-t-il changé*? Annuaire météorologique de France pour 1850, p. 121.

relocation to the south is nowadays merely seen as a symptom of a more refined taste and a better transportation system allowing good quality wine to reach the markets at lower prices and over larger distances.

While Picot, Arago and others, in one way or the other, fully supported the theory of a change in the French climate, others denied it ever having occurred. In a critical assessment Charles Martins, as we have mentioned earlier, rejected Arago's view as being unwarranted. But he also disagreed with the opposite hypothesis and pointed out that Roman records, if used to suggest a climatic improvement, must be read with reservations and can only be looked upon in relative terms—just as the excellent Ideler did 13 years earlier. By examining meteorological observations Schouw[269] and Dove refuted the occurrence of a continuous change in the temperatures of Germany, Denmark, and Scandinavia while Zimmermann[270] later claimed again that the climate in and around Hamburg had cooled off. For Genève Gautier[271] and Plantamour[272] showed that since the middle of the last century not even the trace of a change in temperature could be substantiated.

Arago[273] claimed to have established a pattern of climate change for Tuscany in Italy with cooler summers and warmer winters, which he again attributed to increasing deforestation. But Whitney[274] rightly points out how unreliable the meteorological data collected in the 17th century are, which Arago used. Temperature changes in some other regions of the subtropical hemisphere have been reported as well. It is primarily the fact that in Afghanistan the northern growth line of the date palm tree retreated further south, which was interpreted in this way, as for instance by Bellew[275] and Whitney.[276] Theobald Fischer prefers to link this event to human negligence in cultivation and plantation efforts.[277]

As in France, two very different rivaling hypotheses about climate change are prevalent in North America. At the end of the last century Larochefoucauld-Liancourt[278] supported the assumption that in Canada

[269] Schouw: *Skildring af Vejrligets Tilstand i Danmark* [*Description of the weather in Denmark*, Kjöbenhavn, 1826. Cited and referred to by Ideler, op. cit., p. 428 ff.

[270] Zimmermann, in Poggendorff's Annalen 1856, Volume 98, p. 323.

[271] Gautier, in Bibl. Univ. de Genève, 1843, Janvier, T. XLIII, p. 158.

[272] Plantamour: *Le climat de Genève* [*The climate of Genève*], Genève.

[273] Arago, op. cit., p. 227.

[274] Whitney, op. cit., p. 234.

[275] Bellew: *From the Indus to the Tigris*, London, 1874, p. 239.

[276] Whitney op. cit., p. 233.

[277] Fischer: *Die Dattelpalme*, [*The date palme*] Ergänzungsheft, No. 64 of Petermann's Mittheilungen, Gotha 1881, p. 50.

[278] Larochefoucault-Liancourt: *Voyage dans les État-Unis de l'Amérique septentrionale*, [*A Voyage tothe septentrional United States*], Volume II, p. 207.

summers were getting hotter and winters milder. Much earlier Peter Kalm expressed a similar conviction for parts of the United States, as mentioned by Volney (1803).[279] In contrast, according to S. Williams[280] and Williamson,[281] who were publishing at roughly the same time as the authors just mentioned, and according to general public opinion summers have allegedly become considerably cooler in the New England states, in Virginia (Jefferson), and in Louisiana (Thomassy)[282] and on the whole the climate has become more temperate. The development is similar for Newhaven as Loomis and Newton[283] discovered on the basis of meteorological observations. All these temperature changes are almost unanimously attributed to excessive deforestation or, in the case of Newhaven, to local reforestation. However, these conclusions are conflicting with findings by many researchers who strictly dispute any change in temperature such as Humboldt, Noah Webster, and Forry,[284] and recently Schott[285] backed by a large sample of meteorological data. Draper's[286] objection is based on the observation that the length of time during which the Hudson River is covered with ice in winter remained unchanged.

Apart from all this evidence of a temperature change in historical times there are, beside the ones included above, additional data supporting the idea that temperatures remained constant. A comparison between descriptions of the cultural life and the plant life at antique historical sites and what life is like today led Arago[287] to dispute any change in temperature for Palastine, Syria, Egypt, and Greece. By using the same method E. Biot[288] concluded the same for China; according to him the temperature has not changed in this country in 3300 years. Ideler's and recently Partsch's findings about the climate in Mediterranean countries are along the same lines.[289]

[279] Volney: *Tableau du climat et du sol des États-Unis d'Amérique*, Paris, 1803, T. I. p. 288.
[280] Williams and Jefferson, as referred to by Ideler, see p. 422.
[281] Williamson mentioned by Draper in Zeitschrift für Meteorologie. 1874. p. 240.
[282] Thomassy, mentioned by Dufour, op. cit., p. 364.
[283] E. Loomis and H.A. Newton: *On the Mean Temperature and on the Fluctuations of Temperature at Newhaven*, Transactions of the Connecticut Academy of Arts and Sciences, Volume I, p. 194. (Cited by Whitney.)
[284] Humboldt, etc., mentioned by Draper, op. cit.
[285] Schott: *Tables, Distribution and Variations of Atmospheric Temperature in the United States*, Smithsonian Contributions to Knowledge, Volume XXI, p. 311.
[286] Draper in an article in zeitschrift der österreichischen Gesellschaft für Meteorologie, Volume IX, 1874, p. 239 ff.
[287] Arago, op.cit., p. 222 ff.
[288] E. Biot: *La température ancienne de la Chine*, 1841 (quoted by Dufour).
[289] Partsch: *Verhandlungen des VIII. Deutschen Geographentages zu Berlin*, Berlin, 1889.

Most of all we have the excellent research by L. Dufour about the question of climate change in Switzerland.[290] His critical stance in the discussion about why the Alps' tree lines have dropped to lower altitudes has already been outlined above. He also shows that all the information concerning the past growing of olive trees or grapes in regions where today they do no longer thrive, is in part misunderstood, in part unreliable and in contrast with other information, and can, finally, in part be explained by human interference unrelated to the climate. Generally, all evidence for a climate change in Switzerland that is based on observed changes in the vegetation, is not conclusive because these changes can be explained differently. However, it is nevertheless quite remarkable that these phenomena, as far as they were interpreted as symptoms of climate change, should all point to a deterioration of the climate, to a decrease in temperature. The registered dates for the harvesting of grapes in Switzerland point to the same result. In the sixteenth and early seventeenth century grapes were harvested much earlier in the year than they are today and particularly than they were in the eighteenth century. As interesting as this fact may be, Dufour does not fail to recognise that even the timing of this harvest may depend on reasons which have nothing to do with the climate. He, therefore, leaves the question whether the climate has changed or not, wide open. And in fact A. Angot has recently shown that this delay in the grape harvests of past centuries did not occur in the neighbouring Swiss Jura and in the Département Côte d'Or.[291]

As a result the question whether temperatures changed or remained constant in historical times is today as far from being solved as the question of changes in precipitation. Some say the climate is getting warmer, then again others say it is getting colder. The explanation for the alleged temperature change was sought in completely different causes. Nowadays, of course, no one would relate a cooler climate to increased deforestation, as Arago did. But deforestation is still held responsible for a warming trend. On the whole however, one is definitely more inclined to attribute climate change to general causes. Two hypotheses oppose each other in this regard. Schmick compiles the data favouring a warmer climate and relates this temperature change, which according to him is typical of the northern hemisphere, to the earlier onset of the equinox.[292] In contrast, Whitney talks

[290] Dufour: *Variation du climat* [*Climate variations*], Bull. Soc. Vaudoise des Sc. Nat., X, pp. 359–436.

[291] A. Angot: *Etude sur les vendanges en France*, Annales du Bureau Central Météorologique de France, 1883, Part I, p. [B.] 83.

[292] Schmick: *Die Aralo-kaspische Niederung im Lichte der Lehre von den säcularen Schwankungen des Seespiegels und der Wärmezonen.* [*The Aralo-Caspian Basin in Light of the Theory of Secular Variations of the Lake's Level and the Temperature Zones*], Leipzig, 1874.

about a general global cooling process which manifests itself in historical times all over this earth and is nothing else than the cooling trend which dates back to the beginning of the Tertiary and continued through the earliest geological periods.[293] As in the past, these views are still challenged today by a number of scientists, among them the majority of the meteorologists, who categorically insist: Temperatures remained unchanged in historical times.

In view of the legions of hypotheses about a change in temperature or rainfall the number of attempts made to verify changes in any other aspect of the climate is rather small. Nevertheless, assumptions regarding changes in the wind pattern of certain regions for example have been made, defended and attacked. According to S. Williams and Jefferson, in New England the frequency of westerly winds has decreased while easterly winds occur more frequently.[294] And again this altered state is attributed to deforestation. According to Simony the *Bora* winds have become stronger in such clear-cut areas with *karst* topography.[295] Lespiault attempted to verify a major climate change in France which he ascribes to the growing force of those storms attacking the West Coast of France.[296] Blavier suggests the exact opposite: a decreased wind activity and an increasingly calmer atmosphere combined with more fog which he attributes to a potential change of direction of the Gulf Stream along the French coast.[297] Gruß identifies certain areas in Europe where wind directions have changed in the course of this century.[298] In München, for instance, since 1865 north winds are presumed to blow less often than in previous years, easterly and westerly winds, however, increased; in the city of Leipzig westerly and northerly winds appeared to have increased slightly; in Berlin easterly and north-westerly winds increased as well, while in Lund it is the northerly and easterly winds which occur less often in this century as opposed to the last.

We have arrived at the end of our tour. We have walked through a veritable labyrinth without the benefit of Ariadne's clew. Again and again we encountered the same insoluble and irreconcilable contradictions. It seems almost like a psychological puzzle, that for one and the same country serious scientists have at every step insisted on climate changes which are

293 Whitney, op. cit.

294 S. Williams and Jefferson, reviewed by Ideler, op. cit., p. 426.

295 F. Simony: *Schutz dem Walde*! [*Protect the Forest!*], Schriften des Vereins zur Verbreitung naturwissenschaftlicher Kenntnisse, Wien, Volume XVII, 1876–77, 1877, p. 456.

296 Lespiault, cited by Günther, in Geophysik, II, p. 289.

297 Blavier: *Changement du climat sur les côtes de la Vendée etc.* L'Astronomie (de Flammarion), 1883, p. 106 ff. Cited by Günther. (The original was not available.)

298 Gruß, in Wetter, 1888, p. 137, and in Meteorologische Zeitschrift, 1888, p. [57] No. (155).

mutually exclusive. Equally puzzling is the fact that over and over again the forest is singled out as the scapegoat for a variety of frequently contradictory changes. Something quite striking emerges from all this: the lack of any progress towards a solution. Opinions are still as directly and diametrically opposed today as they were forty years ago. If the climate would measure up to all these assumptions which have been uttered over the past years and decades, it would have to change back and forth from one direction to the other. When we look back on this welter of different hypotheses, we have to admit that even today we are still far from a definite answer to the question whether the climate changed in historical times or remained constant. And even today we still have to adhere to L. Dufour's statement of some twenty years ago: "The question of a climate change in historical times is still wide open and the claim by the majority of meteorologists that climate does not change has found neither more nor less proof than the claim to the opposite.[299] " If the climate is indeed undergoing a continuous change, it is undoubtedly occurring at a very slow pace, which would explain why even today we have no definite knowledge of its specific aspects.

4.1.3 Meteorological Cycles

Parallel to the described research into climate change in historical times, efforts were going on in the past centuries to find meteorological cycles, or secular climate variations in predetermined time periods, motivated partly by hopes to gain a safe forecasting method for the future. These approaches went into two different directions and neither their methods nor their outcome had anything in common. At one time this turned into an unsystematic hunt for periodical data that led to tabulations of a variety of diverse cycles. Then again it was attempted to prove the connection between periodical phases of the sun and similar phases for meteorological elements on earth. This created the wealth of literature about the influence of sunspots and their 11-year periods on the meteorological conditions on earth.

On earth a change in weather is governed by the cyclical change of annual seasons; it seemed therefore rather obvious to conclude that there might be similar cycles of much longer duration, e.g., resembling an annual period of a higher rank, which show a regular pattern of weather changes from one year to the next. The search for periodic weather patterns has in fact always been a thankless job in meteorology. The method of

[299] L. Dufour, op. cit., p. 420.

establishing and verifying such cycles has most often been totally inadequate. A series of brief reports published by Symons' Monthly Meteorological Magazine in April and May 1886 may serve as an example. Brumham, for instance, claims to have identified a 40-year period of cold and warm winters, but in such a way that several of these periods are concurrent. Hot and dry summers return after 425 years; G. T. Gwilliam establishes a return of every 17 years. In a second publication Brumham, on the other hand, identifies cycles of 8, 10, 12, 19, 29, 39, 40, 68, and 136 years for the occurrence of hot summers and forecasts one for 1886, because that particular year would fall into all those cycles with the exception of the 19-year period. The subsequent cool summer of 1886, it would seem, taught him a lesson. Data on which such an approach is based are less than adequate.

Even some distinguished meteorologists have tried to find a periodicity in weather patterns. But more often than not their arguments do not bear up against serious criticism. The highly regarded director of the meteorological network of New South Wales, H.C. Russel, for instance, advocates a 19-year weather period;[300] that, however, is not supposed to mean that every 19 years rainfalls or temperatures reach a maximum, but rather that each year shows the same characteristics as the corresponding one of 19 years back; unfortunately his evidence is inconclusive. Benj. Gould believed to have discovered an 18-year periodic pattern of the wind velocity for Buenos Aires.

The search for a periodic pattern in the occurrence of cold winters is a frequent endeavour. We have already mentioned one such case. But as early as 1741 the Petersburg scientist Krafft stated categorically that extreme winters tend to occur every 30 to 34 years;[301] he tried to back up his findings with less than complete statistical data about severe winters in historical times. Endeavours like this have been attempted until recently, i.e., in 1876 by Chavanne who, based on data about the ice cover of the polar sea, identified periodic reoccurrences at a multiple of three, i.e., every 3, 6, 9, 12, 15, 18, 21, etc., years.[302]

All such attempts have one thing in common: they cling too much to the single year and see the weather as a process developing with an almost

[300] H. C. Russel: *History of Floods in the River Darling*, Journal and Proc. R. Society New-South-Wales for 1886, Sydney, 1887, p. 156 ff.

[301] Krafft: *Ausführliche und umständliche Beschreibung des im Januar-Monat 1740 in St. Petersburg errichteten Eispalastes*, etc. [*Detailed and lengthy description of the 'Ice Palace' constructed in St. Petersburg in January 1740*], St. Petersburg, Academy of Science, 1741, in Russian, pp. 23–29.

[302] Chavanne: *Die Eisverhältnisse im arktischen Polarmeere und ihre periodischen Veränderungen* [*The ice conditions in the arctic polar sea and their periodic variations*], Petermann's Mittheilungen, 1876, p. 254 ff.

mathematical precision. Deviations, of course, do occur, which are declared random occurrences; yet it is not checked why these deviations are in fact lower in number than they ought to be, if the arrangement of years was left to chance. As soon as the random rule is applied, the alleged regular pattern collapses.

A more scientific approach of verifying a periodic pattern is based on multi-year averages. But even then fallacies are easily committed if the number of deviations is not carefully examined in the above manner. This is still evident in a note by A. Duponchel published in the Comptes Rendus in 1888;[303] he determined the 12-year averages of the annual temperature of Paris starting with the year 1804, e.g., from 1804 to 1815, 1816 to 1827, etc., and was surprised to find out that two successive time periods each, "in general" deviate from the multi-year average under opposite signs; indeed, based on this 24-year pattern he even feels inclined to predict a cold winter for 1896/97 and a very mild one for 1908/09!

In examining observational data from large meteorological networks scientists occasionally claimed to have discovered certain cycles here and there. Wild advocated that St. Petersburg's temperature changed in 23-year cycles. Later he also found a cycle of approximately 40 years for the rainfall measured at Russian stations.[304]

A multi-year period of cold winters was found by E. Renou,[305] but not in the same sense as did the authors mentioned above. This cycle implies that a particularly severe winter occurs every 41 years surrounded by a 20-year period with numerous winters which are not quite as severe but still cold, while the remaining 20-year period shows relatively few hard winters. In essence, he assumes that periods with a great number of cold winters alternate with periods with a lot of warm winters. Similarly, Köppen suggests a 45-year period of severe winters for the past two centuries (18th and 19th) and a 130-year period for earlier centuries.[306]

From observations in Praha, Milano, Wien and München Hornstein believed to have identified a 70-year period regarding air pressure. This period coincides with a period of sunspots. A 22-year period of temperature changes, as well as a 7-year one in particular, was identified by Schott for

[303] Duponchel, in Paris Comptes Rendus, 1888, 2^{e} semestre, p. 427.

[304] Wild, *Temperaturverhältnisse des Russischen Reiches* [*Temperature conditions of the Russian Empire*], Supplementary Volume to Report. f. Met. St. Petersburg, 1881, p. 279; also Wild, *Regenverhältnisse des Russischen Reiches* [*Rain Conditions of the Russian Empire*], Supplementary Volume to Report. f. Met. St. Petersburg, 1887, p. 80.

[305] Renou: *Périodicité des grands hivers*, Annuaire de la Société Météorologique de France 1861, p. 19 ff.

[306] Köppen, in Zeitschrift für Meteorologie, 1881, p. 183 ff.

the United States. However, the length of this period varies a little as the case may be.[307]

Of all those meteorological cycles, which occur in a more or less established pattern, none has been treated with the same versatile and scientific interest as the incidental 11-year period of meteorological phenomena that was assumed to have been the result of an 11-year period of sunspot activity.[308]

Soon after Schwabe and R. Wolf had identified the periodical re-occurrence of sunspots, the search for an 11-year temperature cycle started. Previously, Gautier had attempted to establish a 10-year temperature cycle for a number of stations in connection with the 10-year period of sun spot activity discovered by Schwabe in 1843. For the first time in 1853 Fritsch established an 11-year temperature cycle for seven European stations. According to both scientists, years of minimal sunspot activity are marked by very warm temperatures. Subsequently, in numerous publications several other researchers have tried to investigate this link further, i.e., Zimmermann, Piazzi, Smith, Stone, Hill, Hahn, who came up with positive results, while Celoria for Milano, Hann for Wien and Schott for the United States failed to discover a connection between sunspots and temperature; Baxendell, Weilenmann, Blanford, Roscoe and B. Stewart claimed to have found a correlation, in which, however, periods are swapped. According to their theory the temperature development is not inverse to the sun spot activity but runs parallel to it.

Among all related research, Köppen's inquiries into multi-year weather cycles are the most outstanding. Even today, Köppen's findings have not been surpassed. According to these results, during the time span from 1816–1860 sunspot activity and temperatures correspond globally like mirror images as Gautier and Fritsch had asserted, but prior to 1816 or after 1860 this accord is sporadic.

Just as numerous are the attempts to establish an 11-year cycle for the rainfall. In 1872 Meldrum was the first to refer to a periodical pattern of

[307] Schott: *Tables, etc. of the Atmospheric Temperature in the United States*, Smithsonian Contributions, Volume XXI, No. 277, p. 314.

[308] Hahn gave a summary of the position on this question: *Über die Beziehungen der Sonnenflecken zu meteorologischen Erscheinungen.* [*About the Link Between Sunspots and Meteorological Phenomena*], Leipzig, 1877. H. Fritz: *Die Beziehungen der Sonnenflecken zu den magnetischen und meteorologischen Erscheinungen der Erde,* [*The Link Between Sunspots and Magnetic and Meteorological Phenomena on Earth*], Haarlem, 1878; v. Czerny: *Die Veränderlichkeit des Klimas etc* [*The Variability of the Climate etc.*], Wien, Pest, Leipzig, 1881, p. 9 ff.; the latest and most detailed summary was given by v. Bebber in his *Handbuch der ausübenden Witterungskunde* [*Manual of Applied Meteorology*], Volume I, Stuttgart, 1885, pp. 199–257. A well-documented bibliography is included. We can therefore refer to names without quotations.

cyclones in the Indian Ocean south of the equator whose maxima and minima correspond with those of the sunspot activity. As a sequence he later tried to establish that the rainfall is slightly heavier during times of maximum sunspot activity than during times of a minimum. Lockyer, R. Wolf, Symons, Hunter, Brocklesby and H. Fritz agreed with Meldrum's theory while Celoria, B. Stewart, Strachey and Whipple, and in part also Jelinek were unable to verify the correlation from the data they had worked with. In general, however, it is quite likely that rainfall increases with an increase in sunspot activity. A most significant result was achieved by Hill and E. Douglas Archibald who independently established that the summer and the winter rains in India show an entirely different correlation. The summer rains reach maximum amounts at the time of a sunspot minimum, the winter rains at a sunspot maximum in accordance with the continental and dry regions on earth.

It has also been attempted to relate the remaining meteorological phenomena to sunspots. Hornstein, Forssmann, Hahn, S. A. Hill, H. F. Blanford; F. Chambers, Douglas Archibald and J. Allen Broun have found a periodic pattern of barometric pressure corresponding with the 11-year cycle of sunspots. The correlation appears to have been established for Southeast Asia implying that the higher barometric readings are in accordance with the maxima—the lower with the minima. However, where the process of compensation takes place, which is required due to the static nature of the total air mass, has not yet been examined. The frequency of cyclones, it appears, increases with that of the sunspots, complying with the correlation established by Meldrum and defended by A. Poey. The same increases occur in wind velocity, according to Rühlmann, while the influence on wind direction has not yet been verified. The same applies to clouds, thunderstorm activity and hailstorms.

Some went even further and tried to establish the influence of an assumed 11-year weather cycle on hydrographical and even on economical aspects. Reiß, for instance, insisted on a link between periodic floods and sunspots. An 11-year cycle had been claimed earlier, though partly with little success, by Dawson for the oscillations of the American Great Lakes, and by Fritz for the fluctuations of rivers in general and even the sizes of the European glaciers. Hunter talks about an 11-year cycle of famines in India, Javons about one pertaining to commercial crises, etc.

In looking at the current status of the sunspot issue it cannot be denied that there are indeed close links between the different meteorological phenomena and periods of sunspot activity. However, the physical aspect of the argumentation for this correlation is in part in a sad state, as, for instance, regarding rainfall and even in respect to temperature. For, even the simple question whether a sun radiates more heat with or without spots, has not been solved. It remains equally unexplained why a meteorological phenomenon appears to concur with the phase of sunspots in one location and not in others, and why such a concurrence continues for some time in one place and then again disappears. In short, the issue still remains in the dark and unsolved. This can be said of the 11-year cycle of sunspots and much more so of the longer 55-year cycle; its influence on our atmosphere is at this point quite hypothetical apart from the occurrence of Northern Lights and of magnetic phenomena which are not part of this study.

It was the search for a clearly defined cycle which governed the theories thus described and which, in the case of the sunspot cycle at least, led to results. All other theories of cycles have in part no scientific basis whatsoever. Entirely different are the views we would like to address now and which in their majority were produced during the past years.

The whole question of climate change in historical times entered a new phase when it was no longer the continuous change in one direction, be it of the rainfall or the temperature, which was sought or the short weather periods of a certain duration, but when meteorological data were instead examined with regard to secular ups and downs of the climate over long periods of time.

This was initiated by the unusual variations in the size of glaciers. Meteorological conditions had to be the sole cause because the very existence of glaciers is directly linked to certain climatic conditions; their change must necessarily cause a change in the size of glaciers; there could be no doubt about that. For a long time glaciers had therefore been looked upon as a kind of thermometer or measuring device of weather conditions in general. The more striking was the fact that this obvious relationship between glacier size and weather variations could not be verified by meteorological observations until in 1858 v. Sonklar[309] spoke the relieving words and regrettably no one listened. In his excellent report he provided proof of the parallel development of glacial variations and variations in temperature and rainfall over a time span of two centuries. For the more recent time period he used the meteorological observations of stations in Milano and Hohenpeißenberg; for the earlier years he collected as many data as he could obtain about weather conditions of individual seasons as far as they pertained to the territory of the Alps and their surrounding area. Since general weather accounts as well as precise observations were available to him regarding the last 100 years, he tried to quantify such general terms as "cold", "very cold", etc. with the help of a clever, yet

[309] v. Sonklar: *Ueber den Zusammenhang der Gletscherschwankungen mit den meteorologischen Verhältnissen*, [*About the relationship between glacial variations and meteorological conditions*], Sitzungsberichte der Wiener Akademie, Volume 32, 1858, pp. 169–206. Including a diagram displaying curves.

somewhat controversial method. He obtained proportional figures which increase and decrease according to changes in the two factors mainly responsible for the formation of glaciers: cold temperatures and moisture. von Sonklar recognised quite clearly that it is not the weather pattern of a single year that determines glacial variations but rather the consecutive weather patterns of many years. Consequently, he deemed it necessary to adjust the meteorological data in order to be able to demonstrate the parallel development of glacier variations and weather changes. He was able to achieve this by ignoring individual years and their quite irregular and random weather changes and by concentrating instead on their five- and ten-year averages. In neutralising the influence of individual years in this manner, the path was set which generally had to lead to success.

von Sonklar's results are clear and exact. Based on both, the general weather accounts as well as the measurements from observational instruments, he determined that in the Alps the glacial expansions around 1770, from 1810 to 1820, and in the forties of the 19th century coincided with periods of wet and cool weather, whereas the reduction of the glacial cover by the end of the 18th century and in the twenties, thirties, and fifties of the 19th century coincided with periods of dry and warm weather.

But v. Sonklar's work went unnoticed and the question about the causes of glacial changes continued to be looked upon as unanswered until the publication of Forel's research in 1881, which at least prepared the ground for a possible solution. Independently of v. Sonklar, but using a similar method of neutralising the swings of the annual weather patterns Forel achieved the same results. From his observations in Genève he was able to verify that those periods when glaciers expand are indeed immediately preceded by periods of low summer temperatures and heavier precipitation, whereas glaciers diminish in size when the weather is quite warm and dry.[310]

Two years later the same route was taken by Eduard Richter in his accomplished case study of the Obersulzbach glacier.[311] He made use of the rainfall measurements of Klagenfurth for his purposes by determining the 5-year averages in an attempt to show the general tendency of this climate factor. His conclusion was that the rainy periods of 1842 to 1852 caused the glacial expansion in the fifties, while the dry periods of 1852 to 1872 were the cause of the unusually large latest reduction. He finds it remarkable though that the significant increase in rainfall from 1872 to 1878 has not yet shown any effect on the glacial cover.

[310] Forel: *Variations périodiques des glaciers*, Archives de sc. phys. et nat. Genève, 1881, 3. Pér. T. VI, p. 22 and p. 451.

[311] E. Richter, in Zeitschrift des deutschen und Österreichischen Alpenvereins, 1883, p. 75 ff.

These findings were confirmed in a more general aspect by C. Lang, who, by adhering strictly to the methods chosen by Forel, extended the study to all of the surrounding regions of the Alps using rainfall measurements from nine and temperature measurements from five stations primarily located at the foot of the Alps and in the Alps' foreland.[312] Here again, the link between rainfall, temperature and glacial variations showed a definite parallel tendency.

Given the importance of the research by Forel, Richter and Lang for the issue which we will explore later on, we may be permitted to tabulate their results below. Von Sonklar's findings could not be included because he does not separate between temperature and rainfall. Maximum, respectively minimum precipitation occurs in the following years:

		Maximum Precipitation			Minimum Precipitation		
Genève	(Forel)[313]		1842/57	1878/80	1835/41		1858/77
Klagenfurth	(Richter)		1842/52	1872/78			1852/72
Milano	(Lang)	1810/14	1840/49	1880/84	1825/29		1865/74
Praha	"	1815/19	1845/49		1820/24		
Wien	"			1875/79			
München	"		1850/54	1880/84			1870/74
Hohenpbg.	"	1805/09	1835/44		1820/29		
Reichenhall	"		1845/49			1855/59	
Stuttgart	"		1845/54	1880/84		1860/64	
Chioggia	"	1800/09					
Genève	(Forel)[314]			1826/35	1856/75		1836/55

		Maximum Temperature			Minimum Temperature		
Milano	(Lang)		1790/94	1820/29	1845/49	1810/19	1835/44
Stuttgart	"			1825/34	1860/64		1835/44
Regensburg	"	1775/79	1795/99	1825/34		1815/19	1835/44
München	"			1830/34	1865/69		1840/64
Hohenpbg.	"		1790/94	1820/24		1810/19	1835/39

[312] C. Lang: *Der säculare verlauf der Witterung als Ursache der Gletscherschwankunegn in den Alpen* [*The secular weather pattern as a cause of glacial variations in the Alps*], Zeitschrift für Meteorologie, 1885, p. 443 ff.

[313] It must be kept in mind that the annual figures determined by Forel are not directly comparable with Richter's and Lang's findings. The figures for the first period of heavy rainfalls from 1842–57 must be interpreted in such a way that the 10-year averages of 1833–42, 34–43, ..., 48–57 were higher than the multi-year average. The correct way of putting it would therefore be: heavy rainfall from 1837 to 1852, if the first 10-year average of 1833–42 is connected to the middle year 1837, the last of 1848–57 to the middle year 1852.

[314] Only according to summer temperatures.

In general the variations of the Alps' glacial cover correspond with the weather pattern of their surroundings: glaciers decrease when precipitation is too low and increase when it is too heavy; in both instances the cause precedes the effect a little.

Forel and Lang established the same for the temperature.

Undoubtedly the region of the Alps and its immediate surrounding area experiences long-term variations of precipitation and temperature that are reflected by changes in the glacial formations. These are typical signs of secular variations of the weather as Lang called the long-term changes between wet and dry periods, as well as cool and warm ones.

This view received an unexpected confirmation by a small, but, as far as the methodology of this entire research area is concerned, significant study carried out by A. Swarowsky, Wien, about the water levels of the Neusiedler See.[315] He demonstrated that the changes in the water levels of this lake with no outlet and in the glacial areas exhibit a partly striking parallel pattern. He thereby delivered proof that even lakes without run-offs are excellent indicators of secular weather changes—a fact that however had frequently been assumed *a priori.*

The variations of the weather over long periods of time as demonstrated for the Alps and the neighbouring region of the Neusiedler See are in all respects surprising. They seem to be more pronounced than the variations of rainfall and temperature following the 11-year period of sunspot activity; for they are followed by glacial changes. While an 11-year period of glacial variations has been claimed by Fritz, to the unbiased observer it does not exist.

We are still in the dark about these variations; we do not know if they are limited to the regions of the Alps or extend beyond them to the rest of Europe perhaps, or to even more substantial parts of the globe. We can make the assumption that something like this occurs because we know that glaciers do not remain unchanged but often vary in size. But whether such variations all occur at the same time, or with delays in some regions and ahead of time in others—we do not know. And finally, we have no knowledge whatsoever of their cause.

These were the questions which motivated the research recounted on the following pages. I could have started by collecting all the data about glacial variations in European and Non-European mountain ranges. But I was deterred by the consideration that glacial changes are but an inadequate and sluggish measuring device for secular variations of weather patterns. It is first of all necessary to compile the statistical data from observations at many different glaciers of the same mountain range in order to determine

[315] Swarowsky in his report about the XIIth year of the Association of Geographers of the University of Vienna, Wien, 1886, p. 18.

the average periods of increase and decrease because neighbouring glaciers may often react quite differently due to the peculiar feature of their bed and their location. Therefore, we have only recently through Forel's work obtained a more detailed picture of the variations of the Alps' glaciers during the current century. To achieve the same at present for non-European mountain ranges, had *a priori* to be considered hopeless. On the other hand solving this problem seemed easier and more reliable by applying Swarowsky's method, e.g., by comparing changes in the water levels of lakes without run-offs with changes in the rainfall and the temperature of the regions of the Alps. We will begin with the largest of the lakes without a constant outlet—the Caspian Sea.

[1] See Chapter 4b.

4.2 PERIODICITY OF CLIMATIC VARIATIONS DERIVED FROM OBSERVATIONS OF ICE CONDITIONS ON RIVERS, THE DATE OF GRAPE HARVEST AND THE FREQUENCY OF SEVERE WINTERS*

I. Secular variations of ice conditions on rivers. The significance of the ice cover during Russian winters. Factors affecting the timing of freezing and breaking. References, comments, and tables concerning the length of ice-free seasons and the dates of the break-up. Forming of groups. Applying the results about temperature fluctuations to Russia and Siberia and to the time dating back to 1700, in part even back to 1560. Increase in the amplitude of the variations of ice conditions when progressing in a westerly direction, solely explained by the value of the periodic temperature variation at the time of freezing and breaking-up. —**II. Secular variations of the start of the grape harvest.** Angot's discourse. Additional data in the form of hand-written notes by Angot, Forel, and Wehrli. Non-meteorological factors which might influence the grape harvest. Applying the method of finite differences. Data and tables for 29 stations in France, SW Germany, and Switzerland 1391–1888. Group average and the overall average for all data series. Interpretation of the tables. Angot was unaware of variations caused by climatic oscillations. These oscillations are paralleled by variations in temperature and rainfall. —**III. Table of variations in the frequency of severe winters 800–1775.** Comparison with the variations in ice conditions and the start of the grape harvest. —**IV. The average periodicity of climatic variations.** Table of climatic variations from 1000 to 1880. Average periodicity of 34.8 (±0.7) years. Climatic variations in central Europe as a local aspect of global climate variations since 1000.

4.2.1 Secular Variations of the River Ice

Records of the opening and closing of bodies of water date back partly to the year 1700. It was not only the majestic sight of the ice built-up in a mighty river with its ice floes slowly gliding downstream at one time, and crashing, shoving, and swirling at another, which inevitably led to the

* [Chapter 8: *Die Periodizität der Klimaschwankungen, abgeleitet auf Grund der Beobachtungen über die Eisverhältnisse der Flüsse, über das Datum der Weinernte und die Häufigkeit strenger Winter.*]

desire to keep track of this phenomenon. It was above all invaluable for trade and commerce to know exactly when a river would freeze-over in the winter and when it would open up again. How important the ice cover of the rivers is varies quite substantially from country to country. In the oceanic climate of Western Europe rivers freeze rarely, and if at all, then in abnormally severe winters and for a short period of time only. The further we move toward the Continental East, the longer the ice cover will last and the more significant it will be. In the European part of Russia and in Siberia ice provides a safe passage across the river in winter and at the same time serves as a smooth road for sleighs commuting up and down the river between settlements. Of course, it restricts shipping and the transportation of bulk goods to the warmer season. Thus the river's role as a means of transportation is twofold. The river is released from this role in the fall, when its new and fragile ice cover interrupts normal boat traffic and is not yet able to support sleighs and horses, or in the spring when the ice cracks and the floes start moving. As much as the ice used to facilitate traffic, it now prevents it. The river communities that during winter and summer were able to enjoy close neighbourly contacts, have now, at the beginning of the freeze-up and even more so at the beginning of the melting, become inaccessible. At such times traffic comes to a halt for large regions of Russia.

Before we start discussing the observations, we should address the question of the causes for the onset of the freeze-up and the beginning of the break-up of the ice. Should both occurrences indeed be regarded as the direct result of a change in air temperature? This would only hold true to a certain degree. A river begins to freeze when the temperature of its top water level drops below zero degrees. The timing of this occurrence depends not only on the air temperature but also on the river water itself. The entire water mass must have cooled to 4°C, i.e., to its maximum density temperature, before the surface temperature can drop that low. The time this process takes, depends largely on the initial water temperature and also substantially on the water volume. The same river will freeze sooner, when its water levels are low than it would when they are high at otherwise completely comparable conditions.

Water volume has a similar effect on the timing of the break-up of the winter ice cover. Very seldom is melting the only cause. A rising water level is in almost all instances a contributing factor. Caused by the melting of snow and ice upstream, it pushes up against the ice cover cracking it and breaking it up and, as a result, starting the formation of ice floes. The faster and stronger the river-water rises, the earlier this occurs. The rise depends almost entirely on the intensity of the snow's melting process, i.e., indirectly on the temperature in the upstream part of the river's catchment.

The amount of snow fallen during the winter, which provides the melt-water, as well as the thickness of the ice cover which is determined by the winter temperatures and the protective layer of snow, are secondary factors only.

Although the point of freezing and melting of the ice cover as well as its duration is, in this sense, not a true function of the temperature conditions, the latter play a major role. Therefore, the condition of the river ice must primarily reflect the conditions of the air temperature. This does not imply that changes in the timing and duration of the ice cover from one year to the next follow precisely the same pattern as the annual temperature averages. The average temperature, as indicated by the condition of the river ice, is arrived at through a completely different method from the one used to determine our annual average where each month carries the same weight. Winter temperatures and even more so summer temperatures are of little significance in the determination of this average value as opposed to spring and fall temperatures—as pointed out by Wild.[316] Secular variations in the duration of the ice cover as well as in the start of the drift of the ice floes nonetheless follow the same pattern as the secular variations of the annual temperature averages. This is, with the exception of the time span after 1840, demonstrated in Wild's study as well as in our table below which shows that on the whole the duration of the winter ice and the average temperature vary in a parallel pattern. Consequently, data about the freezing and melting of the rivers can be used to substantiate secular temperature variations where actual temperature readings are unavailable.

As an indicator of the temperature, river ice has the advantage over the thermometer in that it is almost completely free of human interference. Factors such as faulty instruments and errors in their positioning which can easily distort the homogeneity of an observation series, do not apply in this case. Not until the most recent decades has it become possible to arbitrarily alter the dates of closure and re-opening of rivers with steamship traffic in the attempt to keep these rivers open for shipping by using machinery, as shown in the case of the Elbe River. Under these circumstances the homogeneity of the data appears to be ensured apart, of course, from writing and printing errors which may have crept into the original publications.

The following sources were used for the tables below:

1. Rykatschew: *Über den Auf- und Zugang der Gewässer des russischen Reiches* [*About the opening and closing of the rivers in the Russian Empire*] II. Supplemental volume to Repertorium für Meteorologie. St.

[316] Wild: *Temperaturverhältnisse des Russischen Reiches* [*Temperature conditions in the Russian Empire*], Supplementary Volume to the Repertorium für Meteorologie, St. Petersburg, 1881, p. 285 f.

Petersburg 1887. Numerous printing errors were corrected prior to incorporating the numerals into the tables, according to the list of misprints included in the publication and above all according to Woeikof's detailed review in the Iswestija of the Imperial Russian Geographical Society, Volume XXIII.

2. For the Donau, Annalen der Hydrographie und Maritimen Meteorologie, Volume 1880, p. 477.
3. Heydenreich: *Klimatische Verhältnisse von Lithauen im Regierungsbezirk Gumbinnen nach den 50jährigen Beobachtungen zu Tilsit,* [*The climatic conditions of Lithuania in the District of Gumbinnen according to the observations in Tilsit over a period of 50 years*], Tilsit 1870.
4. Draper: *Über die Eisbedeckung des Hudson,* [*The Ice Cover of the Hudson*], Zeitschrift für Meteorologie, 1874, p. 240.

Included in the tables are data from 32 stations about the average annual length of time during which the rivers are ice-free, and from 12 stations about the average opening dates of ice-locked rivers, i.e., for each lustrum. Incomplete lustra are marked by a period symbol.

I paid special attention to the degree of accuracy to which the lustra averages reflect secular variations of the ice conditions. For this particular purpose I examined the ice conditions of two rivers, the Newa at St. Petersburg and the Dwina at Archangelsk, and their variations from one year to the next by determining for each year the average of its neighbouring ten years. The final result turned out in favour of the method of lustra averages; the latter give indeed precise indications of the variations. It would take up too much space to reproduce the entire table as originally planned; since Rykatchev published the 10-year-averages for the river Dwina in his extensive study, I may be permitted to revert to a small excerpt pertaining to the river Newa for the years 1801–50. The decades are mentioned in the heading of the table. The left margin shows the individual years. The 10-year-averages for each individual year were arrived at by determining the average of this particular year as well as of the previous five and the successive four years. The lustra-averages are given for their middle year.

Information about missing years is supplied in the account below. Stations are categorized according to their geographical location from East to West, as in the tables.

Lena 1855; Sselenga 1869 f.; Angara 1795, 1803, 54, 55; Ob (Barnaul) 1836, 37, 40, 41, 43, 45, 46, 49, 65; Irtysch 1811, 43–48, 50; Bogoslowsk (Teich) 1846, 55, 74 ff.; Kama (Dedjuchin) 1811, 23, 38–45, 55; Kama (Perm) 1804–24, 27, 29, 35, 36, 44, 61, 80; Belaja 1833, 44, 65; Ai (Slatoust) 1863, 64, 70, 74, 75; Ssyssola 1816, 43, 55–57, 79, 80; Wyt-

Table 5B.1. Variations of the Ice-free Periods of the Newa at St. Petersburg According to 10-year Averages and Lustra Averages with Deviations from the Mean 1816/80 (219.5 days)

	1800		1810		1820		1830		1840	
	10-y. Mean	Lustra	10-y. Mean	Lustra	10-y. Mean	Lustra	10-y. Mean	Lustra	10-y. Mean	Lustra
1	–4.2		–9.1		11.0		5.3		–7.7	
2	–5.3		–8.9		14.6		4.3		–7.7	
3	–5.4	–4.9	–9.5	–2.3	14.4	21.1	3.9	1.3	–5.3	–2.9
4	–8.2		–9.5		16.8		–0.1		–3.3	
5	–10.8		–7.1		18.4		–1.7		–3.5	
6	–14.6		–3.7		18.6		–2.6		–2.1	
7	–18.8		4.6		11.8		–4.0		–4.9	
8	–16.6	–16.7	2.9	–11.8	11.4	15.7	–5.7	–2.7	–4.0	–2.1
9	–12.9		4.1		8.6		–7.3		–3.2	
10	–9.5		4.7		8.5		–6.2		–4.6	

schegda 1837, 38, 41, 42, 53, 56, 57, 61, 80; Ssuchona 1793, 94, 1803, 13, 16–18, 20, 22–24, 26, 28, 29, 32–35, 55–58, 61, 80; Wologda 1829 to 1835, 37, 43, 45–54, 56–58 (replaced by 55), 61, 79 f., Dwina 1880; Onega 1800–11, 79 f.; Newa 1709, 12; Düna 1813, 25, 29, 33; Wjatka Slobodskoj) 1879 f.: Wjatka (Wjatka) 1816, 55–58, 62, 64, 65, 79 f.: Kama (Jelabuga) 1826, 28, 33, 34, 38, 39, 41, 55, 80; Wolga (Ssimbirsk) 1831, 79 f.; Wolga (Ssaratow) 1761, 62, 1806, 50, 57, 58, 80; Wolga (Astrachan) 1814–27, 39, 42, 44, 80; Dnjepr (Jekaterinoslaw) 1847, 61, 80; Dnjepr (Kijew) 1845–55, 79 f; Weichsel 1833, 34, 80; Memel 1870 ff.; Donau 1836, 80; Hudson 1816, 17, 60. —Jenissei 1796, 1803, 04, 9–13, 15, 19, 84, 85; Tschussowaja 1855–61, 65, 66; Kama 1786, 1804–17, 24, 35, 36, 58, 61, 80; Ssyssola 1771, 72, 82–84, 87–89, 91, 93–95, 97–99, 1843, 55–57, 80; Waga (Welsk) 1806, 32, 33, 55, 79 f.: Waga (Werchowashskij Possad) 1801 (replaced by 1800), 04, 07, 32–35, 39 f.; Lickscha 1827, 30, 31, 47, 48, 57, 64; Kyro 1741, 43, 1851; Åbo 1780, 1832, 40; Kumo 1828, 71, 74; Düna 1559–61, 69, 70, 73–75, 99, 1600, 03–08, 10, 11, 13, 14, 20, 24, 25, between 1640 and 1700 only the following years are available: 1643, 49, 50, 53, 59, 62, 67, 87, 89, and 92.

The lustra-averages of the ice-free periods were expressed as deviations from the mean of 1816–80 whereas the ones of the rivers' opening dates were expressed by the difference between the observed lustrum-average and the mean value of 1816–80. As a result, a negative symbol (–) indicates that temperatures are too low, a positive symbol (+) that they are too high. In cases where the 65-year average of 1816–80 could not be obtained directly it was determined by reduction according to neighbouring stations of the same group (see below). The use of this method is endorsed by

Woeikof, whereas Rykatschew opposes it.[317] All figures for the break-up of the ice in Finland's rivers refer to the 65-year average of 1781–1845. The date is determined according to the new style. Rykatschew mentions that when he switched from the old to the new style, he neglected to take into consideration that in the 18th century the difference was not 12 days as is the case in the 19th century, but only 11. I have ignored this minor deviation; strictly speaking, however, the times of the opening of rivers in the past century ought to be reduced by one day, and the smoothed figures in the table below to be increased by one day. For the 16th and 17th century—only data from Riga can be traced back that far—this correction would amount to +3.0 and +2 days respectively. This, of course, has no bearing on the length of the ice-free period.

Length of Ice-free Time at 32 River Stations
Given as Deviations (Days) from the Mean 1816–80

River	Lena	Sselenga	Angara	Jenissei	Ob	Irtysch
Station	Kirensk	Sselenginsk	Irkutsk	Jenisseisk	Barnaul	Tobolsk
N. Latitude	57.8	51.1	52.3	58.4	53.3	58.2
E. Longitude	108.1	106.9	104.3	92.1	83.8	68.2
Mean 1816–80	163.9	207.4	279.7	196.8	198.7	189.7
1736–40	—	—	–10.3*	—	—	—
41–45	—	—	–6.9	—	—	—
46–50	—	—	2.3	—	—	—
51–55	—	—	**12.7**	—	—	—
56–60	—	—	3.7	—	9.5	—
61–65	—	—	–3.1	—	–8.5*	—
66–70	—	—	4.1	—	–6.3	—
71–75	—	—	–1.7	—	**12.3**	—
76–80	—	—	—	—	–2.7	—
81–85	—	—	—	—	2.1	—
86–90	—	—	–7.5*	—	–4.3*	—
91–95	—	—	**5.5**	—	–3.3	—
96–1800	—	—	–5.7	—	7.3	—
1801–05	—	—	–2.9	—	—	—
06–10	—	—	–0.5	—	—	—
11–15	—	—	–24.9*	—	—	–12.9*
16–20	–2.3	—	–4.1	—	—	–8.5
21–25	–1.9	—	7.1	—	—	5.3
1826–30	–3.5	—	**12.3**	5.6	—	2.5
31–35	–7.9*	—	7.5	–5.6	—	0.5
36–40	–0.5	—	–15.5	–10.0*	}–12.9*	–1.5*
41–45	**2.3**	—	–5.9	0.8		} 2.0
46–50	–5.9	–11.2*	–4.3	–3.6	–9.0	
51–55	–2.1	0.4	–18.7*	–2.6	–10.1	—

[317] Woeikof in the Iswestija of the Imperial Russian Geographical Society Volume XXIII.

River	Lena	Sselenga	Angara	Jenissei	Ob	Irtysch
Station	Kirensk	Sselenginsk	Irkutsk	Jenisseisk	Barnaul	Tobolsk
56–60	—	**6.8**	—	0.6	**4.3**	**4.9**
61–65	—	2.4	—	–1.2	–5.9*	—
66–70	—	6.6	—	**12.0**	0.3	—
71–75	—	—	—	0.2	–7.5	—
76–80	—	—	—	1.4	—	—

River	Ob	Teich	Kama	Kama	Belaja	Ai
Station	Obdorsk	Bogoslowsk	Dedjuchin	Perm	Ufa	Slatoust
N. Latitude	66.5	59.8	59.4	58.0	54.7	55.2
E. Longitude	66.6	60.0	56.6	56.3	56.0	59.7
Mean 1816–80	148.5	162.0	187.0	208.1	210.8	174.1
1736–40	—	—	—	—	—	—
41–45	—	—	—	—	—	—
46–50	—	—	—	—	—	—
51–60	—	—	—	—	—	—
56–60	—	—	—	—	—	—
61–65	—	—	—	—	—	—
66–70	—	—	—	—	—	—
71–75	—	—	—	—	—	—
76–80	—	—	—	—	—	—
81–85	—	—	—	—	–18.0*	—
86–90	—	—	—	—	–4.4	—
91–95	—	—	—	—	–6.4	—
96–00	—	—	—	–3.3	**2.6**	—
1801–05	—	—	—	**7.9**	–5.4	—
06–10	—	—	—	—	–5.4	—
11–15	—	—	0.2	—	–18.8*	—
16–20	—	—	–5.6	—	–3.6	—
21–25	—	—	**10.9**	—	–8.6	—
26–30	—	—	–3.6	**6.9**	**3.4**	—
31–35	—	—	–6.6	1.1	–4.8*	—
36–40	–2.1	—	—	–6.3*	–4.2	—
41–45	–4.5	–6.8*	—	–4.9	0.2	–12.5*
46–50	–13.6*	–3.5	–1.0	**–2.7**	**1.2**	–12.5
51–55	1.9	3.5	–1.8	–4.9	–0.8	1.9
56–60	**3.1**	0.0	—	—	–7.3*	2.7
61–65	—	—	—	–16.6*	–1.8	–11.8*
66–70	—	3.2	—	0.3	—	**13.4**
71–75	—	**10.3**	—	–0.1	—	10.2
76–80	—	—	—	–4.1	—	—

River	Ssyssola	Wytschegda	Sschona	Wologda	Dwina	Onega
Station	Ust-Ssyssolsk	Ssolwyt-schegods	Welikij-Ustjug	Wologda	Arch-angelsk	Onega
N. Latitude	61.7	61.3	60.8	59.2	64.5	63.9
E. Longitude	50.8	46.9	46.3	39.9	40.5	38.1
Mean 1816–80	186.0	192.3	196.3	201.7	179.0	199.0
1736–40	—	—	—	—	–11.4*	—
41–45	—	—	—	—	–3.2	—
46–50	—	—	—	—	–11.0	—
51–55	—	—	—	—	8.8	—
56–60	—	—	—	—	–16.8*	—
61–65	—	—	**10.1**	—	–1.0	—
66–70	—	—	9.3	—	–0.2	—

River	Ssyssola	Wytschegda	Sschona	Wologda	Dwina	Onega
Station	Ust-Ssyssolsk	Ssolwyt-schegods	Welikij-Ustjug	Wologda	Arch-angelsk	Onega
N. Latitude	61.7	61.3	60.8	59.2	64.5	63.9
E. Longitude	50.8	46.9	46.3	39.9	40.5	38.1
71–75	—	—	–0.1*	—	**16.8**	—
76–80	—	—	2.5	—	–5.2	—
81–85	—	—	16.9	—	–0.0	–8.6
86–90	—	—	1.5	—	–4.4	–3.8
91–95	—	—	4.4	—	–10.4	–13.8*
96–1800	—	—	—	—	–2.0	–11.2
1801–05	—	—	–3.3	—	–5.4	—
06–10	—	—	–10.1*	–14.7*	–14.8*	—
11–15	—	—	–8.1	—	–13.8	–11.2
16–20	–18.0*	—	} –5.4	—	–12.4	–13.6*
21–25	–0.6	—		—	1.4	6.8
26–30	–1.6	—	} –3.0	**16.6**	**5.2**	**16.8**
31–35	3.8	—		—	1.8	–0.6
36–40	**5.4**	–4.3	–1.7	0.3	2.0	4.4
41–45	–5.5*	–11.6*	–11.5*	–13.0*	–10.2*	–20.4*
46–50	6.6	4.9	2.7	—	1.6	–3.4
51–55	–4.2	**7.7**	2.2	—	–0.2	**3.8**
56–60	**5.0**	0.7	—	**5.3**	**2.6**	–0.8
61–65	4.4	1.7	0.9	–1.9	–1.8	–0.2
66–70	4.0	2.1	**6.3**	–0.5	–4.4*	2.0
71–75	1.2	3.3	–5.1*	–7.1*	–0.8	–6.6*
76–80	–0.2*	–0.5*	7.2	1.3	8.4	11.3

River	Newa	Düna	Wjatka	Wjatka	Kama	Wolga	Wolga
Station	Petersburg	Riga	Slobod-skoj	Wjatka	Jelabuga	Ssimbirsk	Ssaratow
N. Latitude	59.9	56.9	58.7	58.6	55.8	54.3	51.5
E. Long.	30.4	24.1	50.2	49.7	52.1	48.4	46.0
Mean 1816–80	219.5	238.0	195.0	203.2	207.1	230.1	235.3
1761–65	7.3	—	—	—	—	—	9.7
66–70	2.8	—	—	—	—	—	**17.3**
71–75	–8.3	—	—	—	—	—	–6.1
76–80	–5.9	—	—	—	—	—	–9.5*
81–85	–3.5	—	—	—	—	—	6.7
86–90	–12.2*	—	—	—	—	—	–1.9
1791–95	**5.5**	—	—	—	—	—	–0.1
96–1800	0.9	—	—	—	—	—	—
1801–05	–4.9	—	—	**5.2**	—	—	—
06–10	–16.7	—	—	–5.4*	—	—	–5.8
11–15	–2.3	–18.0*	—	–1.8	—	—	–0.1
16–20	–11.8	—	—	–5.4*	–8.5*	—	8.5
21–25	**21.1**	**33.2**	—	**8.0**	–1.3	—	—
26–30	15.7	7.8	—	–7.0	**10.2**	—	—
31–35	1.3	10.3	2.2	–2.2	–10.4*	–19.6*	–17.7*
36–40	–2.7	15.0	0.6	1.6	2.6	–10.5	–13.5
41–45	–2.9	5.6	4.4	–8.2*	–3.9	–4.5	2.3
46–50	2.1	14.2	**10.6**	7.2	–5.1*	–5.5	–0.5
51–55	–9.3*	–19.0*	–1.4	–3.2	2.1	0.5	—
56–60	–7.7	–5.6	6.2	} 1.6	—	3.5	–5.6

River	Newa	Düna	Wjatka	Wjatka	Kama	Wolga	Wolga
Station	Petersburg	Riga	Slobod-skoj	Wjatka	Jelabuga	Ssimbirsk	Ssaratow
61–65	–1.5	0.0	–3.4*		3.3	–0.1	–2.3
66–70	–0.9	–7.6	1.8	2.0	**7.1**	6.9	6.7
71–75	–7.5	0.0	–2.0	–4.4	–0.8	15.9	**10.5**
76–80	4.3	7.4	6.7	11.1	8.9	**20.9**	9.9

River	Wolga	Dnjepr	Dnjepr	Weichsel	Memel	Donau	Hudson
Station	Astrachan	Jekatarinoslaw	Kijew	Warschau	Tilsit	Galatz	Albany
N. Lat.	51.5	48.5	50.5	52.3	55.1	45.4	42.7
E. Long.	46.1	35.1	30.5	21.0	21.9	27.9	–73.8
Mean 1816–80	264.9	277.0	268.7	304.9	264.7	327.8	273.3
1781–85	—	—	—	–3.5	—	—	—
86–90	—	—	—	6.1	—	—	—
91–95	—	—	—	19.7	—	—	—
96–1800	—	—	—	—	—	—	—
1801–05	—	—	–14.5*	—	—	—	—
06–10	2.5	—	–3.9	—	—	—	—
11–15	–17.9*	—	–4.3	—	—	—	—
16–20	—	–6.7	6.5	—	—	—	–18.9*
21–25	—	39.4	**10.3**	**12.5**	**24.5**	—	**18.9**
26–30	–10.9	2.6	4.5	–20.7	–14.9	—	10.7
31–35	–7.5	–11.8	–7.3	–20.4	15.7	—	–5.9
36–40	–2.9	–11.8	–13.9*	–24.1*	–36.3*	–9.4	–7.1*
41–45	6.8	–26.6	5.1	–9.5	11.1	–10.8*	1.5
46–50	–1.3	26.0	—	-8.9	–2.1	–2.8	6.9
51–55	14.1	1.4	—	13.7	–4.3	–27.2	–8.1*
56–60	2.9	—	–13.9*	–13.7	–19.3	4.8	**10.9**
61–65	–10.5*	–71.6	–11.1	–1.1	3.9	–23.8*	–3.1
66–70	7.7	7.8	–4.5	**14.5**	**13.5**	–1.4	—
71–75	2.3	–5.2	**8.1**	2.7	—	–7.2	—
76–80	14.9	13.8	–3.9	12.1	—	7.1	—

Date of Break-up of Ice in Rivers at 12 Stations
Based on the Differences Between Lustra Averages and the Mean of 1816–80

River	Jenissei	Tschussowaja	Kama	Ssyssola	Waga	Waga Werchowashskij
Station	Jeni-sseisk	Ust-Utkin-skajaStanika	Perm	Ust-Ssyssolck	Welsk	Possad
N. Latitude	58.5	57.0	58.0	61.7	61.1	60.7
E. Long.	91.1	59.6	56.3	50.8	42.1	42.0
Mean 1816–80	5. V	21.IV	25.IV	5.V	29. IV	25. IV
1771–75	—	—	—	**10.7**	—	—
76–80	—	—	—	8.8	—	—
81–85	—	—	—	} –1.5*	—	—
86–90	—	—	–2.5		—	—
91–95	—	—	–8.6*	} **6.3**	—	—
96–1800	–4.5	—	–2.0		—	—
1801–05	**2.0**	—	**4.3**	–0.4	—	7.8
06–10	–2.0	—	—	–3.6	–6.0*	3.2
11–15	—	—	—	–5.4*	–3.0	–5.4*
16–20	2.5	—	–1.3	–2.8	–2.4	–2.8
21–25	**2.6**	—	**3.2**	1.4	—	3.2
26–30	0.6	**1.2**	1.0	**4.2**	3.0	1.6
31–35	–1.8	–2.6	–2.5	3.0	11.6 }	4.0

River	Jenissei	Tschussowaja	Kama	Ssyssola	Waga	Waga Werchowashskij
Station	Jeni-sseisk	Ust-Utkin-skajaStanika	Perm	Ust-Ssyssolck	Welsk	Possad
36–40	–5.0*	–3.8	–6.5*	2.4	—	—
41–45	3.8	–4.6*	–5.6	–3.8*	–5.2*	—
46–50	–3.8	–2.4	0.0	–1.6	5.6	—
51–55	–0.2	–1.2	0.0	–1.8	5.8	—
56–60	–1.2	—	**2.8**	**0.3**	—	—
61–65	0.0	**9.3**	–0.2	–1.0	0.2	—
66–70	**3.2**	–1.2	–1.0	–2.4	–0.4	—
71–75	0.4	0.8	0.6	–0.6	2.8	—
76–80	–0.6	-0.8	–3.2*	–4.0*	2.7	—
81–85	–5.3*	—	—	—	—	—

River	Lickscha[1]	Borgå[1]	Kyro[1]	Åbo[1]	Kumo[1]	Düna[2]
Station	Pielis	Borgå	Storkyro	Åbo	Björneborg	Riga
N. Latitude	63.3	60.4	63.0	60.5	61.5	57.0
E. Longitude	30.1	25.7	22.3	22.3	21.8	24.1
Mean 1816–80	3. V	24.IV	26. IV	16. IV	24. IV	7. IV
1741–45		—	–1.7	–8.0	—	–6.6
46–50		—	4.4	–4.8	—	0.0
51–55		—	**9.2**	**1.6**	—	2.6
56–60		—	1.6	-5.0	—	2.4
61–65		—	1.4	0.8	—	**2.6**
1766–70		—	–0.6	—	—	–2.8
71–75		0.3	0.6	–4.4	—	–3.4
76–80		1.3	0.0	–0.2	—	5.0
81–85		–1.0*	2.0	—	—	–7.2
86–90		–0.9	–1.4*	—	—	–8.6*
91–95		**7.9**	**5.8**	—	—	**5.4**
1796–1800		2.1	2.8	—	—	–2.8
1801–05		7.8	4.0	3.8	6.0	–2.8
06–10		–10.7*	–11.4	–11.0*	–11.2*	–13.5*
11–15		–0.7	1.8	–0.6	–0.2	–3.4
16–20		–5.1	–5.5	–4.2	–3.6	–4.2
21–25		**5.6**	2.5	**7.2**	6.4	11.0
26–30	2.7	0.2	1.1	–2.0	0.8	–3.2
31–35	1.0	4.1	**3.9**	3.2	**6.8**	**11.2**
36–40	–1.6	5.8*	0.0	–4.5	–0.2	–6.6*
41–45	–4.2	–1.1	–3.3	—	1.6	–3.8
46–50	8.7	—	0.6	—	2.3	–1.2
51–55	56*	—	–4.2	—	–13.6*	–6.2

[1]) The Finnish stations based on the mean of 1781–1845.
[2]) Previous years see next table.

River	Lickscha[1]	Borgå[1]	Kyro[1]	Åbo[1]	Kumo[1]	Düna[2]
Station	Pielis	Borgå	Storkyro	Åbo	Björneborg	Riga
56–60	5.4	—	—	—	—	2.0
61–65	−4.0	—	—	—	—	**4.8**
66–70	−3.2	—	—	—	—	−0.8
71–75		—	—	—	−0.7	−3.2*
76–80		—	—	—	—	1.2

Date of Ice Break-up of the Düna at Riga [1)] Mean 7. IV						Ice-free Time of the Newa at St. Petersburg [2)] Mean 219.5	
	raw	sm.[+]		raw	sm.[+]	raw	sm.[+]
1556–60	−1.7*	−2.1*	—	—	—	—	—
1561–65	−3.0*	−1.0	1687–92	2.7	—	—	—
1566–70	**3.7**	**1.4**	—	—	—	—	—
1571–75	1.0	1.3	1706–10	—	—	9.1	6.2
1576–80	−0.4	0.4	1701–15	1.6	−0.7	0.5	1.2
1581–85	1.6	−0.4	1716–20	−5.4*	−0.8*	−5.1*	−0.5*
1586–90	−4.4	−3.2	1721–25	5.8	0.2	7.7	5.0
1591–95	−5.6	−5.7*	1726–30	−5.4	1.3	**9.5**	**6.8**
1596–00	−7.0	−5.5	1731–35	**10.2**	**4.7**	0.3	−1.1
1601–05			1736–40	3.8	0.3	−14.5*	−8.1
1606–10	−2.4	−2.6	1741–45	−6.6*	−2.4*	−3.7	−8.8*
1611–15			1746–50	0.0	−1.0	−13.5	−5.5
1616–20	1.5	−1.2	1751–55	**2.6**	1.9	**8.7**	1.4
1621–25	−5.3*	−2.3	1756–60	2.4	**2.5**	−0.7	**6.1**
—	—	—					
1641–50	7.3	—	See tables on pages 110 and 112 for a continuation.				
1651–67	−21.4*	—					

In order to eliminate any locally produced particularities individual observation series were categorised as follows:

For the length of the ice-free period:

1. Siberia: Lena, Sselenga, Angara, River Ob near Barnaul and near Obdorsk and Irtysk.
2. Ural-Region: Reservoir near Bogoslowsk, Kama near Dedjuchin and near Perm, Belaja and Ai.
3. North Russia (the entire catchment area of the White Sea): Ssyssola, Wytschegda, Ssuchona, Wologda, Swina, and Onega.
4. Baltic Provinces: Newa and Düna.

1) Corrections to be added to the lustra averages in order to arrive at the mean value of multiple years.

2) Deviations from the mean of multiple years.

\+ [smoothed]

5. Southeast Russia: Wjatka near Sslobodoskoj and Wjatka, Kama near Jelabuga and Wolga near Ssimbirsk, Ssaratow and Astrachan.
6. Southwest Russia: Dnjepr near Jekaterinoslaw and near Kijew, Weichsel, Memel and Donau.
7. Atlantic States, North America: Hudson.

For the dates of the ice break-up:

8. Siberia: Jenissei.
9. North Russia: Tschussowaja, Kama, Ssyssola, Waga near Welsk and near Werchowashskij Possad.
10. Finland: Lickscha, Borgå, Kyro, Åbo and Kumo.
11. Baltic Provinces: Düna.

Both raw as well as smoothed group averages are included. On page 113 a small table presents the results of the earliest observations of the Newa at St. Petersburg (1706–60) and the Düna at Riga (1556–1760). They form a continuous line of data with the group averages of the categories "ice-free periods in the Baltic provinces" and "dates of the ice break-up in the Baltic provinces", because the latter's earliest lustra are based on the observations at St. Petersburg and Riga alone.

Secular Variations of the Length of Time Rivers Remain Ice-free in Different regions.

Deviations from the Mean (in Days) 1816–80: Raw Mean Values.

Lustrum	Siberia	Ural	Northern Russia	Baltic Provinces
1736–40	–10.3*	—	–11.4*	–14.5*
41–45	–6.9		–3.2	–3.7
46–50	2.3	—	–11.0	–13.5
51–55	7.8	—	8.8	8.7
56–60	6.6	—	–16.8*	–0.7
61–65	–5.8*	—	4.6	7.3
66–70	–1.1	—	4.6	2.8
71–75	5.3	—	8.3	–8.3
76–80	–2.7	—	–1.4	–5.9
81–85	2.1	–18.0*	2.8	–3.5
86–90	–5.9*	–4.4	–2.2	–12.2*
91–95	1.1	–6.4	6.6	5.5
96–1800	0.8	–0.3	–6.6	0.9
1801–05	–2.9	1.2	–4.4	–4.9
06–10	–0.5	–5.4	–13.2*	–16.7*
11–15	–18.9*	–9.0	–11.0	–10.2
16–20	–5.0	–4.6	–9.7	–11.8
21–25	3.5	9.8	3.3	27.2
26–30	4.2	2.2	6.8	11.8
31–35	–1.4	–3.4	0.5	5.8
36–40	–7.1*	–5.2*	1.0	–8.8
41–45	–3.0	–5.0	–12.0*	1.3
46–50	–6.8	–3.7	2.5	8.1
51–55	–5.2	–1.8	1.9	–14.2*
56–60	3.9	–2.3	2.6	–6.6
61–65	–1.6	–10.1*	0.5	–0.7
66–70	6.3	5.6	1.6	–4.2
71–75	–3.6*	6.8	–2.5*	–3.7
76–80	1.4	–4.1	4.6	5.9

Lustrum	SE Russia	SW Russia	Atl. States N. America	Means
1736–40	—	—	—	–12.1*
41–45	—	—	—	–4.6
46–50	—	—	—	–7.4
51–55	—	—	—	8.4
56–60	—	—	—	–3.6
61–65	9.7	—	—	4.4
66–70	17.3	—	—	5.9
71–75	–6.1	—	—	–0.2
76–80	–9.5*	—	—	–4.9
81–85	6.7	–3.5	—	–2.2
86–90	–1.9	6.1	—	–5.1*
91–95	–0.1	19.7	—	4.4
96–1800	—	—	—	–1.3
1801–05	5.2	–14.5*	—	–3.4
06–10	–2.9	–3.9	—	–7.1
11–15	–6.6*	–4.3	—	–10.0*
16-20	–1.8	–0.1	–18.9*	–7.4
1821–25	3.4	15.8	12.9	10.8
26–30	–2.6	–7.1	10.7	3.7
31–35	–9.2*	–6.0	–5.9	–2.8
36–40	–3.7	–19.1*	–7.1	–7.1*
41–45	–2.0	–0.8	1.5	–2.9
46–50	0.9	–4.6	6.9	–0.5
51–55	2.0	9.5	–8.1*	–2.2
56–60	1.7	–10.5	10.9	0.0
61–65	–1.9	–8.5	–3.1	–3.6
66–70	5.4	6.0	—	3.5
71–75	3.6	0.4	—	0.0
76–80	12.1	7.3	—	4.5

Smoothed Mean Values

Lustrum	Siberia	Ural	N. Russia	Baltic Provinces
1736–40	–8.0*	—	–8.7*	–8.1
41–45	–5.5	—	–7.2	–8.8*
46–50	1.4	—	–4.1	–5.5
51–55	6.1	—	–2.6	0.8
56–60	3.8	—	–5.0	3.6
61–65	–1.5	—	–0.8	4.2
66–70	–0.7	—	5.5	1.2
71–75	1.7	—	5.0	–4.9
76–80	0.5	—	2.1	–5.9
81–85	–1.1	–13.5*	0.5	–6.3*
86–90	–2.2*	–8.8	1.2	–5.6
91–95	–0.7	–4.4	1.1	–0.1
96–1800	0.0	–1.4	–2.8	0.6
1801–05	–1.4	–0.8	–7.1	–6.6
06–10	–5.7	–4.7	–10.4	–12.1
11–15	–10.8*	–7.2*	–11.2*	–12.2*

Lustrum	Siberia	Ural	N. Russia	Baltic Provinces
16–20	–6.4	–2.2	–6.8	–1.6
21–25	1.6	4.8	0.9	13.6
26–30	2.6	2.7	4.4	14.2
31–35	–1.4	–2.4	2.2	3.7
36–40	–4.6	–4.7	–2.4	–2.6
41–45	–5.0	–4.8*	–5.1*	0.5
46–50	–5.5*	–3.6	–1.3	0.8
51–55	–3.3	–2.5	2.2	–6.7
56–60	0.3	–4.1	1.9	–7.0*
61–65	1.8	–4.3	1.3	–3.0
66–70	1.8	2.0	0.3	–3.0
71–75	0.1	3.8	0.3	–1.4
76–80	–0.3	–0.5	2.2	2.7

Lustrum	SE Russia	SW Russia	Atl. States N. America	Means
1736–40	—	—	—	–9.6*
41–45	—	—	—	–7.2
46–50	—	—	—	–2.8
51–55	—	—	—	1.4
56–60	—	—	—	1.4
61–65	12.2	—	—	2.8
66–70	9.6	—	—	4.0
71–75	–1.1	—	—	0.2
76–80	–4.6*	—	—	–3.0
81–85	0.5	–0.3	—	–3.6*
86–90	0.7	7.1	—	–2.0
91–95	–0.7	15.2	—	0.6
96–00	—	—	—	–0.4
1801–05	2.5	–11.0*	—	–3.8
06–10	–1.8	–6.6	—	–6.9
11–15	–4.5*	–3.2	—	–8.6*
16–20	–1.7	2.8	–8.3*	–3.5
21–25	0.6	6.1	4.4	4.5
26–30	–2.8	–1.1	7.1	3.8
31–35	–6.2*	–9.6	–2.0	–2.2
36–40	–4.7	–11.2*	–4.6*	–5.0*
41–45	–1.7	–6.3	0.7	–3.1
46–50	0.5	–0.1	1.8	–1.0
51–55	1.8	1.0	0.4	–1.0
56–60	1.0	–5.0	2.6	–1.4
61–65	0.8	–5.4	1.6	–0.9
66–70	3.1	0.8	—	0.8
71–75	6.2	3.1	—	2.0
76–80	9.3	4.7	—	3.0

Variations of the Date of Ice Break-up in Rivers in Different Regions

	Corrections (in Days) Reduced to the Mean 1816–80 Raw Mean Values			
Lustrum	Siberia	N. Russia	Finland[1]	Baltic Provinces
1741–45	—	—	–4.8*	–6.6*
46–50	—	—	–0.2	0.0
51–55	—	—	5.4	2.6
56–60	—	—	–1.7	2.4
61–65	—	—	1.1	2.6
66–70	—	—	–0.6	–2.8
71–75	—	10.7	–1.2*	–3.4
76–80	—	8.8	0.4	5.0
81–85	—	–1.5	0.5	–7.2
86–90	—	–2.0*	1.2	–8.6*
91–95	—	1.2	6.8	5.4
1796–1800	–4.5	2.2	2.4	–2.8
1801–05	2.0	3.9	5.4	–2.8
06–10	–2.0	–2.1	–11.1*	–13.5*
11–15	—	4.6*	0.1	–3.4
16–20	2.5	–2.3	–4.6	–4.2
21–25	2.6	2.6	5.4	11.0
26–30	0.6	2.2	0.6	3.2
31–35	–1.8	2.7	3.8	11.2
36–40	–5.0*	–1.0	–2.4	–6.6*
41–45	3.8	–4.8*	–1.8	–3.8
46–50	–3.8	0.4	3.9	–1.2
51–55	–0.2	0.7	–7.8*	–6.2
56–60	–1.2	1.6	5.4	2.0
61–65	0.0	2.1	–4.0	4.8
66–70	3.2	–1.2	–3.2	–0.8
71–75	0.4	0.9	–0.7	–3.2
76–80	–0.6	–1.3	—	1.2
81–85	–5.3*	—	—	—

	Corrections (in days) reduced to the mean 1816–80 Smoothed mean values			
Lustrum	Siberia	N. Russia	Finland	Baltic Provinces
1741–45	—	—	–3.3	–2.4*
46–50	—	—	0.0	–1.0
51–55	—	—	2.2	1.9
56–60	—	—	0.8	2.5
61–65	—	—	0.0	1.2
66–70	—	—	–0.3	–1.6
71–75	—	10.1	–0.6*	–1.2
76–80	—	6.7	0.0	–0.2
81–85	—	1.0	0.0	–4.5
86–90	—	–1.1*	1.2	–4.8*
91–95	—	0.6	3.7	–0.2
1796–1800	–2.3*	2.4	4.2	–0.8
1801–05	–0.6	2.0	0.5	–5.5
06–10	–0.7	–1.2	–4.2*	–8.3*
11–15	—	–3.4*	–3.9	–6.1
16–20	2.5	–1.6	–0.9	–0.2
21–25	2.1	1.3	1.7	3.6
26–30	0.5	2.4	2.6	4.0
31–35	–2.0	1.6	1.4	3.2

[1] Relative to the mean of 1781–1845.

	Corrections (in days) reduced to the mean 1816–80 Smoothed mean values			
Lustrum	Siberia	N. Russia	Finland	Baltic Provinces
36-40	-2.0*	-1.0	-0.7	-1.4
41-45	-0.3	-2.6*	-0.5	-3.8*
46-50	-1.0	-0.8	-0.4	-3.1
51-55	-1.4	0.8	-1.6*	-2.9
56-60	-0.6	1.5	-0.2	0.6
61-65	0.5	1.2	-1.4	2.7
66-70	-1.7	0.2	-2.8*	0.0
71-75	0.8	-0.2	-1.5	-1.5
76-80	1.5	-0.6	—	-0.3
81-85	-3.7*	—	—	—

The conclusions that can be drawn from our tables go into two directions. In the first place, they confirm the findings about the climate variations of the past century which were based on observations in Western Europe, for the European and Asian parts of Russia as well and provide, in a most welcome way, additional data about the development of temperatures dating back to the year 1700 and even further. It is not until after 1850 that variations are vague and erratic. This is in accordance with the fact emphasized earlier, namely that from that point on temperature variations in general have become much less pronounced and appear to be indistinct. In contrast, the temperature minimum of 1836–50 is evident everywhere, regardless whether we look at the length of the ice-free period or the break-up of the river ice. The latter, in addition, clearly indicates the warm period around 1860 and signifies the temperature decrease towards the year 1880. More important are the results concerning more remote periods of time, for which, generally, very few meteorological observations are available and for Russia none. Observations at the Düna show a minimum temperature around 1560, followed by a maximum around 1570 and a second minimum around 1595, which is then followed by a maximum around 1616/20.* It is note-worthy that from changes in the Alps' glaciers and of Lake Trasim as well as Lake Fucin we were able to identify a wet period around 1600, which corresponds with the cold period observed in Riga at that same time. Unfortunately from 1626 to 1710 the observations at Riga are incomplete so that we cannot draw any definite conclusions regarding the variations of the ice conditions during this time. It seems, however, that there was a cold period during the twenties of the 17th century, followed by a warm period in the forties, and again a cold period in the fifties and sixties. Not until the 18th century is it possible again to give an accurate account of the tempera-ture variations as shown in the following table:

* [See table on page 113]

Cold	Warm
1716–1720	1721–1735
1736–1750	1751–1770
1771–1790	1791–1805
1806–1820	1821–1830
1831–1850	from 1850 on undetermined

This brief presentation corresponds quite well with the one done earlier on temperature variations. It shows an additional warm period from 1721 to 1735 preceded by a cold period, a fact which was not known to us up to now for lack of temperature observations, but which is quite consistent with the water level variations of the lakes. A small inconsistency appears only during the period of 1750–90. Although both the ice conditions of the rivers as well as the temperatures suggest that this period was warm at the beginning and cold at the end, the two factors differ as to the time of this temperature change: according to the thermometer observations it occurs as early as around 1755, according to the ice conditions of the rivers not until around 1770. The latter is exactly the same year that marks the beginning of the last wet period of the past century, thereby removing the only major incongruity between the variations of temperature and rainfall which we encountered earlier. It confirms our assumption that the temperature curve might have been distorted.

In order to deal statistically with the amplitude of the variations in the duration of the ice cover as well as in the date of its break-up, I determined the mean value of the lustra extremes for the different regions based on the lustra in the tables given above. Only the observations prior to 1855 were included, while the indistinct development of the last decades was disregarded.

The figures for the length of time during which rivers were free of ice are again deviations from the mean value, those for the dates of the ice break-up are again differences between mean and actual values, so that, throughout, + (plus) means warm and – (minus) cold.

Maximum and Minimum Lustra Averages and the Amplitudes of the Secular Variations of Ice Conditions on Rivers.

Duration of Ice-free Time	Siberia	Ural	SE Russia	N. Russia	Balt. Prov.	SW Russia	Hud-son
Medium of Minimum	–12.0	–10.8	–8.4	–11.9	–12.5	–16.8	–13.0
Medium of Maximum	4.3	5.5	8.6	8.0	12.6	15.0	10.7
Amplitude	16.3	16.3	17.0	19.1	25.1	31.8	23.7

Date of Ice Break-up	Siberia	N. Russia	Finland	Balt. Prov.
Medium of Minimum	–5.0	–3.8	–6.2	–8.1
Medium of Maximum	2.6	5.7	5.9	7.4
Amplitude	7.6	9.5	12.1	15.5

In Siberia the ice break-up in rivers occurs on the average slightly more than one week later in the coldest lustrum of a secular variation than in the warmest lustrum, and the ice-free period falls short by 16 days per year. These differences further increase towards the West. In the Baltic provinces the rivers open up 15–16 days later in the coldest lustrum than in the warmest and each winter they remain ice-covered for a full 25 days longer than in the warmest lustrum. In South West Russia and Poland this amounts even to a total of 32 days. This geographical trend towards the West is such a general and consistent phenomenon that its factuality cannot be doubted. It would, however, be totally wrong to conclude that towards the West the secular temperature variations increase in size. Rather, the phenomenon has a completely different source. The freezing and melting of the rivers takes place approximately around the time when the annual temperature curve passes the freezing point. In Siberia, this occurs during seasons with rapid periodic temperature changes, i.e., in the fall and the spring. Towards the West this point in time moves further and further into the winter period, during which periodic changes of the temperature are very gradual. A rise or fall of the entire temperature curve by a certain amount will therefore in Siberia cause not more than a slight shift in the times when the freezing point is passed, whereas in West Russia the effect is considerable. This very fact is undoubtedly the base of Rykatschew's findings about a decrease in the variability of the duration of the ice period as well as the freezing and melting dates, which he claimed to have observed "in an increasing distance of continental regions from the ocean and also from South to North."[318] The distance to the ocean and the geographical latitude are of equal importance. Decisive is the extent of the periodic variation in temperature at the time when melting and freezing usually takes place.

In this way, our tables allow us to draw reliable conclusions about the existence of temperature variations yet fail to give any indication of their intensity in different locations. We count 5 1/2 variations from the beginning of the past* century up to the year 1885 and can therefore determine an average duration of 34 years. This result is lower than the one gained from the shorter observation series pertaining to temperature and rainfall. Yet if we ignore the years prior to 1740, we arrive at the previous average duration of 36 years. If we use the observations of the Düna from 1556 on, which are admittedly quite incomplete, we have a total of 9 variations spanning over 325 years and obtaining again an average of 36 years. These results are fully supported by the secular variations in the timing of the grape harvest.

[318] Rykatchev, op. cit., p. 23

* [18th century]

4.2.2. Secular Variations of the Time of the Grape Harvest

There seems to be no phenological event whose records date farther back than the beginning date of the grape harvest. In many regions the individual wine-grower is not at liberty to begin with the harvest at his own choosing. An official decree by the administrative authority, be it the mayor or the local council, is necessary authorizing the start of the harvest and thereby lifting the ban which was imposed on vineyards some time before the ripening of the grapes.

Once the starting of the harvest had become an administrative measure, its execution naturally was put to protocol each year. Records obtained in this manner, though being sparse, render phenological data, which could not be more reliable because the release of the date was decided on the advice of experienced wine-growers.

In fact many old archives and chronicles have already been searched and scientists have repeatedly undertaken the effort to collect the data. Consequently, we have today an important body of critically assessed data at our disposal. Since wine-growing naturally is restricted to certain areas, the data refer to these regions only. The available records are limited to some stations in South West Germany and in Switzerland as well as to a large number of stations in France.

In 1880, when phenological observations were introduced and organized in France universally, the Bureau Central Météorologique de France saw it as one of its duties to collect all the earlier relevant observations. Automatically, so-to-speak, the first step turned out to be the compilation of the data about the start of the grape harvest. To this purpose, the meteorological commissions of the départements were instructed to extract from the old chronicles and archives all the information pertaining to this subject. The reports sent to the Bureau were then carefully researched by A. Angot and their results were published in 1885.[319] From this publication we not only selected the observation series carried out at 19 French stations, the length of which is remarkable, but also data from Kürnbach in Baden and Stuttgart. The French records are available in print up to the year 1879. In addition A. Angot has been kind enough to pass on to me his manuscript notes about the observations of the year 1880, for which I thank him. For Switzerland Ch. Dufour's excellent publication[320] provided four long observation series to which are added two published by Angot. I am also

[319] Annales du Bureau Central Météorologique de France. Année 1883, I. (Paris 1885), p. B.29 to B.120.

[320] *Notes sur le problème de la variation du climat,* [*Note on the problem of climate variations*], Bull. de la Soc. Vaudoise des sc. Naturelles, X., No. 63 (1870), p. 359 ff.

grateful to Prof. Dr. F. A. Forel in Morges for submitting the series of observations done in Pully near Lausanne and to R. Wehrli in Zürich for the one from Altstetten.

Consequently, the following results are based on a total of 29 series. For France I could have easily included 5–10 additional stations from Angot's publication with observations running over a period of 50–60 years. However, the series I used provided such reliable and consistent results, that I saw no further need to increase the amount of data.

Before we enter into a discussion of the tables' contents, we should try to define the significance of our data. Obviously the timing of the grape harvest can not serve as a reliable phenological factor in determining climatic variations unless it is proven that man's interference has little impact and that changes in the cultivation of grapes did not affect the homogeneity of the observation series in any major way over time.

The number of factors which, aside from purely meteorological ones, determine the start of the grape harvest is not exactly small. Those having to do with the location and conditions of the ground are of no concern to us here. These may change from location to location, but stay the same in one site and can therefore never affect the homogeneity of an observation series. We are only concerned with the factors that for that one site have changed over time.

The ripening of grapes depends largely on the age of the vine; the grapes of young vines require a longer time to ripen than those of older vines. But this fact should be of no importance to us because normally the number of young vines is very small and grape picking generally will take place according to the ripening process of grapes from older vines, and not *vice versa*. Of course, if in some places diseases like Phylloxera require the destruction of old vines, which then are replaced by young vines, the homogeneity of a sequence can definitely be disturbed.

Furthermore, the time it takes for grapes to ripen differs from variety to variety. In his "Textbook on Agriculture" DeGasparin, using Odart's system, categorizes all wine varieties into seven groups[321] according to the amount of temperature required for the ripening process. This amount is determined by using the formerly accepted method of summarizing the temperature readings from an actinometer* the results for the seven categories are as follows:

[321] Angot, op. cit., p. 34

* [Schwarzkugelthermometer]

Classification		Grapes	
I.	2264° C	Grapes	
II.	3400	} Medium 3480° C	
III.	3564		
IV.	4133		
V.	4238	} Medium 4250° C	
VI.	4392		
VII.	5000		

As the categories I and VII have no bearing on the wine regions of interest to us, and the remaining are grouped around the values of 3480° and 4250° without any major deviations, Angot suggests to distinguish between two groups only. Actually this is quite sufficient for phenological climate studies.

The difference between the temperature totals which are necessary to ripen the grapes of the two groups is quite substantial, and consequently harvesting takes place at completely different times depending on which variety is grown in one particular location. Therefore, the introduction of a new variety of vines could possibly shift the timing of the grape harvest in a certain region considerably. In the North there is little danger of this ever happening. Along the latitude of Paris only those grapes will reach maturity which do not require high temperatures and the introduction of a more demanding variety is simply impossible. Not so in the South! Here, naturally, varieties requiring different degrees of temperature are able to grow side by side. Preference, however, is given to the varieties for a warmer climate in order to use soil and climate to best advantage. A certain balance has developed: each region cultivates those varieties best suited to its particular climate.

Angot[322] gives evidence of how conservative the wine-growers of a certain region are because of these very reasons. Columellus[323] describes the different grape varieties of Gallium. From his description one is clearly able to distinguish the Pinot grape among other varieties, which even today produces the highest yield in Burgundy. During the time span for which exact written records are available, i.e., since the year 1330 for the region of Beaune and since 1430 for the region of Dijon, viniculture in Burgundy quite obviously has not changed much: the same variety is still grown, the same sites still produce the best wine today as they did in the past. Since the time of Gregor de Tours, i.e., since the VIth century, the same vineyards have always produced the heavy Burgundy wines. And this is a natural course of events; because if a vine is transplanted to a location differing, if only slightly, from the original climatic conditions, the grape loses its distinct characteristics and the wine its bouquet. All in all it is not likely that a region

[322] Angot, op. cit., p.34
[323] Angot, op. cit., p. 83.

ever changed its variety on a large scale, a step which would disrupt the homogeneity of the records for us today.

Cultivation methods are another major factor in the ripening process of the grapes and consequently in the timing of the harvest.[324] One aspect is for instance the distance between the planted vines. The closer together they grow, the more shade there is in the vineyard, the soil remains moist and the ripening process is delayed. Whether the vineyard is fertilized or not is another relevant aspect.[325] Sulphur, which is often used to fight certain diseases of the vine, accelerates the ripening process considerably. As an example, Marés describes[326] the case of the region of Launac, Departément Hérault, where in the years of 1838–54 the vineyards on the average used to start harvesting on September 19, whereas from 1855 to 1872 when sulphur was used regularly, they started as early as September 5. Smoke has the opposite effect, the ripening process is delayed.

In addition to these interferences in connection with various methods of cultivation, another aspect might induce the wine-grower to start harvesting prematurely: his fear of field robbers. In this case, according to Angot,[327] wholesalers have a vested interest in the picking of the grapes before they are fully matured because wine made from these grapes will keep longer and be less susceptible.

Certain weather conditions may be another cause for starting the harvest before the grapes have reached full maturity. In the eastern and northern wine regions of France, for instance, cold weather often forces a premature harvest to avoid a total loss of the crop.[328] There have even been years in which no wine was produced, i.e., in Verdun in 1816, 1821, and 1830.

Last but not least, the timing of the harvest is affected by taste and custom, which have always and everywhere been subject to changes and must naturally reflect in our records. But since they occur gradually, they presumably have no bearing on the rather short-termed variations we are interested in.

Obviously, the potential sources for error are numerous enough and we will have to avoid them in order to base our results on solid ground. We need to take a careful look at our data sequences to ensure their homogeneity and to discover any infractions. It is quite useful in this regard that of all the factors influencing the beginning of the grape harvest the meteorological factors are the only ones which affect large areas in a general way. All that is needed for the discovery of distorting factors is a comparison between the

[324] Angot, op. cit., p.34

[325] Dufour, in Bull. Soc. Vaud. des sc. nat., X., p. 396.

[326] Ladray: Cours de Viticulture et d'Oenologie, Tome I, p. 594. (Citat d'Angot.)

[327] Angot, op. cit., p. 35.

[328] Angot, op. cit., p. 113.

data series of neighbouring regions with regard to any changes in the timing of the grape harvest from one year to the next. Once again it is the method of determining the differences which allows us to distinguish between suitable and unsuitable data. This way it was possible to identify the data series from Perpignan, Pierrefeu near Toulon, Montmorency and Châtillon-sur-Saône as unsuitable.

Having sorted through the data, I formed lustra averages, which are shown in the following table. A lustrum average followed by a period symbol means that this lustrum is incomplete.

Determining the lustra averages was quite simple. In all his tables Angot defines the beginning of the grape harvest in relation to the days of the month of September, e.g., the number 30 means September 30, 31 means October 1, 45 means October 15, –2 means August 29, etc. All I needed to do was to determine the lustra averages of the dates published by Angot in order to arrive at the average starting date expressed through a day of the month of September.

These lustra averages are rendered in full days only. The stations are arranged according to regions and within these according to their geographical latitude.

With regard to the different stations I would like to offer the following preceding comments:

For Verdun the year 1803 is missing (1816, 21, and 30 no grapes were harvested, because they did not reach maturity; the lustra concerned are in brackets and not included in determining the minimum). For Argenteuil 1751, 1803; Foug 1796, 1804–09, 38, 40, 41, 53; Loches 1804, 09; Denainvilliers und Boësses 1740, 60, 74, 75, 92, 96 (between 1741 and 1780 figures refer to the former location, from 1781 on to the latter; Boësses is only 15 kilometres east of Denainvilliers, so that the data from both places can be integrated); Auxerre 1799; Vendôme 1800; Vesoul 1815; Dijon 1391, 95, 96, 1404, 06, 38, 46, 47, 60, 62, 92, 1513, 21, 24, 26, 29, 31, 41–44, 62, 71, 1650, 1720, 94–95, 1814 (longest series, observations start as early as 1366, are however up to the year 1390 so fragmentary, that they seem useless for our purpose); Salins 1563–65, 67, 73, 75, 81, 86, 95, 96, 1605, 10, 15, 18, 19, 29, 34, 35, 53, 56, 57, 73, 74, 1709, 10, 41 (during the years 1794, 1809 and 1843 no grapes were harvested; the lustra averages concerned are formed from 4-year lustra and are in brackets); Lons-le-Saulnier 1669. 74, 81, 84, 88, 89, 91, 94–99, 1702, 03, 06, 14, 15, 17–20, 25–27, 30, 31, 37–40, 42, 77, 79, 91, 95, 1803; Pichon-Longueville (in the Médoc region) 1751, 53, 60, 1793–1800; Médoc (the series represents the entire Médoc region as published by the chamber of commerce of Bordeaux in 1882, comp. Angot, op. cit, p. 41); Tain 1879; Castres 1801, 02, 10.

Southwest Germany and Switzerland

Missing are: For Stuttgart 1768, 1814, 15, 16; Kürnbach 1613, 17, 19, 25, 33, 38–40, 42, 51, 53, 58–50, 63, 73, 74, 78, 89; Altstetten 1611, 12, 16,17, 32, 39, 41, 43, 46–48, 51, 53, 54, 58, 62, 76, 77, 79, 83, 84, 88, 89, 92, 95, 96, 98, 1709, 15, 33 no harvest, 98–99, 1815, 18, 89, 90. For the villages around Lake Genève ranked in a north-east to south-westerly direction: Veytaux 1741, 43, 46, 48, 57, 58, 63, 65, 67–69, 72,77, 79, 1800, 02; Vevey 1800; Lausanne 1500, 01, 08, 16, 24, 25, 27, 32, 34–48, 56, 67, 69, 81, 82, 84, 85, 87, 89, 92, 93, 1596–1603, 05–16, 18, 21–28, 30–32, 34–55, 57, 63–66, 64–74, 76, 79, 80, 82–86, 1720, 21; Pully 1889, 90: Lavaux 1581, 84, 91, 1617–23, 25, 28, 49, 51, 1753, 64–66, 1837, 50, 53, 54, 57, 59, 60: Aubonne 1550, 58, 63, 64, 78, 86–90, 92, 99, 1602, 04, 07, 15, 19, 37, 46, 50, 59, 64, 78, 84–86, 88, 92, 94, 95, 97, 99, 1704, 05, 07, 08, 10, 14, 18, 20, 22, 1857; Rolle 1800–02, 05.19, 27, 29, 30, 38, 54.

To provide a clearer picture of the variations of the harvest times and to enable the combining of observations from several stations into group averages, individual lustrum averages were not reproduced as such but rather in relation to the 65-year mean of 1816–1880. Again, I opted for the correction method; therefore, the table's figures indicate by how much the observed lustra averages have to be corrected in order to obtain the 65-year mean. As a result, a minus sign indicates that the temperature was too low, a plus sign that it was too high, or a plus sign tells us that the harvest occurred too early, a minus sign too late.

Secular Variations in the Timing of the Grape Harvest,
Given as Lustra Averages in Relation to the Mean of 1816–80.
(– = Harvest Too Late, + = Too Early)

Dijon, Côte-D'Or	Raw	Smoothed
Mean 1816–80	29	29
1391–95	0	2.7
96–00	8	7.0
1401–05	**12**	**8.0**
06–10	0*	3.2*
11–15	1	3.5
16–20	12	10.0
21–25	**15**	**12.0**
26–30	6	10.2
31–35	14	7.0
36–40	–6.	2.0
41–45	6	0.0
46–50	–6*	–2.5*
51–55	–4	–2.5
56–60	4.	1.8
61–65	3.	2.5
66–70	0	3.0
71–75	**9**	**4.5**
76–80	0*	2.5
81–85	1	0.8*
86–90	1	1.0
91–95	1	1.5

Secular Variations in the Timing of the Grape Harvest, Given as Lustra Averages in

Relation to the Mean of 1816–80. (– = Harvest Too Late, + = Too Early)

Individual Stations Raw Data	Dijon, Côte-d'Or	Salins Jura	Lausanne Waadt	Lavaux Waadt	Anbonne Waadt
Mean 1816–80	29	43	41	44	44
1496–00	–1	—	7.	—	—
1501–05	2	—	**10.**	—	—
06–10	3	—	7.	—	—
11–15	–6*	—	1*	—	—
16–20	0	—	5.	—	—
21–25	**22**	—	**7.**	—	—
26–30	1.	–4	–2.	—	—
31–35	–1.	6	—	—	—
36–40	4	**15**	—	—	—
41–45	—	–4	—	—	—
46–50	–2	—	—	—	—
51–55	1	—	10	—	–2
56–60	8	—	**11.**	—	**10.**
61–65	–5*	0.	8	—	0.
66–70	1	–6.	–1.*	—	–5*
71–75	0.	**13.**	**6**	—	**–2**
76–80	–2	—	0	—	–6.
81–85	2	1.	—	**17.**	3
86–90	2	2.	**9**	16	—
91–95	–2	–5.	–3.	2.*	–6.
1596–1600	–3*	6*	—	11	–7.*
1601–05	5	—	—	**5.**	—
06–10	1	—	—	–4.	—
11–15	3	—	—	–4.	—
16–20	2	—	—	7.	—
21–25	2	—	—	—	—
26–30	–4*	—	—	–8.*	—
31–35	0	—	—	–3.	—
36–40	**11**	—	—	—	—
41–45	4	—	—	2	—
46–50	3	—	—	–3	—
51–55	7	—	—	2.	—
56–60	1	—	—	**11.**	—
61–65	7	—	—	6	—
66–70	**11**	—	—	—	0
71–75	0*	—	—	–5.*	–16*
76–80	11	—	—	8	**4**
81–85	**16**	—	—	**9**	–3.

Individual Stations Raw Data	Dijon Côte-d'Or	Beaune Côte-d'Or	Volnay Côte-d'Or	Salins Jura	Lons-le Saulnier Jura
Mean 1816–80	29	30	29	43	33
1601–05	5	—	—	**5.**	—
06–10	1	—	—	–4.	—
11–15	3	—	—	–4.	—
16–20	2	—	—	7.	—
21–25	2	—	—	—	—
26–30	–4*	—	—	–8.*	—
31–35	0	—	—	–3.	—
36–40	**11**	—	—	—	—
41–45	4	—	—	2	—
46–50	3	—	—	–3	—

Individual Stations Raw Data	Dijon Côte-d'Or	Beaune Côte-d'Or	Volnay Côte-d'Or	Salins Jura	Lons-le Saulnier Jura
Mean 1816–80	29	30	29	43	33
51–55	7	—	—	2.	—
56–60	1	—	—	**11.**	—
61–65	7	—	—	6	—
66–70	**11**	—	—	—	0
71–75	0*	—	—	–5.*	–16*
76–80	11	—	—	8	**4**
81–85	**16**	—	—	**9**	–3.
86–90	7	—	—	4	–5.
91–95	2	—	2	–3	—
1696–1700	–5*	—	10	–8*	—
1701–05	5	9	4	1	–5
06–10	6	11	6	7.	–1.
11–15	–2	3*	–2*	–3	–7.*
16–20	**9.**	**13**	**13**	10	—
21–25	–2	6	7	0	1
26–30	6	11	12	**12**	—
31–35	1	7	8	4	–5.
1736–1740	–3	2	3	4	—

Individual Stations Raw Data	Kürnbach Baden[329]	Altstetten St. Gallen	Lausanne Waadt	Lavaux Waadt	Aubonne Waadt
Mean 1816–80	45	51	41	44	44
1601–05	—	—	—	**16**	**8.**
06–10	—	—	—	12	–1.
11–15	0.	5	—	13	4.
16–20	5.	0*	–4*	—	2.
21–25	5.	10	—	—	0
26–30	0	3*	—	**20.**	–2*
31–35	–4.*	7	—	12	–1
36–40	—	**21**	—	19	4.
41–45	0.	11	—	7	1
46–50	1	—	—	2	–6.*
51–55	**13.**	—	—	4	2
56–60	—	12	10	**8**	**6.**
61–65	6.	11	—	2	–2.
66–70	9	15	—	4	0
71–75	0.*	2*	—	–8*	–11*
76–80	**10.**	—	—	4	–3
81–85	8	9.	—	0	–12.
86–90	7.	**20**	–4.	2	–10.
91–95	3*	9.	–8	–9	—
1696–1700	7	7.*	–13*	–14	–20.*
1701–05	4	10	–10	–12	–18.
06–10	7	15.	–5	–8	**–4**
11–15	7	7.	–12	–17*	–12.
16–20	**11**	15	–8.	–15	–16.
21–25	4	10	–9.	–12	–16.
26–30	8	**19**	–6	–6	–9
31–35	6	(10)	–11	–13	–15
36–40	8	11	–10	–9	–12

[329] Average according to the Stuttgart station reduced to the mean of 1766–1830.

Individual Stations Raw Data	Verdun	Argenteuil Seine-et-Oise	Fong Meurthe-et-Moselle	Lochses Aube	Les Riceys Aube	Convignon Aube
Mean 1816–80	36	28	41	34	34	32
1741–45	—	—	—	—	—	—
46–50	—	—	—	—	—	—
51–55	—	–3	—	—	—	—
56–60	—	–2	—	—	—	—
61–65	—	**2**	—	—	—	—
66–70	—	–11*	—	—	—	—
71–75	—	–6	—	—	—	—
76–80	—	–4	—	—	—	—
81–85	—	**9**	—	—	—	—
86–90	—	–1	—	—	—	—
91–95	—	–1	—	—	—	—
96–1800	—	–2	1.	—	—	—
1801–05	–4.	2	2.	–1.	–1	—
06–10	–5*	4.	—	3	3	2
11–15	–2	–1	–3	0	0	3
16–20	–1.	–5*	–7*	–8*	–8*	–8*
21–25	**(4)**	0	0	1	1	1
26–30	(0)	2	4	**1**	**1**	0
31–35	0	**5**	1	–1	–1	**4**
36–40	–5	–2	**5.**	–2	–2	–2
41–45	0	–1	1.	–1	–1	2
46–50	–1	5	2	1	1	3
51–55	–8*	–3*	–5*	–7*	–7*	–5*
56–60	–3	1	4	1	1	–1
61–65	**9**	**6**	5	6	6	5
66–70	7	4	**10**	**10**	**10**	**6**
71–75	3	–2	3	–1	–1*	–2
76–80	–3.*	–7*	1*	–6*	2	–7*

Individual Stations Raw Data	Denain-villiers Loiret	Auxerre Yonne	Vendôme Loire-et-Cher	Vesoul Haute-Saône	Dijon Côte-d'Or	Beaune Côte-d'Or
Mean 1816–80	29	32	30	24	29	30
1741–45	–2*	—	—	—	–5*	2*
46–50	1	—	—	—	0	2
51–55	2	–5	—	—	–3	3
56–60	4	–1	—	—	**2**	4
61–65	**5**	**1**	—	—	1	**9**
66–70	–8*	–12*	—	—	–7*	–1*
71–75	–5.	–4	—	—	–1	4
76–80	0	–5	—		–1	5
81–85	Boësses	2	—	—	**5**	**10**
86–90	2.	–6	—	—	–2	2
91–95	1.	0	—	—	1	6
1796–1800	–3	–2	—	—	–1	–3
1801–05	–2	–3	0	5	–2	–4
06–10	1	–3	3	7	–5	–2
11–15	–1	–10*	–1	2	–1	–1
16–20	–7*	-8	–5*	–7*	–11*	–7*
21–25	3	–1	0	4	–7	–1
26–30	–1	–2	2	1	–6	–3
31–35	**3**	**1**	**3**	**4**	–3	**1**

Individual Stations Raw Data	Denain-villiers Loiret	Auxerre Yonne	Vendôme Loire-et-Cher	Vesoul Haute-Saône	Dijon Côte-d'Or	Beaune Côte-d'Or
Mean 1816–80	29	32	30	24	29	30
36–40	–1	–2	–1	–2	**4**	–5*
41–45	–1	–1	–3	0	1	1
46–50	2	2	3	1	2	2
51–55	–6*	–6*	–9*	–8*	–3*	0
56–60	4	3	4	1	3	4
61–65	**7**	6	**9**	**8**	**9**	**9**
66–70	4	**8**	3	7	8	8
71–75	2	4	2	0	1	–6*
76–80	–5*	–3*	–7*	–8*	–3*	2

Individual Stations Raw Data	Volnay Côte-d'Or	Salins Jura	Lons-le Saulnier Jura	Pichon-Longueville Gironde	Médoc Girone	Tain Drôme
Mean 1816–80	29	43	33	23	23	28
1741–45	2*	0*	–2	—	—	—
46–50	2*	5	–3	—	—	—
51–55	3	3	–9*	–10*	—	—
56–60	3	6	**–2**	–9	—	—
61–65	**9**	**7**	–3	**–3**	—	—
66–70	–2*	–6*	–12*	–11*	—	—
71–75	4	0	–4	–5	—	—
76–80	1	8	–1.	–8	—	—
81–85	**10**	**11**	3	**4**	—	—
86–90	1	3	–4	–3	—	—
91–95	6	(10)	**5.**	—	—	—
1796–1800	–2	3	–3	—	–4	—
1801–05	–2	5	–3	2	4	—
06–10	–2	(8)	0	3	3	–2
11–15	–1	3	–1	–16*	–2	–6*
16–20	–8*	–2*	–5*	–6	–8*	–4
21–25	–1	**4**	2	3	1	–2
26–30	–3	2	1	6	4	–2
31–35	**1**	2	**5**	**7**	**5**	0
36–40	–5*	–3	2	–2	–1	**1**
41–45	–1	(2)	2	–5	–2	–7
46–50	2	–1	4	2	1	–1
51–55	–3	–7*	–4*	–7*	–10*	–11*
56–60	4	0	3	–1	–1	–2
61–65	7	**6**	**8**	4	5	11
66–70	**8**	5	6	**10**	**9**	**11**
71–75	2	0	–9	2	3	7
76–80	–6*	–8*	–9*	–7*	–2*	–6*

Individual Stations Raw Data	Castres Tarn	Stuttgart Württemberg	Kürnbach Baden	Altstetten St. Gallen	Veytanx Waadt	Vevey Waadt
Mean 1816–60	28	45[330]	45[331]	51	45	45
1741–45	—	—	4*	7*	–3*	—
46–50	—	—	8	15	5	—
51–55	—	—	4*	8	0	—
56–60	—	—	7	**11**	8	—
61–65	—	—	**8**	10	**14**	—
66–70	—	–1.*	0*	2*	—	—
71–75	—	3	3	6	1	—
76–80	—	2	3	10	1	—
81–85	—	**5**	**5**	**13**	8	—
86–90	—	2	1	10	2	—
91–95	—	4	4	5	**10**	—
1796–1800	–2	3	2	—	3.	—
1801–05	0	–1	–1	2	3	3
06–10	–7	–1	—	4	4	1
11–15	–6	–4.*	—	–1	0	–5
16–20	–11*	–3.	—	–3	–3*	–9*
21–25	0	–3	—	1	1	0
26–30	–2	–2	—	–7*	**2**	**1**
31–35	–1	—	—	**3**	1	1
36–40	–6*	—	—	–5*	–5	–4*
41–45	0	—	—	2	–2	–2
46–50	2	—	—	1	–1	3
51–55	–4	—	—	–3	–6*	–3
56–60	1	—	—	1	–4	1
61–65	7	—	—	**5**	**5**	**7**
66–70	**8**	—	—	4	—	4
71–75	3	—	—	5	—	4
76–80	0*	—	—	–2	—	–4.*
81–85	—	—	—	1	—	—
86–90	—	—	—	–8*	—	—

Individual Stations Raw Data	Lausanne Waadt	Pully prés Lausanne Waadt	Lavaux Waadt	Aubonne Waadt	Rolle Waadt
Mean 1816–80	41	44	44	42	41
1741–45	–16*	—	–13	–20*	—
46–50	–9	—	–13	–16	—
51–55	–14	—	–15.*	–18	—
56–60	–9	—	–7	–13	—
61–65	–6	—	0	–12	—
66–70	–14*	—	–18.*	–16*	—
71–75	–8	—	–9	–12	—
76–80	–2	—	0	–5	—
81–85	–3	—	1	–1	—
86–90	–5	—	–2	–9	—
91–95	7	—	7	3	—
1796–1800	–5	—	–2	–5	—
1801–05	0	—	1	–9	—

[330] mean 1766–1830

[331] Mean according to the Stuttgart station mean of 1766–1830.

Individual Stations Raw Data	Lausanne Waadt	Pully prés Lausanne Waadt	Lavaux Waadt	Aubonne Waadt	Rolle Waadt
Mean 1816–80	41	44	44	42	41
06–10	–1	—	–1	–7	–4
11–15	–7	—	–6*	–10*	–6
16–20	–11.*	–5*	–6	–10	–10.*
21–25	–4	–4	2	–1	4
26–30	–2	–2	0	0	—
31–35	–1	–1	–1	2.	2
36–40	–3	–3	–3*	–3	–1.
41–45	–6*	–6*	2	–4	–1
46–50	4	4	3	0	2
51–55	–5.	–5	–2.	–5*	–6.*
56–60	0	0	—	–2.	3
61–65	9	8	6	9	8
66–70	—	5	—	—	6
71–75	—	3	—	—	4
76–80	—	0	—	—	–6.*
81–85	—	0	—	—	—
86–88	—	–2*	—	—	—

Overall Averages

	Raw	Smoothed		Raw	Smoothed
1496–00	3.0	3.2			
1501–05	6.0	5.0	1601–05	8.5	4.4
06–10	5.0	3.4	06–10	2.0	4.0
11–15	–2.5*	0.6*	11–15	3.5	2.8
16–20	2.5	4.2	16–20	2.0	2.8
21–25	14.5	7.4	21–25	3.8	2.8
26–30	–1.7	3.4	26–30	1.5*	2.2*
31–35	2.5	3.2	31–35	1.8	4.7
36–40	9.5	4.4	36–40	13.8	8.4
41–45	–4*	0.1	41–45	4.2	5.4
46–50	–2	–1.2*	46–50	–0.6*	2.2*
51–55	3.0	3.4	51–55	5.6	4.7
56–60	9.7	5.8	56–60	8.0	6.6
61–65	0.8	2.1	61–65	5.0	6.1
66–70	–0.2*	–0.2*	66–70	6.5	3.2
71–75	4.2	0.7	71–75	–5.4*	0.4*
76–80	–2.7	1.2	76–80	5.7	3.8
81–85	5.8	4.0	81–85	4.0	4.1
86–90	7.2	4.4	86–90	2.6	2.2
91–95	–2.8*	0.1*	91–95	–0.6	–0.8
1596–1600	–1.2	0.9	1696–1700	–4.5*	–2.7*

Raw Data	North France	Central France	SW Germany and Switzerland	Mean
1701–05	—	2.8	–5.2	–1.2
06–10	—	5.8	1.0	3.4
11–15	—	–2.2	–5.4	–3.8
16–20	—	**11.2**	–2.6	3.6
21–25	—	2.4	–4.6	–1.1

Raw Data	North France	Central France	SW Germany and Switzerland	Mean
26–30	—	10.3	**1.2**	**5.2**
31–35	—	3.0	–4.6	–0.8
36–40	—	1.5	–2.4	–0.7
41–45	–2.0*	–0.6	–6.8*	–3.8*
46–50	1.0	1.2	1.7	0.2
51–55	–2.0*	–2.2*	–5.8	–3.6
56–60	0.3	0.7	–0.5	0.1
61–65	**2.7**	**3.3**	**2.3**	**2.8**
66–70	–10.3*	–6.5*	–7.8	–7.8*
71–75	–5.0	0.3	–2.3	–2.1
76–80	–3.0	0.7	1.3	0.9
81–85	**5.5**	**7.2**	**4.0**	**5.5**
86–90	–1.7	–0.5	–0.1	–0.6
91–95	0.0	5.6	5.7	4.5
1796–1800	–1.5	–1.7	–0.7	–1.3
1801–05	–0.9	0.6	–0.3	–0.2
06–10	1.0	3.0	–0.6	0.2
11–15	–1.7	–2.9	–4.9	–3.1
16–20	–6.3*	–6.9*	–6.7*	–6.6*
21–25	1.0	0.3	–0.4	0.3
26–30	0.8	–0.2	–1.2	–0.2
31–35	**1.7**	**2.1**	**0.8**	**1.5**
26–40	–1.3	–1.7	–3.4	–2.1
41–45	–0.6	–0.9	–1.4	–0.9
46–50	2.0	1.4	2.0	1.8
51–55	–6.2*	–5.7*	–4.4*	–5.5*
56–60	1.6	1.2	–0.1	1.0
1861–65	6.6	7.4	**7.1**	7.0
66–70	**6.9**	**8.0**	4.8	**7.0**
71–75	0.9	1.1	4.0	1.5
76–80	–3.9*	–5.5*	–3.0	–4.4
81–85	—	—	0.5	0.5
86–88	—	—	–5.0*	–5.0*

Smoothed data	North France	Central France	SW Germany and Switzerland	Total Average
1701–05	—	3.8	–3.1	–0.9
06–10	—	3.0	–2.2	0.4
11–15	—	3.2	–3.1	–0.2
16–20	—	5.6	–3.8	0.6
21–25	—	**6.6**	–2.6	1.6
26–30	—	6.5	**–1.7**	**2.1**
31–35	—	4.4	–2.6	0.7
36–40	—	1.4	–4.0	–1.5
41–45	–1.0*	0.4	–4.4*	–2.1*
46–50	0.5	–0.1	–4.0	–2.0
51–55	–0.7	–0.6*	–3.4	–1.8
56–60	0.3	0.6	–1.1	–0.2
61–65	–1.2	0.2	–0.9	–0.5
66–70	–5.7	–2.5*	–3.9*	–3.7*
71–75	–5.8*	1.6	–2.8	–2.8
76–80	–1.4	2.1	1.1	1.3
1781–85	**1.6**	**3.6**	**2.3**	**2.8**
86–90	0.5	3.0	2.4	2.2
91–95	–0.8	2.2	2.6	1.8
1796–1800	–1.0	0.7	1.0	0.4
1801–05	–0.6	0.6	–0.5	–0.4
06–10	–0.2	0.9	–1.6	–0.7
11–15	–2.2	–2.4	–4.3	–3.2
16–20	–3.3*	–4.1*	–4.7*	–4.0*

Smoothed data	North France	Central France	SW Germany and Switzerland	Total Average
21–25	–0.9	–1.6	–2.2	–1.6
26–30	**1.1**	0.5	**–0.5**	**0.4**
31–35	0.7	**0.6**	–0.8	0.2
26–40	–0.4	–0.6	–1.8*	–0.9
41–45	–0.1	–0.5	–1.0	–0.5
46–50	–0.7	–1.0	–0.4	–0.7
51–55	–2.2*	–2.2*	–1.7*	–2.0*
56–60	0.9	1.0	0.6	0.9
61–65	**5.4**	6.0	4.7	5.5
66–70	5.3	**6.1**	**5.2**	**5.6**
71–75	1.2	1.2	2.4	1.4
76–80	1.2	–3.3*	–0.4	–1.7
81–85	—	—	–1.8	–2.1
86–1888	—	—	–3.2*	–3.2*

To eliminate any remaining chance occurrences from the lustra averages of individual stations, I grouped regions together as follows and formed averages for these groups for the period from 1701–1885 [on pages 132 and 133].

1. Northern France: Verdun, Argenteuil, Foug, Loches, les Riceys, Couvignon, Denainvilliers, Auxerre und Vendôme. 9 Stations.
2. Southern France Vesoul, Dijon, Beaune, Volnay, Salins. Lons-le-Saulniers, Pichon-Longueville, Médoc, Tain und Castres. 10 Stations.
3. South West Germany and Switzerland: Stuttgart, Kürnbach, Altstetten,Veytaux, Vevey, Lausanne, Pully, Lavaux, Aubonne u. Rolle. 10 Stations.

In addition from the start one single group average was determined for the observations from all stations [on page 132].

The tables show for all group averages raw figures as well as those which were smoothed according to the method I have repeatedly used earlier.

Let us discuss our tables:

Each single observation series tells us that the date of the grape harvest changes not only from year to year but also from lustrum to lustrum. These are not random changes but show a certain pattern; at times the harvest is delayed longer and longer over several lustra, only to start earlier and earlier in the succeeding lustra. There are distinct periods when harvesting started very late and those when the crops were brought home quite early.

It is remarkable that all these variations do indeed occur simultaneously at all stations from northern France to the South, from the West to Württemberg and into Switzerland.

The lustrum of 1876/80 shows that, with one exception, harvesting started later than in the previous lustra at all stations, the same goes for the lustrum of 1851/55. Between these two "minima" lies a clearly pronounced "maximum" for all the stations half of which show it in the lustrum of 1861/65, the other half in 1866/70. During this time grapes were on the average picked 1 1/2 to 2 weeks earlier than in the lustra of 1851/55 and

1876/80. An analogous maximum occurs around the year 1830. Differing slightly from station to station it appears in five instances as early as in the lustrum of 1821/25, in four instances in 1826/30, in 13 in 1831/35 and in three in 1836/40. The years 1836 to 1855 are generally marked by late harvests; but in the second lustrum of the 1840s a slight return to an earlier date emerges which is more pronounced at some stations so that the absolute maximum prior to 1851/55 falls into this lustrum. Its cause appears to be a local one, since it does not show up at all in the smoothed group averages where the lustra of 1826/30 and 1831/35 represent the earliest time of the harvest. — The year 1815 is surrounded by minima again, which at one station appear in the lustrum of 1806/10, at six stations in 1811/15, and at 20 in 1816/20. If we return to the previous century, the lustra from 1781 to 1795 show maxima, 1766/70 a minimum at all stations, 1756–1765 a maximum, 1741/56 again a minimum which seems to have been a general occurrence and was preceded by a vaguely defined maximum in the years 1716–1730. The minimum of 1691–1700, clearly identifiable at five stations in Switzerland and neighbouring France, shifts at four other stations to the lustrum of 1711/15. Moving further back in time, variations turn irregular. However, some periods of fairly regular patterns still occur: minima around 1671–75, 1626–35, 1591–1600, 1561–70; maxima 1651–60, 1601–05, 1581–90 and 1556–60. Dijon, Salins, Lausanne are the only stations providing records of the time prior to 1550; for the time before 1495 observations are only available from Dijon. These observation series seem to indicate that the lustra of 1521/25, 1501/10, 1471/75 and 1421/25 carry maxima, the lustra of 1541/50, 1511/15, 1496/1500 or 1476/80, 1446/50 and 1406/10 minima. However, the variations before 1550 should be considered unreliable due to the low number of stations.

These distinct oscillations have clearly escaped Angot's as well as Dufour's attention despite the fact that both examined the data series of their publications with regard to climatic variations. The number of years that they combined into averages was too big. By using periodicities of 10 or even 25 years they could not possibly detect variations of under 30 to 40 years. They did however come across variations of the harvest date that occurred in the course of centuries. For example, from 1550 to 1670 the harvests at Aubonne took place on the average around October 13, and again around that same date in the years after 1780. But in the years 1670–1780 the average date was several days after October 20. Lausanne and Veytaux show analogous and parallel variations over very long periods of time. The data series from Salins and Dijon again indicate such variations, which however neither seem to be in accord with each other nor with the variations in Switzerland. Angot[332] comes to the conclusion that they may not be

[332] Angot, see B83.

attributable to climate variations but rather to changes instigated by human arbitrariness, be it a change in taste or in the methods of cultivation. To what extent this conclusion is justified will have to be discussed later on.

As opposed to long-term variations our variations over shorter periods of time have the advantage of occurring universally. They reflect the climate variations that we have already observed many times before.

In the last part of his essay Angot has dealt in detail with the question of the meteorological factors which mainly determine the date of the grape harvest. In order to investigate this question for France, he concentrated on the years 1811, 1822, 1834, 1846, 1865, and 1868 showing very early harvest dates, and the years 1816, 1821 and 1879 with very late harvests. Based on the Paris observations he determined the extent to which temperature and rainfall deviate from the standard during the vine's growing period. His results are summarised below:

Deviations from the Paris Standard:

	Temperature April to August	Temperature-sum	Date of the Grape Harvest
Early Harvest	+ 1.39° C	+212° C	17 days too early
Late Harvest	–1.52	–223	20 days too late

Negative temperature deviations during the vegetation period obviously cause a delay of the harvest, whereas positive deviations make early harvests possible. During the years in question, however, rainfall shows only small deviations from the standard amount.

With this result we are already in a position to link those changing harvest dates to climate variations. The variations in the average temperature during the vines' growing season have without doubt caused those changes in the harvest dates. Average temperatures for the months April to September are not at my disposal in order to further support this conclusion. Yet, these monthly averages are included in the annual averages, the variations of which we have already examined earlier. In the following table the variations of the harvest dates are therefore compared with changes of the annual temperature in Western Central Europe and of the rainfall in Southern Germany and France (averages of the observation series from Southern Germany, Northern , Southwest, and Central France).

Congruency between the developments of the three factors—date of the grape harvest, temperature, and rainfall—is excellent during the present century; apart from slight shifts in timing, the variations correspond in such a way that an early harvest coincides with high temperatures and low rainfall, and *vice versa*). For the past century, congruency between temperature and grape harvest is again quite satisfactory while rainfall shows deviations in the period of 1766–1790: its minimum in 1766/70, as well as its maximum in 1786/90 are not congruent.

I do not want to attach too much importance to this result. If we, for instance, replace the less than reliable rainfall observations with those of the Alps' glaciers, the Neusiedler See and the Caspian Sea, it is evident that even during the time period in question the changes in the harvest dates must have been accompanied by corresponding changes in the rainfall. The parallelism between the variations of the harvest dates and the temperature is clearly established for the period 1756 to 1875, e.g., for a total of 120 years, and between harvest dates and rainfall for the period 1731 to 1880, e.g., for 150 years.

This parallelism is so complete that the two curves are interchangeable. The variations of the grape harvest dates reflect the climate variations as we have confirmed them, just as well as temperature and rainfall. This finding is of the greatest importance to us because evidently the curve of the grape harvest dates can now serve as an indicator of climatic variations for even those periods reaching far back in time when temperature and rainfall observations were not available. We are able to trace climatic variations back to the year 1400 thanks to the many extensive listings of the harvest dates in the wine-growing regions of France and Switzerland.

For those time periods dating back even further, a somewhat different kind of data presents itself, which may not be quite as exact but will nevertheless supply useful results: the records about cold winters

Secular Variations of the Grape Harvest Dates, Compared with the Variations of Temperature and Rainfall

Lustrum	Grape-harvest Days	Temperature (°C)	Rainfall (%)
1731–35	–0.8	—	7
36–40	0.7	—	14
41–45	–3.8*	—	–3
46–50	–0.2	—	4
51–55	–3.6	—	–12*
56–60	0.1	–0.52	5
61–65	2.8	–0.53	0
66–70	–7.8*	–0.98*	–13*
71–75	–2.1	0.00	—
76–80	0.9	0.36	–5
81–85	5.5	–0.10	–5
86–90	–0.6	–0.09	7
91–95	4.5	0.48	0
1796–1800	–1.3	0.14	–4*
1801–05	–0.2	0.04	2
06–10	0.2	0.07	7
11–15	–3.1	–0.39*	6
16–20	–6.6*	0.36	–3
21–25	0.3	0.52	–7

Lustrum	Grape-harvest Days	Temperature (°C)	Rainfall (%)
26–30	–0.2	–0.23	–1
31–35	1.5	0.35	–8*
36–40	–2.1	–0.30	2
41–45	–0.9	0.09	8
46–50	1.8	0.05	0
51–55	–5.5*	–0.31*	2
56–60	1.0	–0.09	–1
61–65	7.0	0.36	–14*
66–70	7.0	0.31	1
71–75	1.5	–0.14	0
76–80	–4.4*	—	14

Secular Variations of the Frequency of Severe Winters:
The Number of Severe Winters Among a Series of 20 Winters.

800	2	970	1	1140	5	1310	8*	1480	4	1650	9*
05	3*	75	2	45	6	15	7	85	4*	55	8
10	2	80	2	50	7*	20	5	90	4	60	9*
15	2	85	4	55	7*	25	5	95	4	65	8
20	2	90	5*	60	5	30	4	1500	3	70	4
25	2	95	4	65	5	35	5	05	5	75	4
30	2	1000	4	70	5	40	5	10	5*	80	2
35	1	05	3	75	3	45	4	15	5*	85	6
40	1	10	3	80	4*	50	4	20	5	90	7
45	0	15	3	85	3	55	5	25	3	95	6
50	2	20	3	90	1	60	5*	30	3	1700	8*
55	3	25	1	95	2	65	5	35	3	05	4
60	3	30	0	1200	4	70	4	40	4*	10	4
65	4*	35	1	05	6	75	2	45	4*	15	4
70	3	40	3	10	7*	80	3	50	3	20	4
75	3	45	3	15	7*	85	5	55	5	25	5
80	3	50	5*	20	6	90	6*	60	5	30	4
85	3	55	4	25	5	95	5	65	5	35	5*
90	2	60	4	30	5	1400	6	70	6*	40	4
95	1	65	5	35	4	05	4	75	4	45	4
900	1	70	6*	40	4	10	3	80	6	50	3
05	1	75	4	45	3	15	4	85	6	55	2
10	1	80	3	50	4*	20	3	90	5	60	4
15	1	85	2	55	3	25	7	95	8	65	4
20	3	90	3	60	3	30	7	1600	7	70	5
25	2	95	4	65	4	35	9*	05	8	75	8*
30	2	1100	5	70	4	40	8	10	9	80	—
35	4*	05	6	75	6	45	4	15	10*	85	—
40	2	10	7	80	7*	50	4	20	8	90	—
45	2	15	8*	85	7*	55	1	25	6	95	—
50	2	20	8	90	6	60	1	30	5		
55	0	25	8	95	6	65	2	35	3		
60	0	30	5	1300	6	70	2	40	4		
65	1	35	5	05	6	75	3	45	5		

4.2.3. Secular Variations of the Frequency of Cold Winters

Much has been written about the periodicity of severe winters;[333] focussing in most cases more or less on the individual year. I am not convinced that this method will lead to any sound results. The severity of winters depends after all on so many factors that 'chance' has to play a major role in its occurrence, especially when a single year is concerned. It is completely inadmissible to conclude from one severe winter that a cold weather period occurred as we have just done from temperature and ice observations. The situation is different, when instead on the single year attention is turned towards a possible periodicity in the frequency of severe winters. Based on Pilgram's listing of severe winters I have done just that and examined their frequency with regard to any periodic patterns. Pilgram's essay was not available to me in the original but in a hand-written excerpt instead, for which I thank Professor Dr. W. Köppen. From this listing, using five-year lustra I determined the number of severe winters for every 20 years, e.g., for the time spans 1741–60, 1746–65, 1751–70, 1756–75, etc., and assigned these frequencies to the middle year of the corresponding time span, e.g., 1750, 1755, 1760, 1765, etc. This method resulted in the table [on page 138] which applies to Central Europe because the observations refer to that region only.

As far back as we are able to compare the variations in the frequencies of severe winters with those of the temperature or the ice conditions of the rivers, the three factors correspond quite well; the temperature minima around 1770, 1740, 1660 and 1615–20, as well as the maxima around 1755, 1720–30 and 1680 are all reconfirmed. In other time periods congruency is less clear. The maximum cold period at the beginning of the 18th century shows up in 1700 and then again in the years 1711–20, another one in 1615 and 1621–25. A third one around 1525 is not more than indicated in the listing of severe winters, whereas it is quite pronounced at the Düna river. The time periods around 1570 and 1560 coincide again as expected.

In comparing the variations in the grape harvest dates with those in the frequencies of severe winters, as shown in the table below (page 142), congruency again turns out to be satisfactory. I must confess that its high degree was unexpected. The transition from one period to another often shifts by five or even ten years here and there: e.g., according to the harvest dates a cold period is indicated for the years 1436–55, whereas according to the frequencies of cold winters this period falls into the years 1425–55. However, most periods correspond. Of the 385 years, for which observations regarding the frequency of cold winters as well as the grape harvest dates are

[333] See Chapter 4a.

available, not less than 260 years or 68 percent show the two factors to be congruent in indicating a cold or warm period and only 125 years show differences. Only once do the variations of the two factors differ considerably: according to the harvest dates the years 1591 to 1690 are marked by four variations, whereas according to the frequencies of severe winters this number amounts to only two.

In order to decide which of the two findings is the correct one, we will compare the raw lustrum averages of the two factors (tables pp. 132 and 138) for the time span in question. It turns out that the warm period of 1601–10 which shows up clearly in the dates of the grape harvest, is vaguely outlined in the listing of cold winters as well: 1595, 8; 1600, 7; 1605, 8. Therefore, I am inclined to put the emphasis on the harvest dates in this case. The cold period of 1666–75 is a different matter. It seems to emerge from the harvest dates, yet stretches over only 10 years in the smoothed data, while being restricted to the one lustrum of 1671–75 in the raw data: 1656–60, 8.0; 1661–65, 5.0; 1666–70, 6.5; 1671–75, –5.4; 1676–80, 5.7; 1681–85, 4.0. If we ignore the lustrum in question, the cold period has disappeared entirely. The data about the frequency of cold winters are, therefore, decisive in this case. As a result the variations' occurrence during the time period under discussion is depicted in the table below under the heading 'corrected'.

According to Date of the Wine Harvest		According to Frequency of Severe Winters		Corrected	
Cold	1591–00	} Cold	1590-1625	Cold	1591–00
Warm	1601–10			Warm	1601–10
Cold	1611–35			Cold	1611–35
Warm	1636–45	Warm	1626-45	Warm	1636–45
Cold	1646–50	} Cold	1645-65	Cold	1646–65
Warm	1651–65				
Cold	1666–75				
Warm	1676–90	Warm	1665-85	Warm	1666-85 resp. 90

Given the high degree of congruency between the variations in the frequency of cold winters and the grape harvest dates during the time period from 1390 to 1775 there can be no doubt that the former factor is the correct indicator for the climatic variations. The same seems likely for the years back to the year 1000. Further back in time, the number of listings for cold winters decreases so rapidly, and in addition the intervals between two maxima of the same temperature increase by so much that I distrust these data. I am inclined to regard those variations as coincidental and will therefore exclude them from further consideration.

4.2.4 Average Periodicity of Climatic Variations

Meteorological observations allowed us to trace climatic variations from not earlier than approximately 1730 on. However, the number of 4 1/2 oscillations which have occurred since then is too low in order to define the average periodicity. But since we have collected a larger amount of data consisting, on the one hand, of observations of ice conditions on rivers from 1556 to 1885 and of the grape harvest dates from 1391 to 1888, and on the other hand of listings about the frequency of severe winters from 1000 to 1775, we can approach this task with a better chance of successfully completing it. The following table has been set up for this purpose. Based on the smoothed results of the observations about the start of the grape harvest, and on statistics about cold winters as well as about the ice conditions of the rivers it shows the extent of warm and cold periods. To avoid any arbitrariness in distinguishing between the periods, for the time prior to 1750 I used the method of drawing the line wherever the figures coincide with the mean of the nearest maximum and minimum. For the time after 1750, in addition I took temperature observations into account whenever cases were uncertain. I have also included the data about the changes in lakes with no outlet and glaciers from chapter III.

The table includes the years 1020 to 1890 (or 1888 to be exact). Within this time span of 870 years we count 25 cold and 25 warm periods, that is 25 complete variations. We, therefore, arrive at an average periodicity of 34.8 years, which is slightly lower than previously assumed (36 years).

To find out how significant this figure may be, I determined the length of each variation according to both, the harvest dates and the frequency of cold winters, as shown in the last two columns of the table (p. 142). I arrived at these figures by combining the numbers for the duration of each cold period with those of its neighbouring warm period, that is I summed up the years of the warm periods from 1020–40 and the cold periods from 1040–55 (20 + 15 = 35), furthermore the cold periods from 1040–55 and the warm periods from 1055–65 (15 + 10 = 25), etc.

The length of these periods varies between 20 and 50 years, but is most frequently close to the average value of 34.8 which we established above. The frequency of these periods of varying length is shown in the following tabulation:

Period:	20	25	30	35	40	45	50	years
Frequency:	6	10	12	**13**	12	8	4	cases

This verifies the significance of the average figure of 34.8 years. There are also other methods to verify this.

Secular Variations of the Climate* Represented by the Variations of

	Grape Harvest	Frequency of Cold Winters	Ice Conditions of Rivers
Warm	—	1020–40	—
Cold	—	40–55	—
Warm	—	55–65	—
Cold	—	65–80	—
Warm	—	80–05	—
Cold	—	1105–30	—
Warm	—	30–45	—
Cold	—	45–65	—
Warm	—	65–75	—
Cold	—	75–90	—
Warm	—	90–00	—
Cold	—	1200–30	—
Warm	—	30–45	—
Cold	—	45–55	—
Warm	—	55–70	—
Cold	—	70–90	—
Warm	—	90–10	—
Cold	—	1310–25	—
Warm	—	25–50	—
Cold	—	50–70	—
Warm	—	70–85	—
Cold	1391–15	85–05	—
Warm	1416–35	1405–25	—
Cold	36–55	25–55	—
warm	56–80	55–75	—
cold	81–95	75–95	—
warm	96–10	95–05	—
cold	1511–15	1505–20	—
warm	16–40	20–35	—
cold	41–50	35–45	—
warm	51–60	45–55	—
cold	61–80	55–70	1556–65
warm	81–90	70–90	66–85
cold		1591–00	86–00
warm		1601–10	1601–20
cold		11–35	21–25
warm		36–45	—
cold		45–65	1651–67
warm		65–90 resp.. 85	—
cold	1691–05	1685–05	1702–20
warm	1706–35	1705–30	21–35
cold	36–59	30–50	36–50
warm	56–65	50–65	51–70
cold	66–75	65–75	71–90
warm	76–05	—	91–05
cold	1806–20	—	1806–20
warm	21–35	—	21–30
cold	36–55	—	31–60
warm	56–75	—	}warm
cold	76–90	—	}1861–80

* [This table as well the those on the next three pages was presented by Brueckner in one big table; unfortunately, because of spatial limitation, the table had to be broken into four parts.]

Secular Variations of the Climate*

	Represented by the variations of ice conditions of		Length of periods as given by	
	Lakes	Glaciers	Grape Harvest Dates	Frequency of Cold Winters
Warm	—	—	—	—
Cold	—	—	—	35
Warm	—	—	—	25
Cold	—	—	—	25
Warm	—	—	—	40
Cold	—	—	—	50
Warm	—	—	—	40
Cold	—	—	—	35
Warm	—	—	—	30
Cold	—	—	—	25
Warm	—	—	—	25
Cold	—	—	—	40
Warm	—	—	—	45
Cold	—	—	—	25
Warm	—	—	—	25
Cold	—	—	—	35
Warm	—	—	—	40
Cold	—	—	—	35
Warm	—	—	—	40
Cold	—	—	—	45
Warm	—	—	—	35
Cold	—	—		35
Warm	—	—	45	40
Cold	—	—	40	50
Warm	—	—	45	50
Cold	—	—	40	40
Warm	—	—	39	30
Cold	—	—	20	25
Warm	—	—	30	30
Cold	—	—	35	25
Warm	—	—	20	20
Cold	—	—	30	25
Warm	—	—	30	35
Cold	1600 [1]	1595–10	20	
Warm	—	—	20	
Cold	38 [2]	—	35	
Warm	56 [3]	Increase	35	

[continued on next page]

* [This table was part of a much larger table in the Brueckner's dissertation. It included the table presented on the previous two pages plus the continuation of this one on the next page.]

1) Lake Fucin and Trasimen.

2) Caspian Sea, at high tide, compared with levels of 1715/20.

3) Lake Fucin.

[Continued from previous page: Secular variations of the climate]]

	Represented by the variations of ice conditions of		Length of periods as given by	
	Lakes	Glaciers	Grape Harvest Dates	Frequency of Cold Winters
Cold	74 [4]	1677–81		30
Warm	83 [5]	and		40
Cold	1707–14 [6]	1710–16	30	40
Warm	about 1720	—	45	45
Cold	about 40	50–67	50	45
Warm	about 60	60–86	30	35
Cold	about 80	about 1800	20	25
Warm	about 1800	1800–15	40	-
Cold	about 20	15–30	45	-
Warm	about 35	30–45	30	-
Cold	about 50	45–75	35	-
Warm	about 65	75–90	30	-
Cold	about 80		35	-

In determining the average value of five successive variations we arrive at the following:

For the Time Span	1020–1190	34	years
	1190–1370	36	
	1370–1545	35	
	1545–1715 [1]	34	
	1715–1890	35	

The length of the period remains almost the same irrespective of the time span selected.

Given these findings we can now calculate the probabale error for the mean period of 34.8 years with Fechner's formula. Given the mean deviation of ±7.0 years, this error amounts to ±0.7 years. The uncertainty is ± 5.9 years for each individual period. This figure lets us conclude that it would take 544 cold and warm periods or in other words 272 complete climate variations in order to pinpoint the average length of a period on the basis of lustra averages down to one month only; that would mean approximately 9500 years of observation. Yet even today we can be quite satisfied with our

[4] Neusiedler See.

[5] Lake Fucin.

[6] Lake Zirknitz, Caspian Sea.

[1] The cut-off date is 1715 according to the ice conditions of rivers, the water level variations of lakes, and variations of glaciers; the year 1705 seems to be more precise according to the grape harvest dates and the frequency of cold winters.

result; after all, its margin of error does not amount to more than 2% of the length of a period.

For the time being this evidence applies to central Europe only because the observations of the grape harvest dates and the frequency of cold winters are confined to this region. But we should not disguise the fact that indirectly it may be of universal significance around the globe. The meteorological observations gave evidence that the same climatic variations, which affected central Europe in this century, occurred simultaneously in Inner Asia, North America, Australia, etc., in short, with a few exceptions, all over the globe. It is unlikely that this simultaneous occurrence is merely a coincidence and a peculiar feature of the century. It is far more likely that we are dealing with a regular pattern. Consequently the observations carried out in one country were indicators of climatic variations all around the globe. Those variations in the timing of the grape harvests and in the frequency of severe winters which in central Europe could be traced back to the year 1000 are merely one aspect of the global climate variations occurring since then with a periodicity of 34.8 years, give or take 0.7 years.

We have reached the end of our discourse about climate variations. We have established that their characteristics are identical with those of the variations of temperature, air pressure, and rainfall, and finally we were able to determine their periodicity from observations spanning nearly nine centuries. Are these climate variations indeed so significant that they have a practical impact? To show, that this is very much the case, will be the theme of the next section.*

* [In the following Section 4c.]

4.3 THE SIGNIFICANCE OF CLIMATIC VARIATIONS IN THEORY AND PRACTICE*

Influence of the climatic variations on dimensions of glaciers; as well as on the dimensions and drainage conditions of lakes, on the frequency of floods, and on the water-level of rivers. The effects of these variations as well as those in the duration of ice covers on traffic life. Links between climate variations and agriculture, demonstrated in a tabulation of grape and wheat crops. Anticipating a severe economical crisis in the dry regions of the Great Salt Lake. Link to typhus frequency demonstrated in several tables. Influence of rivers' run-off on the coastal water levels: Coasts of the Baltic Sea and the French channel (table). Declaring some presumed shifts of the beach line as caused by climatic variations (Paschen, Bouquet de la Grye). Sueß goes too far. Significance of climatic variations in determining mean values in climatology, demonstrated at three stations. Prognoses based on climatic variations. List of scientists who anticipated climatic variations: Hann, Schweinfurth, Dove, Zimmermann, Plantamour, Lorenzoni, Kluge, Hagen, Marié Davy, Jevons, J. A. Brown, and perhaps also Fritz. First clearly defined proof by Sonklar, but limited to the Alps. The universal significance of climatic variations has so far not been recognized. Climate variations are reflected in the history of views about climate change.

It is not my intention to describe in detail how important climatic variations are for the multitude of issues pertaining to practical life and to science. I may be permitted a brief outline however. In doing so, I shall refrain as much as possible from generalizations and instead do my utmost to attempt a quantification of this impact. The enormous effect on glaciers is well known. It was, after all, their varying sizes that made us first aware of the occurrence of climate variations. Here is one example for the enormity of these variations. Sonklar measured the glacial area of the Hohe Tauern range [High Tauern] based on surveys which were carried out shortly before and immediately after the glaciers had reached maximum sizes and found it to be 422 sq. km. I repeated his measurement on the new local map of this Austrian area which was put together in the early 1870s, i.e., at a time, when the glaciers had already been receding for almost two decades. The resulting area of 363 sq. km. was 14% smaller than in Sonklar's measurement.[334] As

* [Chapter 9: *Die Bedeutung der Klimaschwankungen für Theorie und Praxis*]

334 Compare this with Brückner: *Die Hohen Tauern und ihre Eisbedeckung* [*The Hohen Tauern mountains and their ice cover*], Zeitschrift des Deutschen und Österreichischen Alpenvereins, 1886. Richter (*Die Gletscher der Ostalpen* [*The Glaciers of the East Alps*]. Handbücher zur Deutschen Länder- und Volkskunde, Volume 3, 1888) determined a glacier area of 381 sq. km in accordance with the same dimensions, resulting in a decrease of 10 percent since Sonklar's time. The difference between this percentage figure and my

glaciers in the Eastern Alps continued to recede until the end of the 1880s, a decrease of about 20% of the area defined by Sonklar is more than likely. In accordance with our climatic variations the size of the glacial area in the Hohen Tauern range changes by approximately this amount. Based on his observations at eight glaciers of the Eastern Alps, Richter determined that during the most recent stage of this decrease glaciers lost an average of 6.17 cubic metres (cbm) of ice per sq. m through melting.[335] For the entire region of the Eastern Alps this amounts to 9 cubic kilometres and for the entire area of the Alps to 25 cubic kilometres, or to the equivalent of an ice cube of nearly 3 kilometres in length, width as well as height. In volume, this cube would in turn be the equivalent to a 15 cm-thick layer of ice covering the entire Alps; the average elevation of the Alps has during the last 20–30 years decreased by 15 cm.

There are other ways in which our climatic variations are able to temporarily modify the geographical picture of a region extensively. In dry regions in particular, where water is scarce to begin with, hydrological conditions change tremendously in the course of variations in rainfall. A map drawn during a dry season will often look totally different from one produced in a wet period. Lakes disappear during dry periods, only to reappear in wet periods, as for instance Lake George in New South Wales, which around 1820 and 1876 and to a lesser degree in 1850 was a magnificent lake, 20–30 km long,[336] 10 kilometres wide and 5–8 meters deep. During the intervening dry periods however the lake vanished entirely from the face of this earth and grass grew in its basin. The neighbouring Lakes Cowal and Bathurst dry out completely during the dry season and reappear during the following wet season. The enormity of this fact is fully comprehensible when we consider that at high water levels Lake George as well as Lake Cowal cover a region comparable to the area of Züricher See. Similar conditions are found at the swampy Lake Hamun in Persia, although this lake does not disappear entirely. The Great Salt Lake, again, undergoes tremendous changes, its surface area increasing by a full 17 percent from its last minimum level in the 1850s to its maximum level in the 1870s, while Lake Fucine's decreased by 19.2 percent between 1816 and 1835. The changes which occurred in the Caspian Sea are smaller in relative terms, yet much more considerable in the absolute. When from 1809 to 1814 until the beginning of the 1840s, the Sea's level fell by 3 meters the overall area was

---

own can partially be explained by the fact that my measurements were based on a special map 1:75,000 whereas Richter used the original maps of 1:25,000.

[335] Richter, op. cit., p. 297.

[336] In my lecture (Verhandlungen des Berliner Geographen-Tages, Berlin 1890) the length was erroneously stated to be 18 km instead of 18 miles.

reduced by 3 percent or 13,000 sq. kms (square kilometres).[337] Climatic changes are demonstrated in another and equally impressive way at several Central African lakes; according to Sieger the Tsad, the Tanganyika and the Nyassa lakes rise at times to such high levels that they continue to run off for as long as a whole decade, then stop doing so completely in the course of a drought. The same applies to the Abistada Lake in Afghanistan and possibly the Goktscha Lake in Armania. These are changes that even a school-atlas must map out.

Also affected are flowing bodies of water. Rivers and creeks dry up for a whole decade; swamps dry out but reappear in the next wet period.

So does the Atrek river. According to G. Sievers the Turkomans claimed that its water levels have dropped considerably over a number of years prior to 1871 and that in the summer it does not even reach the delta area, etc.[338] During the rainy periods destructive floods are again a common occurrence, as for instance in Australia and especially in New South Wales where according to Jevons floods occurred as follows:

Time Span	1799–1821	1822–1841	1842–1858
Number of Floods	14	4	10
Probability	0.61	0.20	0.59

Jevons therefore emphasises the fact that time periods whith a great number of floodings alternate with those marked by extremely low water.[339] Russel later wrongly opposed this statement and replaced it with his own hypothesis of a 19-year period (compare with Chapter 4a). As a result of climatic variations very dry regions suffer from extremes of this kind. But even regions rich in rainfall and water are affected, although somewhat less severely because the variations of the rainfall are less pronounced in these regions.

Through these hydrological phenomena climatic variations have a profound effect on the human environment. River-navigation is highly dependent on the river's water volume which determines the river's depth. As a result, the latter is, in accordance with the climatic variations, either increasing or decreasing. When the five-year average of the water levels of Seine, Donau, Rhein, Weser, Elbe, Oder and Weichsel was half a meter and even lower during the dry seasons around 1830 and 1860 than it was in the rainy periods around 1815, 1850, and 1880, river-navigation had to be severely affected because a drop or rise by this amount makes a big

[337] Defined with the help of the hypsographical curve.

[338] Sievers, in Petermann's Mittheilungen, 1873, p. 292.

[339] Jevons, in Waugh's Australian Almanach 1859, pp. 61–76.

difference. Navigational problems did indeed increase during this time, and soon a huge amount of literary works was produced dealing with the question of the cause for this drop in the river water levels. In most instances increased deforestation was found to be the cause. Nowadays we know better: It is the climate variations that bring about periodic changes affecting river traffic one way or the other.

Varying temperatures affect river traffic in another way through the duration of the river's ice cover.[340] In the middle of a cold period, e.g., the coldest lustrum, navigation is according to the averages of several variations and stations closed down for a period of 16 days longer in Siberia and in the Ural, 20 in Northern Russia, 17 in South East Russia, 25 in the Baltic provinces and 32 even in South West Russia including the Donau and the Weichsel regions than in the middle of a warm period, e.g., warmest lustrum. This influence is evident even in the averages of the cold and warm periods. At Kronstadt, for example, the annual navigation period was as follows:[341]

	1814–21	1822–36	1837–56	1857–63	1864–83
Averages (in Days)	200.5	231.7	204.5	223.4	205.0
Number of Years at Normal Levels[342]	—	11	7	5	7
Number of Years Below Number	8	4	13	2	13
Number of Favourable Years in Percent	0	73	35	71	35

Consequently, the piers of Kronstadt and also the harbour of St. Petersburg remained closed over an average of 3–4 weeks longer during cold periods than during warm periods; that amounts to 1/10 to 1/7 of the entire navigation period. It means that during the cold period those harbours with a more westerly location and shorter closure times receive part of St. Petersburg's shipping traffic that they lose again during the warmer period. Of course the length of the navigation period may also change from year to year. Yet in that case, the following year compensates for the loss of the past year. This is not the case with climatic variations where average values have changed and where either condition, favourable or unfavourable, remains the same on a multi-year average. In the cold periods only one third of all years is favourable, meaning the navigation period is longer than normal, whereas two thirds are unfavourable; in the warm period however two thirds to three quarters of all years are favourable and only one third to one quarter

[340] See Chapter 4b.

[341] Calculated by Rykatschew: *Open and Closure of the Rivers in Russia*, St. Petersburg, 1886, p. 294 (in Russian).

[342] 208 days. The abnormal years 1823 and 1835 are not included in this average value.

unfavourable. The conditions at Helsingfors are quite similar to those at Kronstadt. In the warm lustrum of 1831–35 the pier at Helsingfors was free of ice about one week earlier than in the cold lustra of 1836–55, as were the piers of Arensburg and Pernau. The same phenomenon is to be found on the American continent. Shipping on the canal of Lake Erie in the State of New York was shut down annually as follows: 1828–33, 129 days; 1838–47, 141 days; 1848–57, 135 days. In this way climatic variations go hand in hand with certain changes in trafficability which are by no means unimportant.

Another area where the influence of climatic variations is felt is agriculture. This influence, however, differs considerably according to location. In areas with an abundance of moisture where consequently wet years produce lower crop yields, the dry periods of our climatic variations yield richer crops. In a dry climate, however, where drought and crop failure go hand in hand, it is the dry period in particular in which yields are low. To substantiate this correlation I tabulated the yields for the individual years on the basis of data submitted by Fritz.[343]

A few words about what these figures represent:

The vintages of Volnay, Aargau, Nassau, Württemberg and France are in hectolitres per hectare annually; for Prussia in buckets (of 68.7 l) per Prussian acre (0.26 hectare); for the Domäne Hochburg in Baden and for Hessia again in hectolitres per hectare, for the state of Ohio in gallons per acre.[344] The wheat crop is in bushels per acre.[345]

In the dry and warm years around 1830 vintages are consistently higher in Central Europe. Between 1840 and 1855 they are very low, while around 1860 they increase. Towards 1880 they start to decrease again. It is just the reverse in Ohio; here the maximum rainfall of 1876/80 coincides with a maximum vintage. That conditions in Ohio are reversed in comparison to Central Europe, becomes quite evident when we look at the figures for the wheat crops: crops were poor in the dry 1860s and rich in the moist years around 1880.

Even in earlier years this correlation existed in Central Europe. Low yields were produced in the years 1576–*90*, 1765–74, 1812–17; high yields in 1671–78, *1698*–1708, 1818–28, etc. Grain prices in England were high pointing to, at least in general terms, low crops around 1648, 1700, 1810 and 1855; in Central Europe around *1760*, 1817; in Southern Germany and in Switzerland around 1544, *1587*, 1710, *1795*, 1817, 1847 and 1855. In Central Europe vintages were good in the years 1470, 1534, 1636, 1678,

[343] Fritz: *Periodische Erscheinungen der Meteorologie und Kosmologie* [*Periodical Phenomena of Meteorology and Cosmology*], Internationale wissenschaftliche Bibliothek Volume 68, Leipzig, 1889.

[344] Fritz, op.cit., p. 283, 296; only France according to Angot, op. Cit, p.33.

[345] Fritz, op. cit. ,p. 303.

Secular Variations of Grape and Wheat Crop Deviation from the mean

	Grape Harvest						Wheat Crop
	Volnay, Aargau, Nassau Württemberg	France	Prussia	Domäne Hochburg	Großherzogthum Hessen	State of Ohio	State of Ohio
Average	15	?	7	28	26	70	13.3
1821–25	–6	—	–1	12.	—	—	—
26–30	**6**	—	2	8	—	—	—
31–35	5	—	3	**17.**	—	—	—
36–40	3	—	–2	–6	—	—	—
41–45	–4	—	3*	–15.*	—	—	—
46–50	3	—	2	0	—	—	—
51–55	–7*	–8*	–2	–14	—	—	–0.1
56–60	2	–2	1	–6	—	—	–0.8
61–65	–1	2	—	—	—	—	–2.6*
66–70	**3**	5	—	—	**9**	–4	–1.2
71–75	0	**6**	—	—	–5	–18	0.4
76–80	–7*	3*	—	—	–8*	**14**	**2.3**
81–85	–1	—	—	—	2.	7	2.0

1724, *1784*, *1854*; on the other hand they were low around 1440, 1485, *1605*, 1695, 1765, 1810, 1875; for Southern Germany, the Eastern part of Switzerland, the Central Rhein and the Mosel around 1482 93, 1595-*1610*, 1685-1700, *1755*-74 and *1795*-1820.[346] If we compare these years with the cold and warm or wet and dry periods which we established earlier in connection with the frequency pattern of cold winters and the timing of the grape harvest (Chapter 4b), they fall into the same periodic pattern, with the exception of those years marked in italics. If we simply count the number of exceptions as well as the number of those years which follow the established periodicity, regardless whether they depict a single year or longer time spans, we arrive at 38 for the periodic years, while there are only 11 exceptions. In other words, in Central and Western Europe 77 percent of those years with particularly rich crops fall into the warm and dry periods and the same percentage of years with poor crops falls into the moist and cold periods, which fully supports our findings about the past 60 years.

In dry regions as well as in the tropics in general, aridity is detrimental to crop yields. In the 1860s Mauritius for example suffered tremendously from droughts that were attributed to the rampant practice of deforestation, yet disappeared when rainfall increased after 1865 and especially after 1870.[347] Siberia, too, suffered severe droughts around 1860 and as a result had a

[346] Fritz, op. cit., p. 265, 387.
[347] Compare Köppen in the Annalen der Hydrographie, 1887, p. 280.

number of crop failures. According to Jevon,[348] in New South Wales the years 1799–1821, in general a rainy period, show only three prominent droughts, however the dry period of 1822–41 shows nine and the wet years of 1842–57 again only three. As reported by Darwin,[349] in the years 1827-30 terrible droughts occurred in the provinces of Buenos Aires and Santa Fé at the Parana river killing millions of animals. In 1791 and 1792 East India, the West Indies, and the Cape Verde Islands suffered droughts simultaneously. Both years fall into the middle of two dry periods. Of the droughts in this century, as documented by Fritz[350] for East India, New South Wales, North America and Mauritius, 60% occurred in dry periods, although there have been three rainy periods and only two dry periods since 1800.

Climatic variations are influencing the fertility of Egypt's agricultural land in a very particular way. Agriculture's total dependence on the waters of the Nile is well-known. A rise of the Nile's level by only a few centimetres turns vast, normally infertile regions into moist, fertile and productive areas. According to our table,[*] in 1846–50 the Nile levels rose on an annual average 0.86 m above the level of 1831–35 and 0.38 m above the one of 1856–60, as well as 0.60 m above the 1856–60 level in 1871–75. Such fluctuations are of eminent importance to the productivity of these agricultural regions.

The variations of the climate should turn out to be downright disastrous for the future of the dry regions of Central North America located around the Great Salt Lake. From the early 1860s to the middle 1870s the level of the Great Salt Lake rose by 3 m; its tributaries filled with water which was then used for the irrigation of newly cultivated fields and grasslands.[351] As we have learned earlier, general opinion agrees that the expansion of cultivated land in formerly arid regions has increased rainfall considerably.

In contrast I would like to point out that the improvement of the climate takes place in exactly the same time period in which as a result of climatic variations rainfall increased over more or less all of the global land masses, but particularly over the continental regions. That, on the other hand, similar changes of the climate, to the worse or the better, occurred in earlier time periods as well in America, was established for the current century from observations of the rainfall and the river levels in the neighbouring Mississippi delta, and for far back into the previous century on the basis of temperature observations. These climate variations are identical with those

[348] Jevon, op. cit.

[349] Darwin's *Scientific Explorations*, etc., German by Dieffenbach, 1844, Volume I, pp. 151, 153.

[350] Fritz, op.cit., p. 331.

[*] [in a Chapter not contained in this excerpt]

[351] Compare Gilbert, in Powell: *Report on arid regions*, Washington, 1879, p. 55 ff.

in Europe which could be traced back to the year 1000. The large number of established variations compels us to believe that they will continue to occur in the future. All facts considered it appears most likely that the period of climatic improvement which can be observed at the Great Salt Lake since the early 1860s, will be followed by a period of climatic deterioration. First indications of this development can be detected in the dryer climate in recent years resulting in a lower level of the Great Salt Lake: in 1888 the lake had already reached the previous low of 1864.[352] Should this assumption become fact, these areas will inevitably suffer a severe economic crisis very soon, because the land which used to be cultivatable in the years 1870–80 would quite soon be unproductive as a result of the drought.[353] Then here, too, would become apparent what we have seen happening in Egypt and Siberia: not only will yields vary in size along with these climate changes but also the actual amount of cultivatable land.

Climatic variations again are not without impact on health conditions. I have, however, examined this correlation in one direction only by endeavouring to establish the link between climatic variations and the occurrences of typhus. Weather conditions do not directly affect the frequency of typhus cases but rather by way of the ground water. Even though Pettenkofer's groundwater theory may still have many critics, hardly anyone who has familiarised himself with the evidence will be able to ignore it completely.

The groundwater level falls and rises parallel with climatic variations; it was high around 1850 and around 1880 during the moist and cool periods and low around 1830–1860 during the dry and warm ones.

Secular Variations of the Groundwater (m)

	1856–60	61–65	66–70	71–75	76–80	81–85
München	0.09*	0.27	0.42	0.27	**0.75**	0.35
Salzburg	—	2.97*	3.03	2.99	**3.13**	3.04

Penck was the first to suspect that consequently epidemic outbreaks, too, might be influenced by climate variations.[354] I was able to provide statistical proof of this correlation first for Hamburg and then for other regions as well.[355] Unfortunately, morbidity rates were unavailable, I had to rely on

[352] According to the hand-written curve by G.K. Gilbert provided to me by Dr. R. Sieger.

[353] I mentioned this conclusion in a public lecture entitled *Is our climate changing?*, held March 31, 1888 in the auditorium of the University of Dorpat (compare with article in No. 68 of *Neue Dörptschen Zeitung*).

[354] Penck in the *Münchner Allgemeinen Zeitung*, end of 1887, in an essay about Soyka, *Der Boden* [*The Soil*].

[355] Compare for Hamburg, Brückner: *Grundwaser und Typhus* [*Groundwater and Typhus*, Chapter 1] Verhandlungen der geographischen Gesellschaft in Hamburg, 1887–88,

mortality rates only. These however may give a distorted view of the actual circumstances because they may be affected by medical changes. Even the link between morbidity rates and groundwater levels is of course, particularly in recent decades, no longer a true functional correlation because of the rapid progress in improvements to urban sanitation systems. As a consequence, the figures for Basel are the only ones that show a parallel development of typhus mortality and climate change while at other locations the mortality rate has generally been declining since it was first recorded.

I will begin with the data for Basel, which have been smoothed by forming five-year averages.

In the first half of the twenties the mortality rate is low (below 10 per 10,000 inhabitants), from 1828 to 1849 however it is high with a maximum in 1832 and a break in 1836, low again from 1850 to 1853 (minimum 1853), quite high in 1854-68 (maximum 1864), and once more low from 1869 to 1883 (minimum 1877). If we ignore the first four years the records of which seem unreliable,[356] and determine the mortality rate for those periods which we identified as wet and dry, we get the following results: dry period 1825 to 1840, 11.4; rainy period 1841–55, 12.0; dry period 1856–70, 21.7; rainy period 1871–83, 6.2. The only significant deviation occurs in the first half of the forties, in which the mortality is high despite the fact that these are years with a lot of rainfall.

Annual Typhus Mortalities at Basel per 10,000 Residents.
Five-year Averages

1824	2.4	1839	14.6	1854	11.9	1869	8.8
25	3.2	40	15.1	55	12.5	70	6.3
26	3.0	41	**17.6**	56	18.1	71	5.4
27	3.1	42	17.2	57	20.0	72	5.5
28	11.4	43	14.6	58	23.2	73	5.8
29	12.6	44	13.8	59	23.2	74	5.4
30	13.6	45	13.8	60	24.7	75	6.1
31	16.4	46	12.1	61	19.6	76	6.1
32	**17.8**	47	11.6	62	17.8	77	5.3*
33	10 .0	48	12.3	63	24.8	78	6.1
34	10 .1	49	11.1	64	**30.1**	79	8.3
35	10.5	50	9.6	65	29.9	80	7.2
36	8.6	51	9.4	66	29.4	81	6.7
37	10.1	52	9.3	67	28.2	82	7.7
38	12.9	53	9.1*	68	17.1	83	7.0

Volume 3. The remaining rows have been calculated according to those in Reincke: *Der Typhus in Hamburg* [*The Typhus of Hamburg*], Hamburg, 1890, p. 68.

[356] Small values like this do not occur in later years.

That such variations have not taken place in other cities, is demonstrated by the next table, which contains Basel for comparison as well

It is obvious that, in general, from the year 1860 on, typhus mortality rates are falling; however, this decline has to be seen mainly as a consequence of the steady improvements in the sanitation systems, since all locations show this tendency towards lower figures as the years progress. In my opinion, this improvement since 1860 can at least in part be attributed to rising groundwater levels. Indicative for this correlation is, to me, the mortality decline during the relatively dry lustrum of 1871–75 in München, Frankfurt a.M., Berlin, and Chemnitz, as well as its rise during the dry lustrum of 1856–60 in Frankfurt a.M., Chemnitz, and Basel as compared to the previous wet lustrum. The increase of the rate from 1820 to 1836 in Hamburg conforms again to the then prevailing dry period.

If we approach the issue in a slightly different way and, instead of concentrating on the number of typhus mortalities within a given lustrum, determine the difference from the previous lustrum, the influence of the climatic variations is clearly in evidence. This step has been taken in the next table. The figures show by how much in a given lustrum the typhus mortality per 10,000 inhabitants has increased (+) or decreased (–) in comparison to the previous lustrum.

With a few exceptions, it emerges that the dry periods around 1830 and 1860 show deterioration or at least a complete halt in the general improvement of the typhus mortality rate, whereas the wet periods show a rapid improvement. The table's last column displaying the averages of all other columns shows these variations in the improvement very clearly. The maxima fall exactly into the wet lustra of 1846–50 and 1876–80, the minima into the dry lustra of 1826–30 and 1856–60. These findings underline the probability that climatic variations have a significant impact on the mortality rates of typhus.

Annual Typhus Mortalities Expressed in Lustra Averages per 10,000 Inhabitants.

	Hamburg	München	Würzburg	Augsburg	Frankfurt/M.
1821–25	9.7	—	—	—	—
26–30	13.6	—	—	—	—
31–35	14.2	—	—	—	—
36–40	15.1	—	—	—	—
41–45	**15.9**	—	—	—	—
46–50	12.8	—	13.8	—	—
51–55	10.6	—	14.4	—	8.5
56–60	8.4	**24.0**	11.6	—	**8.8**
61–65	7.7	18.7	13.2	11.6	5.0
66–70	7.2	11.9	**15.7**	**11.9**	5.7
71–75	5.6	15.6	6.9	6.6	6.8
76–80	3.1	7.7	3.1	5.1	2.1
81–85	3.0*	1.7*	1.8*	1.5*	1.2*

	Berlin	Badisches Army-corps	Baden	Chemnitz[357]	Basel
1821–25	—	—	—	—	—
26–30	—	13	—	—	11.2
31–35	—	34	—	—	10.0
36–40	—	50	—	—	12.9
41–45	—	58	—	—	14.6
46–50	—	59	—	—	12.3
51–55	—	27	13.6	6.2	9.1
56–60	10.0	16	10.5	8.0	23.2
61–65	9.6	13*	8,5	5.6	24.8
66–70	8.6	14.	7.5	3.2	17.1
71–75	10.1	—	6.9	4.5	5.8*
76–80	4.6	—	3.8*	1.6*	6.1
81–85	2.6*	—	—	2.5	7.0.

As paradox as this may at first sound to some, it is undeniable that the water levels of the ocean are affected by climatic variations.[358] In his standard work about the currents of the Northern Sea, Mohn has demonstrated how many interdependent factors are involved in determining the sea level at any given moment in time. There is the air pressure, which in its irregular distribution alters the sea level, or the ocean's salt content that near the coasts is greatly affected by any fresh-water influx. There is also the wind's impact, which should neither be under- nor overestimated, and finally there is the temperature which in combination with the salt content determines the water's density.

The distorting effects that all these factors combined produce, are quite severe and reach one meter on Mohn's chart. Of this amount the factors' proportion is as follows: density of the sea water 0.7 m at maximum level, wind 0.8 m, air pressure a mere –0.07 m, i.e., the air pressure cancels out a small portion of the other factors' distorting effect. The sea level is lowest halfway between Iceland and Spitzbergen as well as between Greenland and Norway, and highest close to Greenland and at the estuary of the Baltic Sea.

[358] Mohn: *Nordhavets Dybter, Temperatur og Strominger* [*The Nordic Sea, Temperatures and Currents*], Volume 8 of the essays of Norske Nordhavets-Expedition, 1876–78. Preliminary information which however deviates quite substantially from the final version published as supplement No. 79 to Petermann's Mittheilungen, Gotha, 1885. [The Norwegian term "Nordhav" is today uncommon. It seems that Mohn's suggestion to name the sea between Norway, Faroe Islands, Iceland, Jan Mayen and Spitsbergen "Norwegian Sea" was accepted. (Mohn: *Die Norwegische Nordmeer-Expedition* [*The Norwegain Nordic Sea expedition*] (Petermann's Geogr. Mittheilungen, 1878).]

Change of Typhus Mortalities from one Lustrum to the Next

	Hamburg	München	Würzburg	Augsburg	Frankfurt a.M.
1826–30	**+3.9**	—	—	—	—
31–35	+0.6	—	—	—	—
36–40	+0.9	—	—	—	—
41–45	+0.8	—	—	—	—
46–50	–3.1	—	—	—	—
51–55	–2.2	—	+0.6	—	—
56–60	–2.2	—	–2.8	—	0.3
61–65	–0.7	–5.3	+1.6	—	–3.8
66–70	**–0.5**	–6.8	**+2.5**	**+0.3**	+0.7
71–75	–1.6	**–3.7**	–8.8*	–5.3*	**+1.1**
76–80	–2.5*	–7.9*	–3.7	–1.5	–4.7*
81–85	–0.1	–6.0	–1.3	–3.6	–0.9

	Berlin	Baden Army-Corps	Baden	Chemnitz[359]	Basel	Average
1826–30	—	—	—	—	—	+3.9
31–35	—	**+21**	—	—	–1.2	–0.3
36–40	—	+16	—	—	+2.9	+1.9
41–45	—	+8	—	—	+1.7	+1.2
46–50	—	+1	—	—	–2.3	–2.7*
51–55	—	–32*		—	–3.2*	–1.6
56–60	—	–11	–3.1	**+1.8**	**+14.1**	**+1.4**
61–65	–0.4	–3	–2.0	–2.4	+1.6	–1.4
66–70	–1.0	**+1**	–1.0	–2.4	–7.7	–1.5
71–75	+2.1	—	**–0.6**	+1.3	-11.3*	–2.9
76–80	–5.5*	—	–3.1*	–2.9*	+0.3	–3.6*
81–85	–2.0	—	—	+0.9	+0.9	–1.5

As we partly know already, and can partly *a priori* assume, all these factors reflect the climatic variations. To show this variability for the two factors wind and temperature, is however not possible today because of the lack of adequate observations. Having learnt earlier about variations of the air pressure, those of the salt content can be demonstrated for the Baltic Sea. Its salt content was high from 1869 to 1873, decreasing after that. In the middle of the rainy period, that is 1878–81, off Rügen it was two per thousand lower than around 1870, and in the vicinity of the Little Belt even five to six per thousand. These data appear to point to a decrease in salt content during wet periods and a concentration in dry periods. However these figures must be treated with caution because of the short observation time, the more so since according to Karsten the salt content in the Baltic Sea is mainly influenced by storm tides from the West. Be that as it may, we are in a position to identify fluctuations of the ocean level at different measuring stations, displaying the same periodicity as the climatic variations, as I have stated in 1887 in my lecture at the assembly of the

[359] Hospital admittance in percentages.

German Meteorological Society.[360] Of the series of sea level observations that I used at that time, I will repeat the data of Swinemünde at this point, raw as well as smoothed, and will compare them with the water level observations at the Oder River. In addition, I am including the lustra averages of the water level observations at Brest, Cherbourg and Le Havre, which were compiled by Bouquet de la Grye.[361]

Secular Fluctuations of the Sea Levels Adjusted by Standard Deviations

	Swinemünde (mm)		Oder, Neuglitzen (m)	
	Raw	Smoothed	Raw	Smoothed
1826–30	–7*	–3*	2.1	2.0
31–35	5	–1	1.7	1.9*
36–40	–5	0	2.2	2.0
41–45	6	2	2.0	2.2
46–50	0	**6**	2.5	**2.4**
51–55	**19**	6	**2.7**	2.4
45–60	–14	–10	1.9	2.1
61–65	–33*	–10	1.8*	1.9*
66–70	41	7	2.1	2.0
71–75	–12	–16	2.0	2.1
76–80	47	21	2.4	2.3
81–85	—	—		—

	Raw		
	Le Havre mm	Chrerbourg mm	Brest mm
1826–30	—	—	—
31–35	—	—	—
36–40	—	—	—
41–45	—	—	—
46–50	—	—	—
51–55	—	—	**10**
45–60	—	—	—
61–65 } 66–70 }	–27*	1	–17
		19	–10
71–75	5	–13*	–23*
76–80	**52**	**31**	–6
81–85	–17	21	**7**

Around 1830 and 1860, at Cherbourg and Brest a little later, water levels are low, yet around 1850 and 1880 they are high without exception.

That this conformity in the two periodic patterns of sea level and climate should be merely accidental is out of the question, especially since the rest of

[360] Compare Annalen der Hydrographie, 1888, February issue.

[361] Bouquet de la Grye, in the Comptes Rendus of Paris 1888, II, p. 813.

the stations in the Baltic Sea show a similar pattern. Consequently I have held the opinion that the fluctuations of the Baltic Sea as well as of the Black Sea are the result of a changing influx of river water. They indicate that there is sometimes more, sometimes less water in these seas which are relatively detached from the rest of the ocean..[362] Inquiries, which I carried out later, convinced me however that salt content, too, plays an important role.[363] It varies indeed from lustrum to lustrum according to the amount of river water dispersed into the Baltic Sea in the vicinity of the observation station, and this correlation is more pronounced the closer the station's vicinity to the delta of a big river.

Lorenz had shown[364] that the sea level at a river's estuary forms a very flat cone with its peak pointing into the centre of the estuary. The cause of this phenomenon lies in the fact that in this spot the salt content of the water increases into every direction. The stronger the fresh-water influx, the higher the cone's peak, and vice versa. That is indeed the case along the coasts of the Baltic Sea, as a comparison between sea level observations of stations located near a river's estuary and those further away will confirm. Secular variations of the sea level have different amplitudes, which tend to be most pronounced near a river estuary. Of special interest is the fact, which I will merely allude to at this point, that along with the climate variations during wet periods the drop of the sea level is steeper the further it is away from the fresh-water entrances, and flatter along with the varying climate during dry periods, because this fact helps us to understand the variations in the sea level near the coasts as we see them at Brest, Cherbourg and Le Havre. And indeed the fluctuations at Le Havre at the Seine's estuary are much larger than in Cherbourg and Brest, both far from any river estuaries, namely, 79, 44, and 33 mm.

It follows from all this that the sea level along coastlines and in relatively detached parts of the ocean rises and falls in accordance with climatic variations.

This knowledge sheds some light on the significance of several attempts to draw conclusions from sea level observations as to the rising and falling of the coastline. When Paschen on the basis of his 1849–66 observations at Wismar took a 1520-to-1 bet that the coastline would rise, it is because his observations were carried out at a time when the rainfall variability showed a downward trend. In fact the observations of the following 15 years, when rainfall increased again, have reversed his findings. Of exactly the same value is Bouquet de la Grye's finding, that, judging from the sea level observations mentioned above in condensed form, the ground near Le Havre

[362] Brückner, op. cit.

[363] Brückner, in *Naturforscher,* Tübingen 1887, pp. 291–293.

[364] Lorenz, Sitzungsberichte der Wieder Akademie, 1863, 2, p. 612.

gives way by 2 mm annually, at Cherbourg by 1 mm, and at Brest not at all. If as far as Le Havre and Cherbourg are concerned Bouquet de la Grye had been looking at observations from the early fifties as he did for Brest, he most likely would have come up with the same result for these stations as he did for Brest.

Although I am thus inclined to see some of the alleged changes of the coastline as the result of climatic variations, I would not under any circumstances go as far as Sueß does, who attributes shifts of several feet as they have been observed near water-marks for more than 100 years in Finland and Sweden, to climatic variations.[365]

This discourse leads automatically to the question of the significance of climatic variations and their 35-year periodicity for determining climatological mean values, the so-called standard values. That they must play a role goes without saying. As soon as periodical variations have been established, a mean value will only then approach the standard value if the time periods of opposite characteristics are entered as factors of equal weight, i.e., as quantitatively equal factors, or in other words, if the number of observed years is a multiple of the periodicity of those variations. Otherwise the mean will deviate sometimes more sometimes less from the standard value. As a consequence, 35-, 70-, 105-year, etc., averages are much closer to the standard value than those from time periods which are not a multiple of the 35-year period. The empirical approach confirms this. For the two observation series Praha and Madras, both of which display an extremely regular pattern of rainfall variations, I determined several multi-year averages. Instead of using absolute amounts I relied on the lustra averages expressed in percentage of the 1851-80 averages as shown earlie*. The average value nearest to the standard is in this case the mean of two complete variations, that is at Madras the 1821–85 mean of 102.0% and at Praha the 1816–85 mean of 101.4%. The multi-year averages, dated back from 1885, are as follows:

No. of Years	30	35	40	45	50	55	60	65	70	75	80
Praha 100 +	**1.7**	2.0	3.8	4.0	4.2	2.5	2.0	**1.4**	**1.4**	1.9	1.9
Madras 100 +	**2.2**	2.4	4.9	4.8	5.1	2.4	2.8	**2.0**	3.3	2.6	—

It is evident that indeed the 35-year mean is closer to the standard value than any of the other multi-year means. Its standard deviation is only 0.4% and 0.6% respectively, while the 50-year mean deviates by 2.8% and 3.1% respectively, i.e., by more than five times that amount. The great advantage of the 35-year mean is obvious. It is only surpassed by the 30-year mean

[365] Sueß: *Antlitz der Erde* [*Face of the Earth*], II. Wien, Praha, Leipzig, 1888.

* [in a chapter not contained in this anthology.]

which deviates from the standard by only 0.3% and 0.2% respectively. The reason for this lies in the fact that the last variation of the rainfall occurred over an average of 30 years instead of 34.8.

It is worth noting that all of the multi-year averages cited above are higher than the standard value, none is smaller, that is they are not to both sides of the standard value. It could not be otherwise. In going back in time from the year 1885 we are starting with the end of the last rainy period. This last wet period and the previous dry period are both included in the 30- respectively 35-year mean two wet periods and a dry one in the 50-year mean, etc. If we start off with wet lustra the number of dry periods can never be higher than that of the wet periods. It can at most equal that number. If we had started off with a dry period the multi-year averages would have been below the standard mean with the exception of the 35- or 70-year averages which are nearly equal to the standard mean.

The two stations, which we used in this example, are located in regions where the variations in rainfall are not very pronounced, yet we see a substantial influence on the averages. Naturally, this influence is even stronger where variations are more pronounced as for example in Western Siberia. The observations at the West Siberian station Barnaul are unfortunately limited to one periodic variation, therefore we can demonstrate this influence only under the assumption that the variation between 1846 and 1880 represents a standard variation.[366] If we start with the year 1846 and assume that the variations continue to occur with the same regularity as in the years 1846-1880, each 35-year mean (101.3) represents the "standard value". The same applies to the 70-year, 105-year mean, etc., as long as the number of years is a multiple of 35. Should we, however, want to find this standard value which is based on 35.x + w years of observation at not more than 1%, it would possibly require 440 years of observation.

Consequently there can be no doubt that climatic variations are not to be ignored in determining climatological averages that are supposed to serve as standard values, because averages that do not include a full cycle of variations often vary substantially from the elusive standard value. It is a very fortunate coincidence that Hann suggested the time span of 1851–80 as the standard period for determining climatological mean values and strictly adhered to its application, because it covers almost exactly one complete climate variation.

In view of the great and far reaching significance of climatic variations for so many areas of daily life, one is compelled to ask whether it should not be possible on the basis of what is known about these variations to put up prognoses for the future. In fact there can be no doubt whatsoever that these

[366] The mean values of the lustra between 1846 and 1880 are as follows 111, 101, 87, 66, 76, 114, and 154%.

periodic variations, of which we identified not less than 25 starting with the year 1000, will continue to occur in a similar pattern in the future. With equal certainty we can predict even more. There was every indication that a maximum of rainfall and minimum temperatures would occur around the year 1880; consequently we have to assume that we are presently approaching a minimum of rainfall and maximum temperatures, in other words a warm dry period. But this is all we can predict with certainty. When the minimum rainfall is to be expected, can only be estimated with an error margin of ± 6 years, which may in all likelihood be the error margin of one single periodic variation: the minimum should occur approximately at the turn of the century. All it means, though, is that around this time the dry and the warm years will be more frequent than wet and cool ones. A prognosis of this kind has no impact on Europe where rainfall variations—and these matter most for practical purposes—are moderate. It seems to be of practical importance only for continental regions where rainfall variations are more pronounced: for Siberia, Australia and especially Central North America. No doubt these regions are presently approaching a time of crop failures which according to the latest news about the harvests in the Western States of North America may already have begun. The approaching dry period will most likely destroy thousands if not hundreds of thousands of livelihoods.

One may wonder at first glance why they have escaped scientific scrutiny until now. However, there has been speculation about them. Sporadically, most often in connection with unusual occurrences involving rivers and lakes, the opinion is raised in publications, that the rainfall at certain locations in particular, will for a while deviate to one side of the standard amount and then to the other. Hann, for instance, emphasises in 1869 in a report about various essays dealing with the increasingly dryer climate: "Surely the effects of alternating periods of aridity and heavy rainfall are given too little consideration."[367] To what extent a similar statement by Schweinfurth is to the point, is difficult to decide.[368]

Analogous conclusions were drawn occasionally on the basis of long-term observation series. Dove, who, by the way, opposed the idea of meteorological cycles, recognised in 1838 quite correctly that the years 1808–24 were colder than those from 1797–1804 and 1820–30.[369] Zimmermann makes similar remarks in 1856.[370] Plantamour identified multi-year periods alternating between relatively warm and relatively cool temperatures for Genève. According to his findings the years 1826–34 were mostly too warm, nearly all of the years 1835–60 too cold, and finally the

[367] Hann, in Zeitschrift der Österreichischen Gesellschaft für Meteorologie, 1869, p. 18.
[368] Schweinfurth in his preamble to Baedler *Ägypten* [*Egypt*] I. Th. 1877, p. 79.
[369] Dove, Verhandlungen der Berliner Akademie1838, p. 345 f.
[370] Zimmermann, Poggendorff's Annalen, Volume 68, 1856, p. 318.

years 1861–1875, too warm again. In 1873 Lorenzoni arranged the rain observations of the Padua station according to decades and found the following periodicities:[371] dry periods 1733–46, 1784–95, 1812–44, 1856–71; and wet periods 1747–77, 1795–1811, 1845–55.

Both, Klug (1874) and Hagen discovered that German rivers were at nearly the same or a lower level in the years 1817–35 and 1855–73 compared to 1836–54.[372] On the other hand, the periodic time spans of high and low water levels identified by Marié Davy for the Seine conform only in part with our climate variations.[373]

Such unusual variations were even identified for countries outside of Europe. In 1859, Jevon, for instance, observes that in New South Wales floods were rare and droughts unusually frequent in the time periods prior to 1798 and from 1822–41, whereas in the time spans 1799–1821 as well as from 1842 up to the year of his writing, floods were frequent and droughts rare.[374] In 1877 John Allan Brown demonstrates that at Madras as well as at Trevandrum in India the years 1818–27 and 1843–52 had excessive rainfall, and the years 1828–37 and 1860–69 were remarkably dry. He concludes that it would be of interest to find out if such variations also occur at stations located further north.[375]

Finally, at the beginning of the 1880s, Fritz compiled a number of data from all over the world in order to demonstrate the variability of the volume of the mainland's waters.[376] He concentrates on establishing a link between these varying water levels and sunspots, and, consequently, fails to find out about our climate variations. Whether he should be included in this chapter is therefore debatable. A more likely candidate is Reis who in 1883 held the view that the "periodic reoccurrence of water shortage and drought is linked to sunspot activity" and who determined a primary periodicity of 110–122 years with a secondary one of 56 years. But only a few of the wet and dry periods in his differentiation coincide with the ones we have identified, others not at all.[377] Reis was completely unaware of our variations.

Yet none of these scientists has recognized the variations as precisely and was as determined to verify them for a certain area on the basis of

[371] Lorenzoni, Zeitschrift für Meteorologie, 1874, p. 188.

[372] Kluge, Zeitschrift für Bauwesen, 1874, p. 507. Hagen, Verhandlungen der Berliner Akademie, 1880.

[373] Marié Davy, Zeitschrift für Meteorologie, 1874, p. 146.

[374] Jevons, in Waugh's Almanach, 1859, p. 79.

[375] John Allan Brown, in Nature, Volume 16, 1877, p. 333.

[376] Fritz, in Petermann's Mittheilungen, 1880.

[377] Reis: *Periodische Wiederholungen* [*Periodic Repetition*], etc., Leipzig, 1883. Contrary to Reis: *Lehrbuch der Physik* [*Textbook of Physics*], Issue 7, p. 847, I would like to point out that the fluctuations which I had identified have nothing in common with those mentioned by Reis if he implies a 20-year period; his findings directly contradict facts.

meteorological observations as von Sonklar did in 1858. Not until a quarter of a century later did the works of Forel, Richter and Lang follow who again concentrated on one particular area, the Alps.[378] At no time did they mention the possibility that these variations might occur on a far larger scale as Heim assumes to a certain extent with regard to glacial variations. In addition, the universal aspect of this phenomenon, its simultaneous occurrence and global impact could not have been clearly identified before a large number of meteorological stations had actually experienced and registered the dry period in the 1860s and the wet period around 1880.

Our climate variations are also instrumental in solving a psychological problem which we pointed out at the beginning of our study.

We described the clash of opinions about the issue of climate change, some predicting a dryer or warmer climate, others claiming that it will be moister or colder, and finally a third group that it will not change at all. The identification of climatic variations resolves at least some of these contradictions. In fact, in our opinion, climate does change in one direction for a certain time and then again in the other. Depending on whether the observations of one time period or the other are consulted, the resulting conclusions will have to be contradictory. That this is indeed the case is apparent from the small table below. It shows the number of essays advocating a change of climate (wetter or dryer) for the current century in certain parts of the world according to the time spans of their publication. The account is based on the literature described in the first chapter including the studies about the increase and decrease of river levels, yet excluding those authors who advocate a kind of geological climate change on the basis of many centuries of observations, such as Whitney, Fischer, etc. The numbers of authors, which assessed the different time spans as wet or dry was:

	The Climate Turns	
	Wet	Dry
1790–1805	0	1
1805–1825	1	1
1825–1845	1	6
1845–1860	1	2
1860–1875	0	15
1875–1888	13	7

Even if this categorisation can of course in no way claim to be complete because our bibliographical account was incomplete and I was not even able to find the publication years for all of the cited works, the overall picture is

[378] Heim: *Gletscherkunde* [*Facts of Glaciers*], Stuttgart, 1885, p. 520.

in essence correct.[379] It is quite evident that during each dry period and particularly at the end of it and at the beginning of the succeeding wet period many voices were raised assessing a tendency towards a dryer climate, which grow silent again in the course and at the end of the wet period. They are replaced by authors arguing for a tendency towards a wetter climate. Four of the advocates of a dry climate fall into the lustrum of 1836/40 and eight into the lustrum 1866/70, i.e., into lustra immediately following the height of the dry period. Climatic variations are clearly reflected in the views on the direction of the impending climate change and their timing.

These changes in the climate and in the rivers' water levels are predominantly attributed to changes in the earth's flora. Almost all of the studies confirming the link between deforestation and a decrease in rainfall are produced during dry periods, and those attributing an increase in rainfall to reforestation during wet periods. Blanford's findings belong into this category[380] because his observations concluding an increase in rainfall due to reforestation fall exactly into a time period when rainfall in general increased over global land masses. Furthermore even, paralleling the climatic variations, views about the effects of forestation are changing completely. In the 1830s as well as in the 1860s and at the beginning of the 1870s it is common opinion that deforestation reduces the water levels in rivers, in the 1850s however the view prevails that it is deforestation which increases the water level.[381] The most drastic change of opinion took place in Australia. As common as it was twenty years ago at the end of the last dry period to blame deforestation for the increasing drought, as common was the view in the 1880s, that deforestation in particular was to blame for Australia's wet climate.[382] "Protect the Forest" used to be the slogan; "Down with the Forest" it is today.

In this way, climate variations guide us through at least part of the labyrinth of hypotheses and opinions about climate change that we described at the beginning of our inquiry. They teach us at the same time that those diverse opinions are actually not without merit in so far as they tell us about certain stages in the course of climatic variations. The error lies however in the fact that the conclusions drawn correctly from these observations were not viewed in their restrictive application to a certain time period but were extrapolated into the past and into the future. In addition, all of the explanatory attempts must be judged unsuccessful; the forests' influence on rainfall, in particular, has not at all been established.

[379] because some of them were not accessible to me

[380] See Chapter 4a.

[381] According to Schmid, op.cit., p. 25.

[382] See Chapter 5a.

We have come to the conclusion that in the course of the past nine centuries our globe has experienced climatic variations with a periodicity of about 35 years and an amplitude which we tried to establish for different regions of the globe. We saw their effect on all the hydrological phenomena, the glaciers, the lakes with no outlet, and also the rivers and river-lakes. Through geology we have learnt about the occurrence of climatic variations in the distant past. They were quite similar in character but differed in frequency, perodicity and impact on organic and inorganic forms surpassing by far the climatic variations of historic times. These variations are manifest in the changes between ice ages and inter-glacial ages, the nature of which remains partly in the dark. Our findings about the climate variations of past centuries may perhaps be able to contribute to the elucidation of that darkness.

Chapter 5

About the Influence of Snow Cover on the Climate of the Alps*

Of all the Alps' characteristics, none appeals more to the people of the plains than the everlasting snow covering the highest mountain peaks; nothing attracts more than the distant glowing glaciers. The snow mass lasting throughout the summer appears like a strange world to them because they are used to seeing snow as a winter's companion, only appearing in the fall and disappearing in the spring. And indeed, those glaciers are proof of an unfamiliar climate; mountain peaks rise up to this climate zone which exists everywhere at relatively low elevations above the inhabitants of valleys and plains and only reaches down to the lowlands many thousands of kilometres further north in the polar region. The snow masses of the Alps' mountain peaks most vividly demonstrate the fact that the higher the elevation above sea level, the lower the temperature and the colder the climatic conditions. Of course, forms of organic life, the mountains' flora in particular, are further proof of this. But the ever so gradual change of the flora is not as evident as the contrast between the green vegetation in the lowland and the dazzling white snow fields of the towering mountains, which in its overwhelming splendour fascinates us again and again.

The part of the Alps' surface that is permanently buried under ice and snow is not very large in comparison to the total range of the mountains. Ed. Richter has determined the glacial area of the Eastern Alps to be 1,462 sq. km, drawing an imaginary line reaching east of the Rhine across Mount *Splügen* to Lake *Como*. This is only 1_% of the entire area of the Eastern Alps. In relation to the territory of Switzerland, however, the portion of the glacial area is quite large. The total glacial area in Switzerland is 1,840 sq. km. These 1,840 sq. km make up 4 1/2% of all of Switzerland and 7 1/2% of the Swiss Alps. Exact figures for the French and Italian Alps unfortunately do not yet exist. But on the whole it may be assumed that approximately 2 1/4% of the Alps' area are covered by glaciers.

* *Über den Einfluß der Schneedecke auf das Klima der Alpen*, Zeitschrift des Deutschen und Österreichischen Alpenvereins, 1893.

Given the small size of these areas, one would assume the influence on the climate of these regions is not substantial. But it is by all means noticeable. When we approach a glacier at the end of a valley in midsummer, we can feel a cold wind coming from the direction of the glacier. The air stream is cooled off by coming into contact with the ice. It then flows down the valley, often retaining its low temperature over stretches of several kilometres. This flow of air can at times become quite turbulent. Stübel and Reiss describe such winds in the South American Andes, caused by the difference in the temperature of air above the glacial snow and above the snow-free valleys. These winds are so vehement that they may overthrow horse and rider, making travelling on those days simply impossible. Such is the force with which the cooled air funnels down from the glacial surfaces.

If, as in the above example for the Andes, the eternal ice fields of the Alps exert some influence on the climate of nearby valleys, it can only have a fairly limited impact due to their small size. The snow cover changes in size throughout the year, reaching down from the peak during the cold season and eventually covering the entire mountain range down to its base. As it expands, its influence on the climate increases. This influence is greatest during the winter, not in summer. Therefore, we will mainly limit our investigation to the winter season.

In order to study the influence of the snow cover on the climate of the Alps, it is first of all essential to have a thorough knowledge of the dimension of the snow cover. At present there is not much research to fall back on. Few observations of the various seasonal expansions of the snow cover are available. Only two series of observations do exist which have been carried out over a number of years and involve the altitude levels of the bottom snow line in the mountains and their monthly change. One took place at *Mount Säntis* during the years 1829–51,[383] the other was carried out during the years 1863–78 by Prof. A. von Kerner at the mountain ranges of the *Inn Valley* near Innsbruck. Observations at the Inn valley were done separately for the northern and the southern slope.[384] We have compiled the significant results below, omitting the data for January and February. In these months the snow line sank far below the measuring station to some point in the valley and often could no longer be determined. The month of August was also excluded because during this time the hillsides and peaks often have no snow at all and consequently no snow line could be established.

[383] Results from these observations are published in this Journal, Volume XVII, p. 49.

[384] Compare the publication of F. von Kerner: *Untersuchungen über die Schneegrenze im Gebiet des mittleren Innthales* [*Observation of the Snow Line in the Area of the Central Inn Valley*]. Denkschriften der mathematische naturwissenschaftliche Classe der Wiener Akademie der Wissenschaften. Volume LIV Wien 1887.

Altitude of the Bottom Snow Line over Several Months

	Inn Valley		Mount Säntis
	South Slope	North Slope	
	m	m	m
March	960	720	720
April	1270	1110	910
May	1700	1540	1310
June	2190	2030	1910
July	2680	2470	2530
August	—	—	—
September	3210	2760	2470
October	2150	1890	1740
November	1300	1010	1020
December	740	680	740

If similar observations were available for a number of other locations in the Alps, especially in the West and South, we might have the answer to the interesting question how large the snow-covered areas are each month. We would then have an essential database for the question about the influence of the snow cover on the climate of the Alps.

Of all the official meteorological stations, only those of Bavaria regularly observe the snow cover. The Royal Bavarian Meteorological Centre in München was the first among all the official meteorological centres in Europe to be credited for having implemented mandatory snow cover observations at their stations.[385] This observation is very extensive indeed. The stations do not merely note down daily if there is snow on the ground or not, but take an actual measurement of the depth of the snow cover by using gauges that have been set up at suitable locations. The observations began in the winter of 1886/87. If they are continued for some years to come, it will be possible to determine exactly the average altitude of the snow line of the Bavarian Alps, at least for those months when the snow line extends below the height of Mount *Wendelstein*, i.e., from October until the end of May. In looking at the observations of the stations at *Miesbach* (717 m), *Hochkreuth* (1989 m) and *Wendelstein* (1730 m) done during the five winters of 1886/87–1890/91, I discovered that in the vicinity of *Mount Wendelstein* the bottom snow line averaged at above 1700 m in October, at around 1000 m in November, below 900 m from December to March, at 1000 m in April, and at 1700 m in May. Despite this very brief observation time the results tie in closely with those gained at *Mount Säntis* and in the *Inn Valley*.

Even though we currently know little about the extent of the snow cover during the various months, we are able to draw a number of conclusions regarding its influence on the climatic conditions of the Alps, especially in the

[385] Only the observations of the Society of Science of Katharinenburg, Russia start at an earlier date.

winter. A conclusive inquiry, however, is out of the question because there are not enough data available. Rather, the following remarks are meant to encourage further observations that could then later lead to a more detailed inquiry. For the same reason, in describing the influence of snow I will focus on temperature and humidity only. Before we, however, take this main step, we should look at the physical attributes of snow that impose these conditions.

5.1 THE PHYSICAL CHARACTERISTICS OF SNOW

Air has the characteristic, especially when it is dry, to let warm rays from the sun infiltrate without interference; the immediate warm-up during this process is extremely low. The rays, therefore, reach the ground in nearly full strength being eagerly absorbed by it and warming it up. One can witness this process at any time in the mountains. If, on a mountain peak, air is measured away from direct sunlight, its temperature will always be fairly low, whereas for example the dark rocks and also the dark clothing of the mountain climber will be heated up by the sun rays fairly quickly.

Warm air rises from the surface of the ground up into the air layers immediately above the surface. In many ways the surface functions like the top of an oven. It is therefore of some significance with respect to the temperature of the air whether the surface below heats up or cools down quickly or slowly. Wherever the surface absorbs heat quickly, the air temperature will rise up high; but where the surface heats up slowly and very little, e.g., as in the case of a body of water, air temperatures will remain low. The same relates to the cooling process: the lower the temperature of the surface, the greater the drop of the air temperature.

This close correlation between the air temperature and the temperature of the surface below remains, even when the ground is covered with snow. In this case, due to the physical properties of snow, the temperature conditions of the snow cover differ in many ways from temperature conditions of the bare ground; this of course in turn influences the air temperature.

Snow never reaches a temperature above zero degrees. Once it has reached a temperature of zero degrees and is warmed up further, this additional amount of heat is used up entirely in the melting process. As long as the air temperature remains above zero, snow will always be cooler than air and, consequently, will be trying to cool the air by drawing on its warmer temperature for the melting process. This is obvious and yet, as Woeikof rightly points out in his fundamental work about the influence of a snow cover on ground, climate, and weather[386] which we will have to quote more often, it is still being given too

[386] Penck's *Geograhische Abhandlungen*, Volume III, Issue 2, p. 15, Wien 1889. Woeikof was the first who drew attention to the significance of snow as a climatological factor.

little consideration. This cooling effect of snow continues not only as long as the thermometer indicates above zero degrees but also at temperatures below zero.

To begin with, even at air temperatures below zero the snow cover is not warmed by the sun as much as is bare rock or soil because a good portion of its rays bounces off the snow surface without any warming effect. This has to do with the composition and colour of snow. If fresh-fallen snow is walked on, crystals flash everywhere, as if the surface is covered with diamonds. All these countless crystals reflect sunlight like a mirror, not just in one but in all directions with the same kind of intensity. Reflection decreases as snow ages because the crystals grow smaller during the melting process and snow compacts even more. Even then the degree of reflection remains high. In addition, the whiteness of snow, a result of its high volume of air, enhances reflection. Zenker points out that soil reflects only 1/30 of the infiltrating sun rays while snow reflects 1/6.[387] In my opinion this figure is much too low, particularly for fresh-fallen snow. Just consider the strong effect of the heat and brightness of sunlight reflected by snow, experienced when hiking across sunny snow covered glaciers. Skin tans in no time and burns quickly. As white paper already reflects approximately 40% of light, fresh-fallen snow should at the very least reflect 1/3, that is at least ten times as much as soil. Regrettably, exact tests are non-existent. All reflected sun rays are lost as they are now increasing the temperature of the snow cover. It is for this reason that the amount of heat available for this increase is much smaller in a snow cover than in bare soil.

While, on the one hand, reflection causes the snow cover to warm up more slowly, its enormous capacity of radiating heat, on the other hand, enables it to cool down rapidly.

How much heat a substance is capable of emitting depends, as is well known, on the size of its surface. A surface reduced in size by a high polish gives off the smallest amount of heat; in contrast, the rougher a surface is the larger it becomes and the greater is the loss of heat. The surface of a snow cover can be considered unusually rough even though it levels uneven terrain and, therefore, appears smoother to the naked eye than bare soil. In relation to its volume a single snowflake's surface is rather large with many edges which enhance reflection. The snow cover consists of numerous such flakes; therefore its surface space reflecting heat is many times larger than its actual size. This enables the extraordinary cool-down during clear weather periods, particularly at night. Radiative loss is highest after fresh-fallen snow allowing this type of snow to cool down to the lowest temperature. The older the snow, the smaller the surface becomes because of melting ice tips and compacting snow crystals

[387] *Vertheilung der Wärme an der Erdoberfläche* [*Distribution of temperature on the surface of the earth*], p. 63, Berlin, 1888.

and subsequently its capability to cool down rapidly by giving off heat is reduced.

To a large degree, the cooling effect on the snow surface is increased even further because snow is a poor conductor of heat. As a result the amount of heat lost in the top layer is not replaced adequately enough from below as is the case with bare soil. The low heat conduction of snow is caused by its feather-like structure capable of a large number of air pockets between the ice crystals. Snow contains an inconceivable amount of air. It is easily calculated. One only has to measure how much water a snow cover of a given density loses during the melting process. Not until the past few years have such analyses been carried out on a larger scale. However, there are almost none for the Alps. The most complete series of observations was carried out by Abels at Katharinenburg where snow—like in the Alps—often falls at fairly low temperatures.[388] Abels discovered that the ratio of snow and melt-water from newly fallen snow varies between 1:7.0 and 1:45.0; i.e., a new thick layer of snow of 100 cm yields not more than 2.2 cm to 14.3 cm of water or a compact layer of ice of only 2.4 to 15.9 cm thickness. P. Schreiber found similar data in Chemnitz: 1:6.6 to 1:34.[389] As a result, 83 to 97% of newly fallen snow is made up of atmospheric air and only 3 to 17% of solid ice. The air content is always higher, the lower the temperature and the weaker the wind was during the snowfall. Snow is much more firmly packed in snow drifts where its air content amounts to only 63 to 76%, according to Abels. Over time snow compacts, then compresses. On the one hand it is probably the weight of the snow mass itself that compresses it; on the other hand intervals of melting and freezing obviously play an important role. Air is forced out during compression. Abels discovered this while observing the density of the bottom layer of the snow at the beginning of the winter, measuring it throughout the whole winter. The air content dropped from 84% right after the first snow fall to 67% at the end of the winter before the snow melted completely. The glacial mountain range provides the opportunity of observing this process even further. According to Ratzel, glacial snow of the Tyrol Alps has 55% air and that of lower layers with less air cavities only 33%; the older the glacial layers, the lower the air content, until it is reduced to a minimum fraction when glacial snow has turned into glacial ice.[390]

Air is known to be a poor heat conductor. Consequently, the air content of snow reduces the conduction of snow considerably, meaning that the snow's conductivity improves the less air it contains. Because each layer of snow is apt to lose air after it has fallen and begins to take on the consistency of glacial snow and ice, conductivity increases with age.

[388] Repertorium für Meteorologie, Volume XV.

[389] Meteorologische Zeitschrift, 1889, p. 141.

[390] same journal, p. 433.

The characteristic of snow being a poor heat conductor is well known by every farmer. He has experienced many times that a blanket of snow prevents cold temperatures from reaching the soil and protects the seeds from freezing: The snow cover keeps the warmer temperature in the ground. This has been verified by several measurements of the ground's temperature. The top layer of snow may cool down considerably with no interference by warmer temperatures from below which is evident from the fact that the temperature of the snow increases rapidly from top to bottom layer. Based on observations taken from both Becquerel, Woeikof arrived at an increase of 0.31 to 0.36° per centimetre. Often the increase may even be much higher. On January 18, 1893 I measured the temperature of the snow on the terrace of the Swiss Meteorological Centre in Zürich at 6 o'clock in the evening and got a reading of –20.1° C at the surface, while 12 cm below the temperature was –6.1°. This results in a temperature difference of 14.0° per 12 cm, or 1.2° per centimetre. The fact that the increase was not steady, but more rapid near the surface than down below, leads us to the conclusion that as little as 1 cm below the surface snow may be several degrees warmer than at the surface itself, proving how effectively the surface layer prevents the cold from spreading. This is very important for our investigation.

Let us summarise the properties of snow with regard to its influence on the air temperature.

1. Snow never warms up to more than zero degrees and must therefore have a cooling effect on air as soon as the air's temperature is above zero.
2. The surface of snow reflects a relatively large amount of the sun rays and, consequently, cannot warm up as quickly as bare soil during frost.
3. The surface of snow is extremely well suited for reflecting heat and can, therefore, cool off rapidly.
4. Because the heat conductivity of snow is very low due to its high air content, a loss in temperature cannot be made up for by warmer temperatures from below.
5. The ability to reflect light and heat is strongest at the surface of newly fallen snow and decreases with older snow. The opposite holds true with regards to its ability to conduct heat. Consequently, under the same conditions new-fallen snow will be colder than old snow.

5.2 OBSERVATIONS OF THE SNOW TEMPERATURE AT DAVOS

As might be expected, temperature observations of a snow surface are of great significance to the question of the influence of snow on the climate. Unfortunately, up to this date they are only available for few locations. The most extensive series of measurements was carried out at Sagastyr, a polar station at the mouth of the *Lena River*. Observations in the Alps have been, until now, non-existent. At my suggestion, F. Imhof and C. Mosca from the Meteorological Station at Davos kindly agreed to undertake measurements of this kind for the months February, March, and December 1891 and January and February 1892. The results are highly interesting.

The observations at Davos were carried out three times daily, at 7 o'clock in the morning, 1 o'clock midday, and 9 o'clock in the evening on a softly sloping meadow immediately behind the station,[391] only a few meters from the building and the shelter containing the station's air thermometer. This was located about 3 meters above the snow surface. The meadow had sunshine during the afternoon, but was in the shade at the time of the observations.[392] That is why the measured temperatures reflect only in part the snow conditions of an open area of snow: The l o'clock-readings on clear days are too low in comparison to a snow area exposed to the sun. Thus, in the course of our investigation we have to keep in mind that readings relate to an area of snow in the shade. Morning and evening readings are not affected by this; because as I learned from my experiments on the terrace of the Meteorological Centre in Zürich and in my own garden in Bern the temperature of an area of snow is the same shortly after sunset, regardless whether it has been exposed to shade or bright sun light during the day.

First, I will refer to the monthly averages of the air temperature, the snow surface as well as the difference between the two. The plus-sign indicates that the air temperature was higher than the temperature of the snow surface, the minus-sign indicates the opposite. The average cloud cover is also indicated in tenths of the total area of visible sky.

Throughout the entire observation time the temperature of the snow surface was lower than the air temperature on an average of 3.9°, and in none of the observed months did the air temperature differ by less than 2.2°. The difference was always greater at 9 p.m. than at 7 a.m. and the greatest at 1 p.m. The latter

[391] Unfortunately thislocation was unavoidable for external reasons.

[392] It must be noted that the obtained temperatures of the snow surface represent maximum values, i.e., they are slightly higher than the actual temperatures. This is already indicated by the fact that the thermometer repeatedly shows above-zero-degrees for the temperature of the snow surface, which of course is not possible. Readings may be too high for the following reasons: Firstly, it is not entirely unavoidable that the thermometer sinks into the snow; consequently the instrument does not measure the temperature of the surface but of the top layer, which may be several millimeters in depth. Secondly, the thermometer protects the surrounding snow a bit from losing temperature. In any case, the loss in temperature that the glossy ball of the thermometer undergoes is smaller than the temperature gain of the surrounding snow particles protected by the instrument. Melloni, on the other hand, discovered that glass is nearly as sensitive to temperature as soot. But new experiments by H.A. Hazen, W. Köppen et al. have shown that a thermometer's sensitivity is much lower than that of glass; it is between that of glass and that of metal.

is explained by the fact that the snow area remained in the shade well past 1 p.m. If we omit the 1 o'clock measurement and use only the averages of the 7 a.m. and 9 p.m. measurements,[393] we arrive at an average difference of 3.2° between air and snow temperature.

Monthly Averages of Air Temperature, Snow Temperature, the Difference Between the Two, and of the Cloud Cover at Davos

			7h a.m.				1 h p.m.			
			Temperature			Cloud	Temperature			Cloud
		Days	Air	Snow	Diff.	Cover	Air	Snow	Diff.	Cover
1891	Feb.	28	−12.0	−16.7	4.7	2.2	1.9	−7.2	9.1	1.8
	March	31	−5.3	−7.0	1.7	6.6	2.9	−1.1	4.0	6.4
	Dec.[394]	12	−9.3	−13.9	4.6	4.5	−3.1	−10.1	7.0	4.0
1892	January	31	−9.7	−11.9	2.2	6.2	−1.7	−6.0	4.3	5.6
	Feb.[395]	20	−6.0	−7.5	1.5	6.6	1.2	−3.2	4.4	6.6
	Average	122	−8.5	−11.4	2.9	5.2	0.2	−5.5	−5.7	4.9

			9h p.m.				Average 1/4 (7+1+2x9)			
			Temperature			Cloud	Temperature			Cloud
		Days	Air	Snow	Diff.	Cover	Air	Snow	Diff.	Cover
1891	Feb.	28	−9.1	−14.7	5.6	1.1	−7.1	−13.3	6.2	1.6
	March	31	−4.0	−6.0	2.0	5.5	−2.6	−5.0	2.4	6.0
	Dec.[394]	12	−8.5	−14.0	5.5	2.5	−7.4	−13.0	5.6	3.4
1892	January	31	−8.8	−11.6	2.8	4.3	−7.3	−10.3	3.0	6.6
	Feb.[395]	20	−6.2	−7.7	1.5	4.2	−4.3	−6.5	2.2	5.4
	Average	122	−7.3	−10.8	3.5	3.5	−5.7	−9.6	3.9	4.6

This is a greater difference than has ever been found previously. At Sagastyr in the Lena estuary and at Katharinenburg, the only stations where snow temperature[396] measurements were taken over a number of months, the average difference between snow and air temperature from the 7 a.m. and 9 p.m. measurements was only 1.8° C, i.e., not much more than half of what we found in Davos. Evidently, in alpine valleys such as Davos, the difference between air temperature and snow temperature is much greater than in lower regions. The reason is quite obvious. Temperature is transmitted much stronger at higher altitudes than in the plains because the protective layer of the atmosphere is much thinner and contains less moisture. A second reason will be introduced later.

[393] This figure is of course not the average of the entire day.

[394] The first 19 days of December are missing.

[395] 9 days are missing when thermometer readings were not taken because of a severe snow storm.

[396] The measurements of Christoni at Modena are not applicable because they are not taken from the very top of the snow surface but 2 cm below the surface.

All in all, only 47 individual measurements, i.e., 13% of the total, showed an air temperature equal to or lower than that of the snow surface. All of these 47 measurements were taken on days with snowfall and, as a matter of fact, according to the daily entries in the log book, 43 of them were done while it was snowing. Consequently, at Davos the snow surface is warmer than air only when it snows.

This result prompts us to take a closer look at the difference between air and snow temperature of those measurements taken during a snowfall. This is shown in the following table.

Temperature of Snow Surface and Air during Snowfall

7h a.m. (30 Observations) Temperature			1h p.m. (21 Observations) Temperature			9h p.m. (24 Observations) Temperature		
Air	Snow Surface	Diff.	Air	Snow Surface	Diff.	Air	Snow Surface	Diff.
–6.3	–6.2	–0.1	–2.2	–1.7	–0.5	–5.6	–5.3	–0.3

Even the averages show the temperature of the snow surface to be warmer than air during a snowfall. The result remains the same if a larger sample is taken into account. Of all the measurements during a snowfall the snow surface was warmer than air at 7 a.m. in 53% of the cases, at 1 p.m. in 62%, at 9 p.m. in 60%, and on average in 58%. This leads to the interesting result that at Davos the temperature of falling snow is usually higher than the temperature of the air near the ground it falls upon. As falling snow obviously has the same temperature as the air layers it originates from, the air temperature at the bottom of the Davos valley is colder than air layers high above the valley. This can only be explained by the fact that at the onset of a snowfall cold air stagnates in the Davos valley whilst the warmer west winds above carry the snow. This conclusion awaits of course confirmation by data from other stations.

If we compare the average temperature difference between snow and air for each month as shown above with the average cloud cover, we clearly recognise the strong correlation between this factor and the temperature of the snow surface. We anticipated this because the cloud cover is mainly responsible for the intensity of the sun's influence. This influence is even more evident if we classify the measurements according to the proportion of overcast skies and determine the average temperature differences at those various degrees of cloud cover. This is shown in the following small table. The measurements taken during snowfall, already been summarised above, have been omitted in this table.

		7h a.m. Temperature		
Cloud cover	No. of observ.	Air	Snow surface	Difference.
0	28	–14.3	–20.4	6.1

7h a.m.		Temperature		
Cloud cover	No. of observ.	Air	Snow surface	Difference.
1–4	20	–4.5	–5.9	1.4
5–8	22	–4.3	–6.8	2.5
9–10	20	–4.5	–5.9	1.4

1h p.m.		Temperature		
Cloud cover	No. of observ.	Air	Snow surface	Difference.
0	31	0.0	–10.0	10.0
1–4	19	2.2	–1.0	3.2
5–8	26	1.8	–1.8	3.6
9–10	19	2.2	–1.0	3.2

9h p.m.		Temperature		
Cloud cover	No. of observ.	Air	Snow surface	Difference.
0	49	–9.7	–15.5	5.8
1–4	18	–3.6	–4.9	1.3
5–8	13	–4.3	–6.7	2.4
9–10	18	–3.6	–4.9	1.3

This table emphasises two facts. First, it shows the unusually close relationship between the temperature of the snow surface and the cloud cover. On bright, cloudless days when nothing interferes with radiation, the snow surface is at its coldest; the more overcast the sky becomes, the higher the temperature rises. The temperature of the snow surface takes on the temperature of the air, but not entirely; in fact the temperature difference between the two increases the brighter the sky is. As a result on overcast days the snow surface on an average is only 1–3° colder than the air but on clear days 6–10°. This difference increases at some morning and evening measurements to 12°, around noon even to nearly 14°. I must admit that I had not expected such big differences. They are by far larger than what has ever been measured in the plains.

To explain such enormous temperature differences in bright weather we need to consider the overall atmospheric conditions. If skies are clear above the Alps, that region is generally under the influence of a high-pressure system. This was also the case during the time of our observations. Of the 108 days of observations under a clear sky, during 95 days the Alps were exposed to maximum barometric pressure, or anti-cyclonic conditions, to use another expression. Under such weather conditions we have the strange phenomenon which remained puzzling for a long time and has only recently been explained by Hann, that the temperature increases from the floor of the valley towards considerable altitudes. In winter, high pressure carries warm air to high altitudes and cold air to the valleys. This is linked to the general circulation of air in an anti-cyclonic system and to resulting thermo-dynamic conditions of the ground surface. In an anti-cyclonic system air moves downward: air masses fall from high altitudes down toward the ground where they spread horizontally in all directions. Although they come down from an altitude of many kilometres up where their temperature was very low, they arrive near the ground with their temperature considerably higher because while descending the air is exposed to increasing amounts of pressure and is compressed; of course, compression generates heat. This warming process is important: it amounts to 1° C for a drop of 100 m. One is able to feel this very distinctly on the peaks of the Alps where the air tends to be extraordinarily warm under anti-cyclonic conditions, as Hann has pointed out previously. The same is not felt at all at the floor of the valleys; on the contrary, here an icy cold prevails. This cold is a consequence of the strong radiation from the surface of the ground which is enhanced considerably by a snow cover. This effect is enhanced by dry air and a cloudless sky, both characteristics of anti-cyclonic conditions. The lower the humidity in the air, the less likely is any interference with radiation. The cold temperature of the underground is passed on to the bottom layers of air; for these do not circulate with the rest of the air, particularly when located in secluded valleys—they stagnate. The more stagnant this air becomes the stronger is evidently its ability to adopt the low temperature of the underground—in our case the snow cover—and the smaller is the difference between the two temperatures. This is exactly what happens in Davos. Whenever it is severely cold in the valley, it is considerably warmer in higher altitudes. This is a known fact to locals and visitors despite the lack of meteorological observations for these altitudes.

It is evident from our findings that in bright weather, i.e., under maximum barometric pressure, the air at a certain altitude must be much warmer than the snow surface below. What seems remarkable, however, is that in Davos this is the case at as little as 2–3 m above the snow surface. If air would stagnate completely in this location, the snow surface would have to transmit much more of its low temperature to the bottom layers of the air mass so that there would be only a small difference between air and snow temperature at a height of 2–3 m. As we know from our findings that is not the case. We must therefore conclude that the air in the Davos valley travels, if only slowly, and is thereby replaced by descending warmer air masses. This slow air movement can be well and quite reasonably explained because of Davos' location right next to a very flat water shed with the valley of the *Landwasser River* running to the south-west and that of the *Landquart River* (or a small side arm of this river) to the north-west. And indeed, log book entries confirm that, on days when Davos is under the influence of a maximum high pressure system, a very light stream of air is noticeable travelling steadily at near-zero speed in the direction of the Davos valley. We are able to state: The big difference in air and snow temperature in Davos on bright days and anti-cyclonic weather conditions is the result of an additional supply of warm air descending from higher altitudes and replacing the slow stream of air moving along the valley. That is, under anti-cyclonic weather conditions Davos is comparable to a slope location.

The observed big difference between snow and air temperature at Davos during anti-cyclone weather conditions refutes the statement made by several sources that the cold temperature in deep valleys is, at least to a certain extent, caused by air masses which cool off at night along the slopes and descend to the floor of the valley. If in fact the snow surface along the slopes—and not the surface in the valley—was to be the primary cause of the cold temperature of the air in the valley, this air obviously should not be considerably warmer than the snow surface at the floor of this valley. Consequently, the observations in Davos show that the cause of the cold air at the valley floor is to be found in the low temperature of the snow surface in that location and not in the snow surface along the slopes. I have to admit, it certainly does not quite make sense to me that air along the slopes which is known to be warmer during the day as well as at night than air at the valley floor should carry cold air down into the valley. All it can possibly carry is a warmer temperature because it is warmer to begin with and warms up even further by approximately 1° C per 100 m when descending. From the observations at Davos we can conclude that under anti-cyclonic weather conditions the cause for those extreme cold temperatures at the floor of alpine valleys is a much more localised phenomenon than had been previously assumed.

From the above discussion we may conclude further that under calm anti-cyclonic conditions the temperature difference of air and snow along most slopes will have to be especially pronounced because a steady stream of new warm air flows down from higher altitudes replacing the old air falling to the valley floor. In this case, air has even less time to cool down while moving along the cold snow surface than it does in Davos. Observations supporting this assumption will yet have to be carried out, of course.

It appears as if our explanations would contradict Woeikof's finding that a strong air turbulence reduces the temperature difference between air and snow by increasing the snow temperature. But this only seems to be so: for, slow hardly noticeable movements of air must have a different effect than strong winds. This is simply the result of the lower specific temperature of air. We can distinguish between three different scenarios in which an air mass comes into contact with a snow surface of different temperature. If air is completely stagnant, a gradual transfer of temperature between air and snow must result in a complete adjustment. In fact the air will cool down to the snow temperature because its specific temperature is low. By comparison the difference in temperature will be very small. If air circulates violently above the snow surface, if for instance there is a strong wind, large air masses are rapidly being moved across the snow surface so that the snow almost reaches the air's temperature despite the latter lower specific temperature. Again, in this case the temperature difference is minimal. If however air moves very slowly across the snow, there will not be sufficient time to adjust completely to the snow's temperature; but at the same time, it is incapable of raising the snow's temperature substantially because due to the slow speed the respective air mass being moved across is small. In this case the temperature difference will be big.

Unfortunately, the Davos observations can confirm this thesis only in part because complete air stagnation does not seem to happen in Davos where slow circulation of air appears to prevail all the time. Neither are strong winds likely to occur due to its sheltered location. Observations under a clear sky in the morning generally show a zero wind force. Among the observations in the evening (at 9 p.m.), however, four were made at a wind force of 0 to 1 and at 1.

As small as this number is, I still think it can be averaged and compared with the overall average.

Observations		Temperature °C		
	Number	Air	Snow Surface	Difference
All Observations	49	–9.7	–15.5	5.8
Wind Force 0–1 Observ.	4	–5.6	–9.2	3.6
Difference (in °C)		4.1	6.3	2.2

The temperature difference between snow and air was indeed small in these four observations, i.e., on the average 3.6°, while all 9 p.m. observations under clear skies show an average difference of 5.8°, which is 2.2° higher. This smaller difference is primarily the consequence of the high temperature of the snow surface, which is 6.3° higher than the overall average. Even the air is warmer but only by 4.1°. But this is exactly what is to be expected if we assume that a stronger air turbulence warms up the snow surface. Strong movements of air increase the temperature of the snow surface and reduce the difference to the air temperature. The latter part of this statement corresponds with Woeikof's observations at the *Lena* estuary.

The observations of the snow surface temperature at Davos allow us to approach another question that has been raised recently, i.e., the possible influence of the snow surface on the humidity of air.

It is generally assumed that, in the mountains, evaporation of snow is high. That snow layers lose depth as they age is generally seen as proof of this view, however this is not altogether correct. This loss is at least in part nothing more than the result of compression when air is forced out during melting and refreezing. There is no evidence at all that high evaporation is the only factor in reducing a snow cover and I am quite sure that such a generalised assumption is incorrect. Much rather, the reaction of snow to the gas molecules of moisture in the air is not always the same but alternates indeed. At times moisture is extracted from the air and deposited as ice on the snow surface drying the air, so to speak. At other times water is lost through evaporation raising the moisture content of the air. The question is which of the two processes outweighs the other. Woeikof believes that evaporation plays a greater role than condensation and P.A. Müller had tried to prove it for Katharinenburg. In the Alps, at least in the valleys, the situation is different as we have learned from the observations at Davos.

This question can be solved by determining the dew-point of the air and comparing it with the temperature at the snow surface.

If an air mass that contains a certain amount of moisture cools down gradually, water will suddenly condense in liquid or solid form when a certain temperature is reached—condensation occurs. This temperature is called the dew-point. The more moisture is in the air, the higher will be the dew-point, and the lower the moisture content the lower the dew-point. If we expose a substance with a temperature lower than the dew-point of the air mass under examination to this air, moisture will condense on the surface of the colder substance, i.e., the substance will be covered with small dew-like water drops or small ice crystals. If the temperature of the substance is known and also the dew-point of the air, which is easily determined from the observed moisture, we can predict whether condensation at the surface of that substance will or will not occur. The question whether a snow surface at a given moment condenses moisture taken from the air or gives off evaporation, can, therefore, be put this way: Is the snow surface temperature below or above the dew-point of air? If it is below, condensation takes place, if it is above, evaporation.

For each single measurement of the snow temperature I have, based on observations of the moisture level in the air at Davos, determined the equivalent dew-point in two different ways: Firstly, according to the old tables of dew-points as they have been in use exclusively so far, and secondly, on the basis of those figures which Friedrich and Ekholm point out as indicators of the dew-point of ice.[397] Limited text space does not allow for a reprint of these long tables in their entirety; I shall contend myself with the averages for the individual months and the observation times in the table below. It contains the average snow temperature as well as the difference between dew point and snow temperature. A positive difference means condensation at the snow surface, a negative—evaporation.

[397] The dew point of ice is slightly higher than the dew point of water.

Relation between the dew-point of air and the snow temperature

7h a.m.

	Number of Days	Snow Temperature	Difference Against the Dew Point: Above Ice	Difference Against the Dew Point: Above Water
1891 Feb.	28	–16.7	3.8	3.2
March	31	–7.0	0.1	–0.01
Dec.	12	–13.9	4.0	3.6
1892 Jan.	31	11.9	1.3	0.9
Feb.	20	7.5	–0.7	–1.1
Avg.	12	11.4	1.3	1.0

1h p.m.

	Number of Days	Snow Temperature	Difference against the Dew Point: Above Ice	Difference against the Dew Point: Above Water
1891 Feb.	28	–7.2	2.0	1.8
March	31	–1.1	–2.6	–2.8
Dec.	12	–10.1	32	2.9
1892 Jan.	31	–6.0	0.6	0.4
Feb.	20	–3.2	–1.8	–2.0
Avg.	12	–5.5	0.1	0.0

9h p.m.

	Number Of Days	Snow Temperature	Difference against the Dew Point: Above Ice	Difference against the Dew Point: Above Water
1891 Feb.	28	–14.7	4.5	4.1
March	31	–6.0	0.0	–0.2
Dec.	12	–14.0	4.8	4.4
1892 Jan.	31	–11.6	1.9	1.4
Feb.	20	–7.7	–0.9	–1.2
Avg.	12	–7.3	1.9	1.5

The table shows that the dew point of air, whether it was determined in one way or the other, is on the average higher than the temperature of the snow surface, meaning that condensation occurs more often at the surface than evaporation. This is particularly true in the morning and in the evening. Negative differences, i.e., evaporation predominates, are found in only two months, March 1891 and February 1892. Both months had heavy snowfalls and generally cloudy skies. In February and December 1891 in contrast, when snow temperatures were below the dew-point by an especially high margin, weather was consistently bright as happens quite often in the Alps during the winter when they are under the influence of a high-pressure system.

These results are further confirmed by the following small tabulation indicating how frequently evaporation occurred during the different months and times of the day. I restricted my calculations to the dew-point under freezing conditions.

Frequency of Snow Evaporation

		Days	7 a.m.	1 p.m.	9 p.m.	Average
1891	Feb	28	11%	11%	7%	10%
	March	31	61	81	55	66
	Dec.	12	25	33	8	19
1892	Jan.	31	32	57	27	39
	Feb.	20	60	79	59	65
	Average	122	39	53	31	41

Difference between the Dew Point of Air and the Snow Temperature at Various Degrees of Cloud Cover

7h a.m.				
Overcast Cloud Formations	No. of Observations	Snow Temp.	Difference against the Dew Point	
			Above Ice	Above Water
0	28	–20.4	5.3	4.5
1–4	22	–15.7	3.1	2.5
5–8	22	–6.8	0.5	0.0
9–10	20	–5.9	–0.2	–0.4
Snowfall	30	–6.3	–1.1	–1.4
1h p.m.				
Overcast Cloud Formations	No. of Observations	Snow Temp.	Difference against the Dew Point	
			Above Ice	Above Water
0	31	–10.0	4.1	3.9
1–4	23	–8.1	2.6	2.5
5–8	26	–1.8	–3.3	–3.5
9–10	19	–1.0	–2.3	–2.5
Snowfall	21	–1.7	–3.0	–3.1
9h p.m.				
Overcast Cloud Formations	No. of Observations	Snow Temp.	Difference against the Dew Point	
			Above Ice	Above Water
0	49	–15.5	4.9	4.3
1–4	25	9.6	1.1	0.9
5–8	13	6.7	0.2	0.0
9–10	8	–4.9	–0.5	–0.7
Snowfall	24	–5.3	–1.3	–1.5

Frequency of Evaporation of the Snow Surface Determined on the Basis of the Dew-point

Overcast	7h a.m.		1h p.m.		9h. a.m.	
	above ice	above water	above ice	above water	above ice	above water
0	4%	4%	6%	6%	0%	2%
1–4	5	5	16	22	28	28
5–8	45	55	89	92	46	46
9–10	55	65	84	84	62	62
Snowfall	80	87	90	90	75	79

In February 1891, from a total of 100 observations, only 10 showed evaporation and a total of 90 condensation, in December only 19 and 81 respectively. In March 1891 and in February 1892 in contrast evaporation was noticed during 2/3 of all observations. On the average, in the morning and in the evening the number of cases with condensation was twice as high as those with evaporation, while at noon this number was almost equal.

Of all observations combined, an average of about 41% indicates evaporation and 59% condensation. If the dew-point of water is used instead of ice, these figures vary a little. The number of observations relating to evaporation increases to 44% and the one relating to condensation decreases to 56%. Again, according to these numbers, condensation prevails.

As we have seen from the monthly averages, the cloud cover is a significant factor in how the snow surface relates to moisture in the air. The following data make this even more evident showing the difference between dew-point and snow temperature on the one hand and the percentages of those instances with evaporation on the other hand in relation to different degrees of cloud cover.

As long as less than half of the sky is overcast, condensation outweighs evaporation considerably and even more so the brighter the weather is. At a cloud cover between 5 and 8, the two processes are counterbalanced. At completely overcast skies (clouds 9 to 10) evaporation outweighs condensation quite considerably, but even more so during snowfall.

We are at the end of our discussion of the observations relating to Davos. Let us once again summarise briefly the most significant results:

1. The temperature of the snow surface at Davos was consistently lower than the temperature of the air. Only when it snowed was it the same or higher.
2. The smaller the cloud cover, the lower is the temperature of the snow surface. In addition, the difference in temperature between air and snow increases as the cloud cover decreases.
3. As a consequence the snow surface has a strong cooling effect on the air temperature which increases with a diminishing cloud cover. Thus this influence is most intense under anti-cyclonic weather conditions and lessens with increasing wind levels when the snow surface adjusts to the temperature of the swiftly passing air masses above.
4. The low temperature of snow often initiates condensation of moisture at its surface in the form of frost crystals. This condensation is most frequent in bright weather while under overcast skies evaporation is predominant.

The results from the Davos observations primarily apply to the alpine valley of Davos. However, they are partly of such a universal nature that they undoubtedly apply to other alpine valleys as well—such as the *Engadin*. What these relationships are like at the bottom of the mountain or at the peak on the other hand cannot be predicted. Observations are the answer to that question.

There can be no doubt, of course, that the cooling effect of the snow surface on the air temperature will again be a factor in those locations.

5.3 INFLUENCE OF THE SNOW COVER ON WEATHER CONDITIONS IN THE BAVARIAN ALPS

From the Davos observations we learned about the quality of the snow cover's influence on the temperature. Regrettably they can not tell us about its quantity. This is only possible by comparing the temperature with and without a snow cover present under otherwise equal conditions. It is not possible in Davos because there is snow on the ground each winter, and winter months without snow are virtually non-existent. In order to make such a comparison, we have to go to lower altitudes. The only location for a comparison of this kind are the Bavarian Alps because observations of the snow cover are not being carried out anywhere else.

Obviously a comparison can be made in two different ways. First, one could compare two temperature readings of a particular location, one taken at a time when snow was on the ground, the other when under similar weather conditions and in the same season the ground was snow-free; or one could compare the temperatures of two different locations simultaneously, where one location has snow and the other doesn't. We will look at both methods.

In December 1888 and in January 1889, the region of the Alps, Bavaria and Switzerland in particular, was under the influence of anti-cyclonic weather conditions, i.e., the air pressure was high and strong air turbulence was absent. The valleys were quite often fogged in, while higher elevations had clear skies. The air pressure decreased in February and weather conditions changed bringing plenty of precipitation. In December 1890 the weather pattern was quite similar. During most of the month the northern foot of the Alps was at the centre or at the periphery of the anti-cyclonic system. This is also true for January 1891 when weather conditions were only temporarily disrupted by a short melting period between the 21st and 25th. Even more pronounced were the anti-cyclonic conditions in February 1891. However, in one point both winter seasons differed considerably—in the amount of snow on the ground. In the winter of 1888/89 until the end of January only a few spots in the valleys were covered with snow. A permanent snow cover did not materialise until February. In the winter of 1890/91, on the contrary, the ground was completely covered with snow from the end of November onward. This development is best reflected in the data for the beginning and the end of a permanent snow cover in Bavaria as shown below.

Duration of Permanent Snow Cover of the Bavarian Alps

Station	Height Above Sea Level (m)	Winter 1888/89 Beginning	End	Days
München	526	21.II	10.III	18
Rosenheim	446	3.II	24.III	50
Miesbach	717	3.II	12.IV	69
Oberstdorf	842	3.II	20.IV	77
Peissenberg	994	3.II	21.IV	78
Hochkreuth	989	3.II	16.IV	73
Wendelstein	1730	3.II	22.IV	79
			1890/91	
Station	Sea Level (m)	Beginning	End	Days
München	526	25.XI	22.II	90
Rosenheim	446	25.XI	5.III	101
Miesbach	717	25.XI	10.III	106
Oberstdorf	842	24.XI	22.IV	150
Peissenberg	994	25.XI	17.III	113
Hochkreuth	989	25.XI	9.III	150
Wendelstein	1730	17.X	8.V	204

In 1888/89 there had been an occasional snow cover prior to the recorded months but snow was not very deep and disappeared quickly as the following figures demonstrate.

Number of Days with Snow and the Depth of the Snow (cm)

			December		January		February	
			Days	Depth	Days	Depth	Days	Depth
München	526	88/89	0	0	13	6	24	13
		90/91	31	11	31	36	22	21
Rosenheim	446	88/89	1	1	17	10	26	25
		90/91	31	7	31	40	28	41
Miesbach	717	88/89	12	2	16	9	26	62
		90/91	31	15	31	57	28	67
Oberstdorf	842	88/89	0	0	12	11	27	85
		90/91	31	17	31	73	28	67
Peissenberg	994	88/89	6	2	14	11	26	92
		90/91	31	33	31	60	28	51
Hochkreuth	989	88/89	20	10	17	17	27	100
		90/91	31	22	31	70	28	84
Wendelstein	1730	88/89	20	14	20	25	27	72
		90/91	31	16	31	148	28	90

Differences in snow covers are quite extreme in December and January and do not even out until February. What were the temperatures like during that time? The following small table has the answer.

Average Monthly Temperature in the Winter 1888/89 and 1891/92

		December	January	February
	1888/89	−1.8	−3.5	−2.8
München	90/91	−7.1	−6.2	−3.4
	Difference	−5.3	−2.7	−0.6
	1888/89	−6.8	−4.6	−3.8
Rosenheim	90/91	−7.1	−7.6	−4.1
	Difference	−6.3	−3.0	−0.3
	1888/89	−3.5	−5.0	−4.5
Miesbach	90/91	−9.1	−8.2	−5.3
	Difference	−5.6	−3.2	−0.8
	1888/89	−1.4	−5.8	−5.3
Oberstdorf	90/91	−10.2	−9.8	−5.9
	Difference	−8.8	−4.0	−0.6
	1888/89	−1.4	−4.6	−6.0
Peissenberg	90/91	−7.8	−7.1	−2.9
	Difference	−8.7	−2.5	+3.1
	1888/89	−0.7	−5.4	−10.0
Wendelstein	90/91	−5.5	−9.9	−4.6
	Difference	−4.8	−4.5	+5.4

Clearly, the winter with heavy snowfall shows a lower average temperature in December and January than the winter with less snow. This confirms our finding that a snow cover reduces the temperature considerably. The months of February 1889 and February 1891 cannot be compared directly because the first one was stormy or "cyclonic", the second one "anti-cyclonic". The temperature appears to be the same. However in February of 1889 it is the result of high morning and high evening temperatures and in February of 1891 of high temperatures at noon. If we exclude the latter and consider only the night temperatures, i.e., the daily minima as shown below, February 1891 with a snow base in the valley and in the plains was considerably colder than even January 1889 with hardly any snow, although normally February tends to be warmer than January.

Average Temperature Minimum

	January 1889	February 1891
München	−5.5	−6.6
Rosenheim	−6.7	−10.5
Miesbach	−8.5	−10.2
Oberstdorf	−11.5	−11.7

Conditions in Switzerland were similar; unfortunately no records about the snow cover are being kept there. Only Zürich and Basel report the monthly number of days with snow. In the winter of 1888/89 in the time period from December 1 to February 1 Basel had a light blanket of snow on January 7 only, and Zürich from January 10 to 31. In Zürich however the snow cover was permanent in all of December of 1890 and January 1891, in Basel from December 1 to January 24.

Temperatures in these months were as follows:

	December			January		
	1888	1890	Difference	1889	1891	Difference
Basel	–0.7	–4.5	–3.8	–1.4	–4.5	–3.1
Zürich	–1.5	–5.8	–3.3	–2.5	–5.3	–2.2

The differences are more pronounced in December because during this month in 1888 there was no snow at all, as opposed to the permanent snow cover of 1890, while Zürich had only a partial snow cover in January 1889 and on the other hand Basel was snow-free towards the end of January 1891. In any case, these figures again demonstrate quite impressively the cooling effect of snow.

But even when we compare the temperatures of two different locations, one with snow and the other without, as they occur at a particular time of day we clearly notice the cooling effect of snow. However, it is not easy to find two such comparable locations, especially in the winter, a time when the region of the Alps is normally under a permanent snow cover. Spring, however, offers ample opportunities for comparison because wherever snow remains on the ground, the temperature does not rise substantially above 0. Since the mountains get large amounts of snow, it often stays longer than at comparable altitudes in the plains delaying the onset of spring. At a time when Bern has no snow anymore and temperatures rise well above zero, the floor of the alpine *Aar* valley at the same altitude is still buried deep in snow and the temperatures are crisp.

This cooling effect of the snow cover is emphasised quite well by observations of the Bavarian stations. For comparison I chose two pairs of stations, at Lindau and Kempten and at Rosenheim and Traunstein. In Lindau at 399 m sea level, the ground is often free of snow, while in Kempten located 300 m higher up (696 m) the snow still keeps. In the same way Rosenheim (446 m) is often free of snow, while Traunstein (597 m) still has snow. The temperatures of the paired stations cannot be compared directly because the locations are at different sea levels and therefore, with few exceptions Kempten will always be colder than Lindau, and Rosenheim colder than Traunstein. If snow has a cooling effect, the temperature difference between Lindau and Kempten and Rosenheim and Traunstein respectively must obviously be particularly large as soon as the lower-level stations are snow-free and the upper locations are not. However the difference will be smaller if both locations either do or do not have snow at the same time. This is indeed the case. Based on the snow observations of the five winters from 1886/87 to 1890/91, I was able to identify those days when snow conditions at both locations were equal, and those days when only the upper locations had snow. The temperatures for those particular days were provided by the data published annually by the Bavarian Central Registry for those 4 stations and I came up with the following small table.[398]

Temperature Difference Between Kempten and Lindau	Number of Days	Temperature Difference	Number of Days	Temperature Difference
	December		January	
When L. was snow-free, K. had snow	25	3.6 °C	25	3.6 °C
Both locations had or did not have snow	100	3.1	110	3.2 °C
Difference		0.5		0.4
	February		March	
When L. was snow-free, K. had snow	30	4.1 °C	40	2.2
Both locations had or did not have snow	80	3.0	45	2.0
Difference		1.1		0.2

Temperature Difference Between Rosenheim and Traunstein	Number of Days	Temperature Difference	Number of Days	Temperature Difference
	December		January	
When R. was mainly free of snow, T. had snow	15	1.2 °C	30	1.2 °C
Both locations had or did not have snow	130	0.5	105	–0.4
Difference		0.7		1.8

[398] The analysis was made with pentade [5 days] averages of temperature. Pentades of equal type were considered those pentades, when at both the upper and lower location, snow was present or absent for at least 4 days. Pentades of opposite type were defined for Lindau and Kempten as those when Lindau had no snow at all and Kempten all the time; for Rosenheim and Traunstein, when opposing conditions prevailed for at least three days. This was required, because the number of pentades of opposite type would have been too few (only 8) for Rosenheim and Traunstein.

	February		March	
When R. was mainly free of snow, T. had snow	20	1.5 °C	30	1.7 °C
Both locations had or did not have snow	130	0.6	9.5	1.0
Difference		0.9		0.7

During the 25 days in December when the snow cover in Kempten remained, while it had already disappeared in Lindau, Kempten was colder by 0.5°C, during the 25 days of January by 0.4°C, and during the 30 days of February even by 1.1°C. These conditions were similar for Rosenheim and Traunstein.

When in spring the snow cover has disappeared from the deep valleys and starts retreating to high elevations, it still maintains its winter temperatures by repeatedly cooling the air, especially at night, down to 0° and below. This process intensifies the contrast between the snow-free valley down below where spring has arrived already, and the wintry mountain valley. At that time the difference in temperature between the two locations is at its highest. Woeikof has demonstrated this nicely by citing a number of examples from the Swiss Alps.[399] In April the bottom of the snow line is at an altitude of approximately 1000 m. The city of Chur (603 m) is already snow-free, while nearby Churwalden (1213 m) or at least its neighbouring slopes still have snow which does not disappear before May. Consequently, in April Churwalden is colder than Chur by 4.6°C, in March by only 4.2°C and in May by only 4.3°C. In June the snow line has moved up toward the 2000 m range and covers the area between *Sils* (1810 m) and the *Julierpass* (2244 m). As a result, the temperature difference between both locations is highest in June: in May 2.5°C, June 3.7°C, July 2.6°C. We can make the general statement that the temperature difference between two locations diminishes most rapidly with increasing altitudes in the month in which the lower location is snow-free while the upper location still has some winter snow. This is especially true for higher elevated valleys, but not for mountain peaks. These are towering in atmospheric conditions where the contact between air and snow surface is very brief, where air is quickly replaced and cannot cool down significantly. On mountain peaks as well as on slopes, the temperature can, therefore, rise far above zero despite the presence of snow; in valleys where air circulation is severely hampered, the average temperature does not stray very far from zero degrees as long as there is snow.

This rapid temperature decrease in early summer is in some respect of importance. Woeikof pointed out quite correctly that it may be one of the main causes for the high number of heavy thunderstorms in early summer because the rapidly falling temperature induces air to circulate upward carrying the war-

[399] Woeikof, op. cit., p. 98.

mer moisture-laden air masses from the valleys. This causes heavy condensation and thunderstorms.

We are at the end of our discussion. In its course we had to be satisfied with demonstrating the influence of the snow cover on the climate by giving at least a few examples. Furthermore, these examples were taken from a fairly small region, mainly between the *Inn* and the upper *Salzach* in the south, the *Rhine* in the west and the lower *Salzach* in the east because the required observation records were only available for this region. Although examples are of good use whenever a problem needs to be illustrated to a larger audience—this, however, is no way to find a definite solution to the problem. At each step we were confronted with new questions which could not be answered because the necessary data are lacking. This means, observational data have to be collected for all the regions of the Alps.

This is desirable in three directions. Firstly, it is required to determine the annual vertical movement of the bottom line of a snow cover as has been done at Mount *Säntis* and near Innsbruck. Observations of this kind would be of particular importance for the farthest East of the Alps, such as the lower *Tauern* range, as well as for the southern valleys of the Alps. They can easily be carried out by anyone, but have to be done daily or almost daily on a continuous basis over a number of years in order to be meaningful. Hillsides that are not too steep are best suited for this. The elevation of the snow line can easily be estimated daily from a standpoint in the valley by using certain points on the slope of which the elevation is known, for visual orientation. Prof. v. Kerner's observations in the *Innthal* can serve as an example.

Secondly, it is necessary to make as many observations as possible at locations with meteorological stations using the same method of daily snow observations which proved so successful in Bavaria, i.e., recording day after day if snow is on the ground or not, and possibly measuring its depth. This depth can be determined simply by using measuring poles (*Schneepegel*) once daily in snow as far away as possible from buildings, walls, trees, in short from all obstacles which could cause snow piles. Again, such observations are very simple and require very little effort and material. A manual explaining this method has been published by the Central Meteorological Station in Bavaria.[400]

Thirdly and finally, it seems desirable that observations of the temperature and the snow surface like the ones in Davos should be carried out in some other locations, particularly in areas with a variety of different elevations. Of particular interest would be observations in a location at the foot of the Alps, in one of the steep main valleys as well as at a mountain observatory. These observations require more effort and attention to detail as well as a certain skill in handling sensitive thermometers. They should therefore best be undertaken by meteorological stations.

[400] Published in Meteorologische Zeitschrift, 1887, p. 15.

Once observations have been carried out in all of these three directions and published, one will be able to take the next step examining the influence of the snow cover on the climate of the Alps in its entirety as we have outlined in our remarks. They have served their purpose already if they entice one or the other person to carry out observations of this kind.

Chapter 6

Influence of Climate Variability on Harvest and Grain Prices in Europe*

Over the past 15–20 years the German farming industry went through a severe crisis. A critical situation emerged that was unlike any other before—caused by the stiff competition from American and particularly Russian grain on the German market. This competition evolved within a few decades, for not until the 1870s and 1880s did Russia's grain export increase so enormously. Several events seem to be responsible for this boom in Russian export: during the second half of the 1860s the North American War of Independence paralysed America, Russia's biggest competitor. Expansion of the German and especially the Russian railway system reduced transportation costs and instantly brought the rich grain fields of South and Southeast Russia closer to the West. At the same time, Russia's agriculture expanded. In addition, the exchange rate of the Rouble dropped offering a tremendous incentive for export. Because of the low exchange rate while prices of goods stayed the same abroad, the Russian trader had more money in his hands than before; but above all the demand for grain in Central and Western Europe increased with a rising industry and growing population. These are the major reasons that have been held responsible for Russia's increasing export. They can be attributed to man and his struggle for survival. Another instance, however, which is completely beyond human control, should not be omitted but rather be included in this number of causes despite its being called secondary: It is the fact that Russia had a series of exceptionally rich crops in the 1860s and 1880s up to and including the year 1888 and only very few poor harvests, whereas the West suffered numerous crop failures.

This aspect has been neglected so far. One has become used to the fact that poor harvests are considered chance occurrences whose economical impact on a country may change the grain price and grain trade considerably from year to year, but not regularly. In other words: those factors which have

* *Der Einfluß der Klimaschwankungen auf die Ernteerträge und Getreidepreise in Europa.* Geographische Zeitschrift, 1895.

nothing to do with human intervention are considered to have no lasting effect on the outcome of crops as soon as longer time spans, such as decades, are concerned and not the individual year. There is, however, no justification for this assumption.

I.

The outcome of crops depends to a large extend on the climate of the region because it is the climate factor and not the soil which ultimately decides whether a plant grows or not. The plant life on this earth develops therefore in close dependence on the climate. Both, temperature and precipitation are determining factors; the former is mainly responsible for defining the polar boundaries of plants; the latter often determines their latitudinal expansion. If weather conditions would remain the same year after year, the habitat of a plant species would, *ceteris paribus,* be defined by the same isotherm and the same isothete all over the globe. In reality this is not the case because for plants to disseminate, more important than the average temperature and the average precipitation is the variability of both elements. For several years a tree may have had barely sufficient moisture to grow in a particular location, it and its seedlings will perish if in the following year it is denied even this bare minimum. This is a common fact. In a series of good years a plant will perhaps extend its habitat; yet, the first bad year which does not supply the warmth and moisture required for its survival will force it back into its old boundaries. Plants are only able to survive in the long run in those places where even in bad years the required minimum amount of warmth and moisture is supplied. Obviously, weather conditions of bad years in particular are responsible for limited plant dissemination. This primarily refers to the natural habitats of plants.

Where man interferes, conditions are in part different. It can often be to his advantage to grow a plant that prospers in good years only, but refuses to bear fruit in bad years or perishes entirely. The yield of one good year may compensate for many bad ones. For example, in medieval times and later there used to be vine growers in many regions of northern Germany and northern France, where viniculture is non-existent today, even though they did not produce at all in some years. Because of the high freight costs it was more advantageous to accept poor crops than to import grapes from the south. As infrastructure improved later on and transportation became cheaper and taste may have become more refined, the viniculture retreated back to the south and vineyards were left to some other cultivation more suitable for that climate.

A similar development is seen happening today in the area of grain production.

Temperature conditions determine the polar boundary of the various types of grain cultivation. Temperature restricts wheat growing in Europe primarily to regions south of the 60th or 61st parallel, yet permits the more fastidious types of grain like barley and oats to advance almost up to the Northern Cape and to the White Sea. Nonetheless in the main agricultural regions of Europe the temperature's impact is limited to those regions and therefore its effect on the outcome of the crop is only secondary. Only near the grain's most northerly line of growth, e.g., in Scandinavia and Northern Russia, crop is often damaged due to frost. Too much heat does not destroy grain at all as long as enough water is available, e.g., wheat grows well in the tropical part of India's subcontinent!

Moisture is much more important for growing wheat. Without water there would be no agriculture, but with too much water none either. This fact comes to mind when we look at the extent of agriculturally used land on this earth or rather, examine the cause of crop failures. In all those regions where rainfall is sparse, droughts and poor harvests go hand in hand. In regions with excessive rainfall, bad harvests are mainly inflicted by rainy years. Of course, it is not the absolute quantity of rainwater that tips the scale, but the ratio between the amount of water and evaporation. The same amount of water that inevitably drowns the crop in a cold climate may just be sufficient enough for the irrigation of grain in a hot climate. In contrast, a small amount of rain that could cause a drought in a hot country due to heavy evaporation is often entirely sufficient in a cool climate. In Europe the wet coastlines of the North Atlantic Ocean and Central Europe respond in a way exactly opposite to the continent's dry interior. Southern Russia and Great Britain including Ireland are extremes in this regard. Drought has to be the cause of the crop failure in 1891 in the central and eastern provinces of Russia. From August to October 1890, drought damaged the winter grain crop; the following winter was dry with little snow so that the seedlings froze in part. Due to the small amount of snow that melted early, the soil did not get the usual amount of moisture. The frosts of April 1891 were not met by a protective blanket of snow and damaged the seeds. Consequently winter grain was destroyed. The summer grain however fell victim to drought and hot winds in May, June, and July: crop failure was the result.[401] Such events are not too unusual in Russia, although they, fortunately, rarely reach such enormous dimensions as in 1891.

In England, the exact opposite happened when numerous crops failed, followed by high costs of living at the end of the thirties, during the forties, and at the beginning of the fifties, as well as when harvests turned out poorly

[401] Woeikof, in Meteorologische Zeitschrift, 1892, p. 40.

in 1872, 1875, 1877, and 1879; all of which were excessively wet years. Complaints were raised that because of extreme wetness in the fall, parts of the agricultural lands could not be cultivated and in the wet summer grain was flattened and rotted in the fields.[402]

Central Europe and the German Empire in particular are in between these two extremes. Dry years are usually good years for a number of agricultural activities, especially for wine and fruit growing, but also for grain, whereas they are bad for the grasslands and cattle raising. Although there are some years when crops are damaged by dryness, however, more common are bad harvests due to too much wetness. We are reminded of the poor harvests of the wet years around 1880.

In Southern Europe and to a larger extent also in the tropics, at least as far as the production of grain is concerned, conditions are similar to Southern Russia. Years of famine in the East Indies coincide with dry years.

It would be of interest, at least for Europe, to determine the borders of regions where grain growing is primarily damaged because of too much rain, and those regions where damage is mainly caused by drought. A precise line cannot be drawn however, because there is a transitional zone where damages by too much rain are just as often as those by drought. In addition, this borderline may well be different for each variety of grain. On the whole, though, this line can be drawn from interior Russia to the Southwest toward the Carpathian mountains and continuing along their range to the Alps all the way to Southern France. In the East it still runs through the forest belt but parallel to the steppes; in the West it almost coincides with the border of the subtropical region. The area to the North suffers more from excessive rain, the one to the South more from drought.

In those European countries influenced predominantly by ocean climate, damages to grain cultivation by too much precipitation are so significant that, given the present high wages and low freight costs inviting foreign grain imports, grain production has to deal with the greatest difficulties. Farmers have partially abandoned wheat growing in favour of grassland and cattle raising. This can be said of Great Britain and Ireland, Holland, Denmark, Scandinavia, and the provinces of Schleswig–Holstein and West Prussia, also Switzerland, and, in general, the regions on the northern slopes of the Alps which enjoy a lot of irrigation and not too much warmth. The grasslands need water; it has to rain a lot before it becomes too much. Wet years which would destroy the grain indigenous to the steppe are quite often good years for raising cattle and vice versa. This has happened in the dry summer of 1893: pastures, especially in the plains and foothills, in

[402] Numerous examples are mentioned by Tooke and Newmarch, *Geschichte und Bestimmung der Preise 1723–1857* [*History and Definition of Prices 1723–1857*], in German by Asher, Vol. I and II.

Switzerland and Southern Germany were dried out, whereas the wheat fields had a good yield. The regions of Western Europe with cool summers are indeed predestined for raising cattle and for grass cultivation. Here, this is a natural line of production and the transition to cattle raising which is taking place under pressure from competition by huge grain suppliers such as America, Russia, and Hungary, is simply a more climate-conscientious way of making use of the land.

If we move Southeast from our borderline the danger of poor growth because of drought increases steadily. In particular, the regions east of the lower and middle part of the Volga River suffer quite often from famines as a result of drought. Here again, we have an outpost of grain growing regions that are presently still viable because labour as well as land is cheap and wheat cultivation therefore survives even multiple poor crops, but mainly because a more natural and at the same time more advantageous use of the land still has to be established, as has been done in the west combining cattle raising and grassland cultivation.

My comments have been sketchy so far; a geography of the causes of crop failures still has to be written. Presently, sufficient data are not yet available. Data about harvest damages are not yet systematically collected for all regions, as for example has been done in Prussia. Only some general outlines could be mentioned here. Because of the variability of the climate one weather system may be the most frequent cause of crop failures in one location and another in some other location. Indeed, complete shifts in agricultural production occur under the influence of climate, although they are instigated by man-made constructions--by the system of modern transportation: That the agricultural use of land has been given up in exposed areas close to the ocean, is to a large extent the result of an adverse climate. This inclemency has always existed, but in former times it was more reasonable to cope with it, today it is more advantageous to accommodate.

Just as the climate can facilitate a local shift in agricultural production, it can influence the timing of economical factors, especially the harvest and the grain price. I will explain this briefly.[403]

[403] A detailed explanation of this question must be left to a later larger publication because my data collection is far from complete. Even here, I can only publish a portion of what I have so far accumulated to avoid being too explicit. The following questions have been addressed in several of my publications: Ed. Brückner: *Inwieweit ist das Klima konstant*? [*How Constant is Today's Climate?*—Chapter 3 in this collection], Verhandlungen des VIII. Deutschen Geographentages zu Berlin, 1889, p. 101. —Brückner, *Klimaschwankungen seit 1700* [*Climate Variability since 1700*—excerpts are given in Chapter 4 of this collection], Wien, Hölzel, 1890, p. 275–279. —Brückner, *Über die praktische Bedeutung der Klimaschwankungen.* [*About the Practical Importance of Climate Variability*], Compte rendu du Vme Congrès international des Sciences géographiques. Bern 1892, p. 616. —Brückner, *Rußlands Zukunft als Getreidelieferant*

II.

In view of the climate's strong influence on the outcome of the crop, one is obviously only allowed to exclude the weather factor when explaining the economical conditions and their change from one decade to the next, if it can be proven that the average weather condition or, in short, the climate remains constant. Only then the probability of good harvests in a region will remain steady year after year.

Until recently, it was believed that the global climate is indeed constant. New research has however shown that this is not the case; in fact climate undergoes many changes which vary around a middle range. I do not refer to the 11-year weather cycle that has often been cited but not yet verified, which from time to time disappears entirely and is caused by the 11-year cycle of sunspots, but rather to the far more important 35-year climate variations. The highly regarded director of the Bavarian Meteorological Institute, C. Lang, who, regrettably, passed away before his time, has established their occurrence for the region of the Alps in 1885 in his usual clear and precise way. I had the privilege of applying his findings globally.[404] Meanwhile the obtained results have been confirmed by various sources, particularly by R. Sieger, E. Richter, and Heintz whilst Fritsch, Sonklar, Forel and Ellis had already confirmed their likelihood for certain specific regions earlier.[405] Nowadays, climate variability has to be considered a fact.

Climate variations involve changes in temperature, air pressure and rainfall that occur globally at the same time. The length of these variations, i.e., the time that elapses between two extremes of the same kind, is 35 years on an average, sometimes more sometimes less. The temperature is the one factor in this process on which all the others depend upon. These temperature variations are common to almost all the countries around the globe. All of them experience cold periods as well as warm periods simultaneously. The change in temperature averages 1 °C. That is quite a lot, meaning nothing else but that, for instance, during the five years around 1840, the average yearly temperature of Berlin was a full degree lower than during the five years around 1825 which would be the same as if Berlin had moved by 3 degrees latitude to the north.

Temperature variations have an impact on the distribution of air pressure. The transfer of moist maritime air from the ocean to the continent seems to

---

[*Russia's Future as a Grain Supplier*], Supplement to Münchener Allgemeine Zeitung, November 19, 1894. Some sentences have been transferred from the latter two publications with permission from the respective editors.

404 Compare my book: *Klimaschwankungen seit 1700* [*Climate Variability since 1700*—see also Chapter 4 of this book], Wien, Hölzel, 1890.

405 The complete literature about climate variability, summarized by myself, is to be found in Geographisches Jahrbuch, XV, 439 and XVII, 348.

be hampered during warm periods, and easier during cold periods. This in turn will influence the amount of rainfall in that region. Over most of the land masses rainfall varies to such an extent that the cool periods are at the same time moist and the warm ones dry. In our latitudes the amount of rainfall during the wet period is 20% higher than during the dry period, in Russia around 25–30% and in West Siberia even more than 100%.

Over the last two centuries cold periods with a lot of precipitation on the continent seem to centre in the years 1705, 1740, 1775, 1815, 1850, and 1880; warm periods with dryness on the continent in the years 1720, 1760, 1790, 1830, and 1860.

Because of the wide range of these climate variations, they will have an impact on the economy as well. That is actually the case; their influence on the yields in agriculture—largely through rainfall—is significant. The quantity and quality of vintages in France, Germany and Switzerland, by the way, are better in dry and warm periods than in wet. It is more significant that grain crops show the same impact.

Unfortunately, the available amount of data proving this impact is rather limited and lacking in quality. Long-time observations of the yields of field-produce in kilogram per hectare would be most suitable. However, data like this are available only for the past few years. For Russia in particular, the giant among grain producers, they do not go further back than 1883. The situation is better in Prussia where the so-called *Erdrusch*-table is available since 1859 accumulating these data for each province and each year. If one would have to rely exclusively on such harvest tables, there would not be much point in comparing crop results and climate variability. Fortunately a different indirect approach leads to results, at least for earlier decades—I am talking about a closer look at grain prices.

Prior to a world trade in grain the price of grain was, first and foremost, decided by the outcome of the crop in the producer country. Years with poor crops have always been years of high food prices and vice versa. Therefore, the annual changes in grain prices reflected changing crop yields. Though the price was occasionally influenced by other events such as war or pestilence, these factors were merely reducing the crop outcome's impact, not obliterating it; the two kinds of impacts simply added up. We are, therefore, entitled to draw conclusions from the development of the grain prices of those earlier years to the crop yields. Lately, of course, this holds no longer true. Nowadays grain prices are no longer solely affected by the turn-out of a crop in the producer country but also by that of American and Russian crops. Today, we have a world price for grain, which strongly influences prices in individual countries. Therefore, only until the years 1850 or 1860, in England even up to 1840, can the grain prices in central and western Europe serve as an indicator for crop yields in the producer country. With this restriction in mind, we can use these prices for our purpose.

The method of categorising the data is the same I have previously used to determine the variability of meteorological factors. Five years each were combined to a lustrum average, e.g., 1801–05, 1806–10, 1811–15, etc. This way, random influences caused by irregular weather patterns from one year to the next, were eliminated in most cases, yet long-term patterns remained unchanged. Numbers obtained in this way are shown in the tables at the end of this essay. They each contain, in addition, the amount of rainfall in percentages of a multiple-year average. Where rainfall observations were unavailable, I have included the dates of the grape harvest. I have previously proven that these dates primarily follow the climate variations as they change from lustrum to lustrum. The lustra averages of the rainfall as well as of the dates for the harvesting of grapes are expressed in their difference to the multi-year average; the minus (–) symbol indicates that the rainfall's percentage was too low and the onset of the grape harvest too early by the respective number of days. Therefore, the minus symbol generally indicates a dry period. All relevant explanations as well as sources are included in the tables' footnotes. The average yield and price for the wet and dry periods are mentioned at the end of each table. These figures were arrived at by determining firstly the time span of wet and dry periods for each country based on the average rainfall per lustrum and then by determining the average value for these time spans.

In addition, some of these numbers are represented graphically. Prior to this, they were mathematically adjusted in order to exclude even more of those random occurrences and, in doing so, to show their line of development much better. I used the same method of adjustment as in my research on climate variations according to the $(a + 2b + c)/4$ = ordinate of b, where a, b, and c are neighbouring lustra averages. For the first respectively last link, the formula $(2a + b)/3$ = ordinate of a respectively $(r + 2)/3s$ = ordinate of s was applied.

The time spans marked by the average year of the lustrum at the top and at the bottom, e.g., 1803 represents the lustrum 1801–05, were expressed as abscissa and the figures for grain prices, yield, or rainfall as ordinates.

The rainfall curve rises and falls according to the amount of rainfall. An up or down of the curve by one increment represents a change in rainfall by 2 1/2% of the multiple-year mean. The curves for the times of the grape harvest, the prices, and the yields as well as Russia's rye exports again rise and fall with the quantities they represent—but the scale is different.

Let us discuss these tables[406] and curves.

[406] Tables are placed at the end of this essay.

The influence of climate variations on the harvest as well as on the price of grain is obvious, but varies considerably in different regions. The same difference between West and Central Europe on the one hand and East Europe on the other, which we discovered earlier with regard to the cause of crop failures, can be established again. During dry periods regions with ocean-climate have high yields and consequently low grain prices because their crop failures are mainly caused by too much rainfall, as in England, France, Belgium, Denmark, Germany, and Austria. Regions with a continental climate, as for instance Ohio and Russia, have low yields and high prices during those periods. In contrast, the yields of oceanic countries are low during wet periods and prices high, whereas continental agricultural regions enjoy good harvests and low prices.

Changes in yields are quite pronounced in Prussia. Whenever the amount of rainfall increases (Figure 1, Table I [page 213]) yields decrease—in fact, for all types of grain. Rainfall around the year 1880 was very heavy and the yield very small, while during the sixties and at the beginning of the seventies rainfall was low and, therefore, the yield was high: one curve is the exact mirror image of the other. In addition, variations are quite pronounced; during the wet lustrum of 1881–85 the crop of wheat was down compared to the dry lustrum of 1861–65 by 225 kg per hectare or 16%, and that of rye by even 325 kg or 26%. Even if one combines several lustra, the influence is still evident. Throughout the entire wet period of 1876–93 10% less wheat and 19% less rye per hectare was harvested than in the dry period of 1859–75. Consider the meaning of this—over a period of 15 years a loss of 10 respectively 19 percent! Exactly the same results turn up when the average harvest is estimated.[407]

In view of this obvious influence of climate variations on the yield, it comes as no surprise that grain prices suffer (compare Figure 2 and 4, Table II, III, and IV [on pages 214-216]). The development of prices does not show the same parallel pattern as is evident between the changes in rainfall and crop yields. There are times when suddenly a country's political situation, the competitive influence, changing exchange rates, etc., interfere with this correlation.

The dry period of 1821 to 1840 has a steady strong impact in the ocean-regions. In all of them grain prices are considerably lower than what they had previously been. With the beginning of the wet period in the forties prices start to rise again; a maximum is reached almost everywhere in 1851–55, at the same time or immediately after the maximum rainfall. This

[407] With the exception of the lustrum of 1846/50.

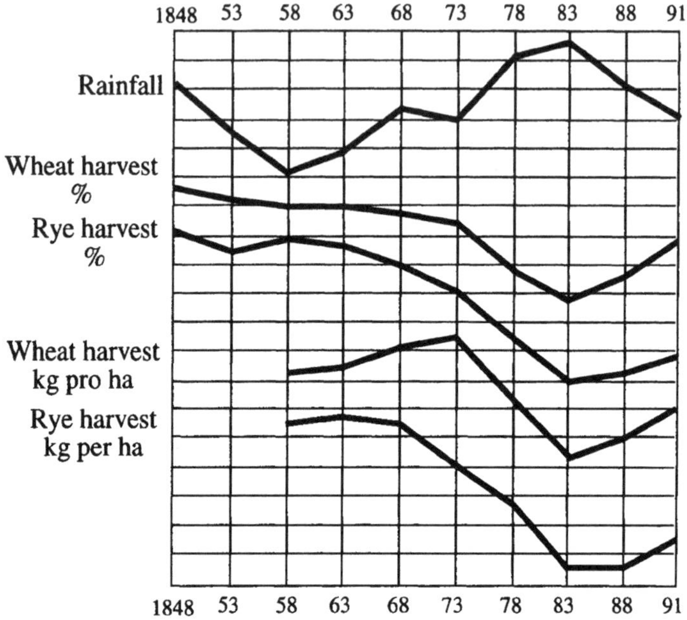

Figure 6.1. Fluctuations of Rainfall and Grain Crop in Prussia. Regenfall = Rainfall; Weizenernte = Wheat Crop; Roggenernte = Rye Crop.

development is quite evident in all countries of Western Europe, except to a relatively lesser degree in England. In all of these countries prices then drop slightly in accordance with less rainfall until 1861–65, Austria being the only exception. Again this period is followed by a price increase as rainfall increases as well. It is, however, of short duration and the price already peaks in the years 1871 to 1875, that is 5 to 10 years prior to the maximum amount of rainfall, only to fall considerably after. These prices drop despite the fact that crops in the producer country continue to worsen and do not begin to improve until the second half of the eighties (compare with Prussia). Consequently, it has nothing to do with the outcome of the crop at home and is the result of massive grain imports from America and Russia.

In order to determine the extent of the price changes due to climate variations, it is advisable to ignore absolute values and to simply deal with the maximum lustra expressed in percentages of the minimum lustra. This has been done in the following small table, i.e., for those lustra showing maxima and minima in the majority of countries.

	Wheat England	Wheat France	Wheat Belgium	Rye Denmark
1816/20 to 1831/35	117	113	115	—
1851/55 to 1831/35	111	112	115	118
1851/55 to 1861/65	112	111	112	112
Average	113	112	114	115

	Wheat Prussia	Wheat München	Wheat Zürich	Wheat Austria
1816/20 to 1831/35	115	118	115	115
1851/55 to 1831/35	116	117	112	115
1851/55 to 1861/65	111	—	112	—
Average	114	117	113	115

On the average the price during the wettest lustra is higher by 13% than during the driest ones — certainly this is a substantial amount.

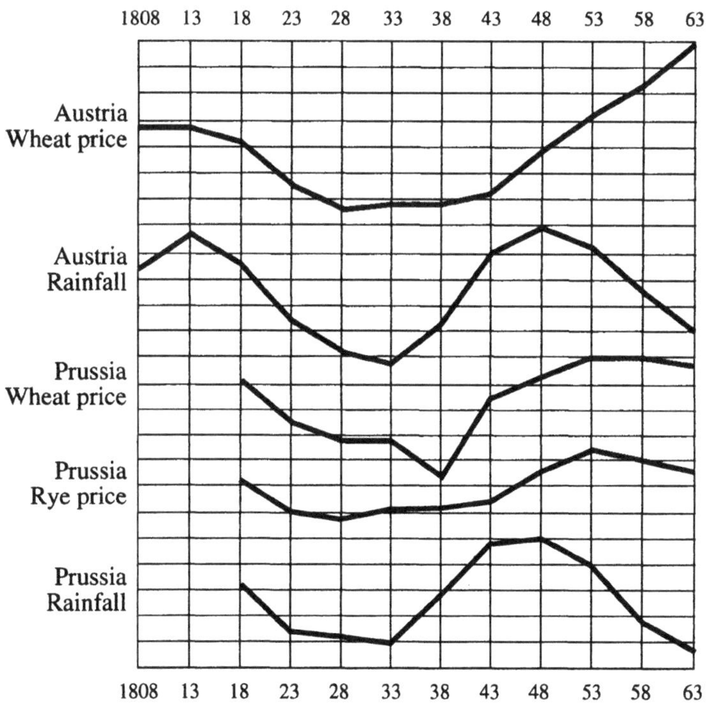

Figure 6.2. Fluctuations of Rainfall and Grain Prices in Austria and Prussia

The situation as we have just described it for West and Central Europe is quite the opposite in countries with a continental climate as far as the correlation between climate variations on the one hand and the harvest and grain prices on the other hand is concerned. There is a very strong influence of climate variability on the wheat crop in America (Ohio, Figure 3, Table V [page 217]). In this case, a decrease in rainfall coincides most visibly with lower crop yields up to 1861–65; after that both increase. In the wet lustrum of 1876–80 the crop per acre was a full 50% higher than in the dryer lustrum of 1861–65. After 1880 a small reduction in the yields reflects the slightly lower rainfall. In general, grain prices react accordingly all over the United States. In the sixties (1862–1870) during years with small amounts of rain—i.e., 3% below the average—the price for a bushel of wheat was 143 cents, in the next 5 years with a rainfall of 6% above the medium—only 100 cents.[408)] However, I would put less weight on price development and more on the yields because at the same time the introduction of farm machinery reduced production costs in America enormously.

The situation is similar in Russia. Unfortunately this cannot be verified as clearly as for the United States because in Russia the official crop statistics were not kept until 1883. For earlier years, only sporadic vague estimates for certain areas are available, which are not conclusive for the situation in the whole empire. However it is a fact that Russia had a number of excellent harvests in the seventies and eighties up to and including 1888 which coincides with the last period of high precipitation (see Table V [on page 217]). Although bad harvests did occur, such as the crops of 1879 and 1885, these were exceptions. However, after 1888 crop yields decreased tremendously as official statistics show.

Prices are the only indicator for those earlier years. In Moscow (Table 5) rye prices were high during the dry lustrum in 1831–35, obviously because yields were low due to droughts. The price fell and remained low in the following wet period of 1836–55, then rose again. Another drop as was to be expected from the good yields in the wet period of 1871–85 did not take place. Undoubtedly, this is related to the extreme devaluation of the Rouble. Despite the relatively low grain prices abroad, Russian traders still earned a lot of money, which had consequences for the price at home.[409]

408 Calculated according to Francke's figures in Zeitschrift des preußischen statistischen Bureaus, 1887, p. 125. The rain is the mean of four groups averages, given in *Klimaschwankungen* [*Climate Change;* Chapter 4b in this collection], without the Atlantic states.

409 Exchange rate fluctuations are also the cause that Hungarian grain prices do not reflect any influence from climatic fluctuations.

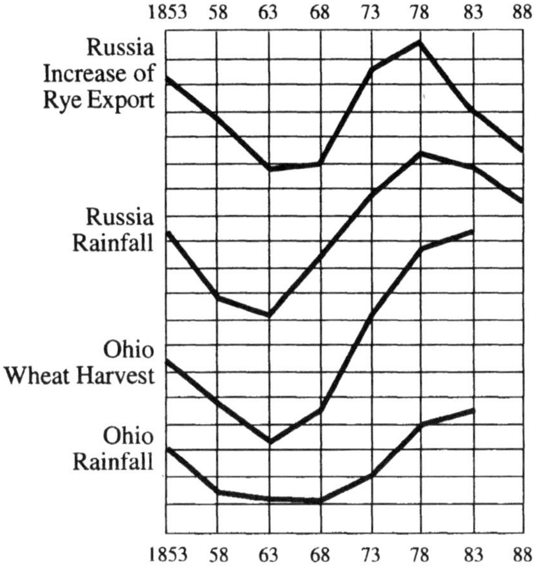

Figure 6.3. Rainfall fluctuations and increase of rye export in Russia and of wheat crop in Ohio

Let us take another look at climate variability and grain prices in the past century (Table II [page 214], Figure 4). The interrelations are the same as in the current century, as far as observations are available. In England rainfall and wheat prices show exactly the same development from 1701 to 1790. There is a slightly lower congruency between wheat prices and the harvest dates for grapes in France as shown in Table II and during the first half of the century in Germany. This is not surprising since both products may be to a great extent affected by the amount of rainfall, but not exclusively. A general trend is obvious though. For example, during the wet period of 1770 prices rise significantly everywhere, which is superseded as well as followed by periods of low prices and less rain. The years 1710 and 1730 in England, and 1715 and 1740 in Germany stand out as times when costs of living were high. This time of high prices at the beginning of the forties does coincide with war-times in Germany; but since at the same time heavy rainfalls were recorded, I am inclined to attribute the high costs of living, at least in part, to the latter. A strong increase in prices and their continuous high level can be observed everywhere from 1790 to 1820. In Germany this development could, at least in part, be the result of the climate (compare Figure 4). Prices again are not noticeably higher than later in the forties. In England, however, rainfall does not show maximum amounts at this time but some time later.

As a consequence, it may be assumed that the high costs of living are mainly the result of political tension: All of Europe was heavily armed.[410]

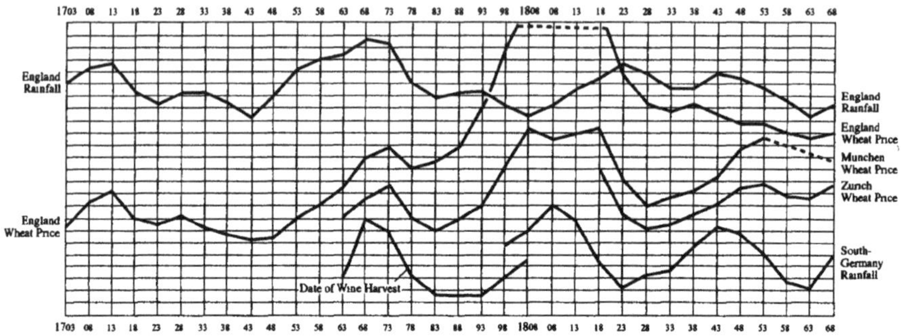

Figure 6.4. Fluctuations of rainfall and grain prices in England, Southern Germany, and Switzerland.

III.

If grain crops and grain prices are influenced by the amount of rainfall, it should also be possible to link climate variations to trade and trade policies. I do not have all the data necessary to verify the correlation for all countries, a task that would exceed the purpose of this essay anyway. Nevertheless, I would like to show in what way climate variability reflects directly on trade policy by citing the example of the grain trade between Russia and Western Europe.[411]

Primarily two forces rule the export of grain from Russia to Central and Western Europe: the demand in the West, and the surplus in the East. All other factors such as the construction of railroads, improvements in agriculture, trade policies in general can never be considered as prime causes; they merely serve the purpose of responding to demand or of disposing surplus. We must expect higher exports if the demand increases in the West and the East has a large surplus, i.e., when crops are bad in the West due to a lot of precipitation and for the same reason crops are good in the East. Lower exports must be expected if in the West demand goes down because dry

[410] According to Geering when reviewing my essay about Russia's future as a grain supplier in Neue Zürcher Zeitung, January 17, 1895.

[411] I will mainly use the data from Laves, *Getreideproduktion und –handel im Europäischen Russland* [*Grain production and trade in European Russia*], Jahrbuch für Gesetzgebung,. 2c (Schmoller) N.F. V, p. 293.

weather produces good crops and in the East the surplus is reduced because of bad harvests.

It is quite interesting to be able to recognise this phenomenon clearly in the development of Russian grain exports. At the beginning of our century, during the period of high precipitation, which lasted in some countries up to 1820 in others up to 1815, England and Central Europe had crop failures, and Russia exported relatively large amounts of grain, i.e., from 1800 to 1813, 4 million hectolitres annually. It is during this time that rapid settlement in Southern Russia's steppe takes place. As early as 1816 the port of Odessa exported 7 to 8 million metric hundredweight, an amount that was not reached again until 1878 and surpassed in 1879. The West now begins to enjoy good harvests during the dry period. England prohibits the import of wheat due to its own surplus. Throughout central and western Europe grain prices are low, demand is low because home production is self-sufficient, and by 1825 to 1840 Russia's export goes down to 3,4 million hectolitres annually because of drought. Again another moist period follows. England abolishes the grain tax in 1846, bad harvests increase the price of grain. Transportation and certain laws are improved which benefit grain imports in Western Europe, because demand increases, and success is not far behind: in the years 1844–53. Russia exports 11 1/2 million hectolitres grain annually, almost four times as much as in the thirties. Export decreased during the Crimean Wars, but even in the years 1856–64 it was slightly below the amount of 1844–53. This backdrop was explained by the fact that so many people lost their lives during the Crimean Wars. However, how much do 100,000–200,000 people count, and the number of dead was certainly lower, compared to a population of more than 50 million of which 12 million were men in the work force? Crop results seem to give a much more plausible explanation for the decrease of the export: in the West good harvests are the result of fairly dry weather, while in Russia the same weather produces bad harvests. Since the middle of the sixties Russian export experiences an unprecedented big increase coinciding with wet weather conditions, during which harvests in Russia are extremely good, but in Western Europe extremely bad. As a result Russian grain exports clearly reflect the variability of the climate.

To convert this more precisely into numbers, the columns 7, 8, and 9 were created in Table V [page 217]. The upward trend of the grain export continues due to a steadily growing population in the West as well as to improvements in transportation. It is, however, briefly interrupted or at least slowed down because of climate variations. In order to isolate the climate's influence, the general upward trend had to be eliminated. I achieved this by drawing a line through all the export figures of each lustrum according to the method of least squares. For each lustrum I then determined the differences between the figures gained through this method and the actual export figures. Whenever these differences were positive, the export's increase had accelerated; were they negative, the increase had decelerated. The figures of column nine in Table V relating to the rye export were arrived at in this way. These figures were adjusted and graphically depicted as shown in Figure 3. A comparison with the variations in the amount of rainfall clearly confirms that the increase of grain export is accelerated in wet periods, while it slows down in dry ones, even a slight decrease takes place. Since this correlation has been verified for the entire 19th century, chance is obviously out of the question: climate variability has a clear impact on the grain export of Russia. This does not mean that we deny any influence by various other factors such as competition, political constellations, customs and tax regulations, etc. It seems, however, that none of these factors can fully hide the influence of climate variations.

Finally, a further question comes to mind: What will weather conditions in the next decades be like as a result of climate variability and how will they influence trade?

From the year 1000 on, not less than 25 full climate variations have been identified and consequently there can be no doubt that the 26th will not fail to occur. All indications are that the last wet period peaked in the year 1880; from then on or even at least in part from the year 1885 on, rainfall decreased significantly in all parts of the globe. Our tables offer examples. Lately Heinz mentioned this reduction in rainfall for Russia from 1881–85 to 1890.[412] It has progressed further since then. I predicted this development in the year 1888 and mentioned at the time[413] that we are approaching a dry period, which can be expected to peak around the turn of the century. I added furthermore that the coming dry period will create major economical crises in the continental regions in particular, and will destroy the livelihoods of thousands, if not hundreds of thousands. Events have already confirmed this. The Russian crop of 1888 which was not exceptional, yet might still be regarded as good, was followed by the bad harvest of 1889, the mediocre one of 1890, and finally by the complete crop failure of 1891. The harvest of 1892 was again quite mediocre. In 1893, however, Russia was blessed with a good harvest.[414] This should not be surprising. For, periods of dryness are not characterised by a series of bad harvests but rather by their increased frequency and a decrease in the frequency of good harvests. In short,

[412] Repertorium für Meteorologie XVII, No. 2.

[413] In my inauguration lecture in the small auditorium of the University of Bern in May 1888. Compare also Verhandlungen des VIII Deutschen Geographentages zu Berlin, Berlin 1889.

[414] According to the official harvest reports of the Central Committee for Statistics of the Russian Ministry of the Interior.

indications are that Russia is on the brink of a period with harvests of inferior quality. This fact alone should have an adverse effect on export. It is, of course, quite certain that Russia alone would be in a position to satisfy all of Europe's demand for grain, even in bad years, simply by intensifying her agricultural methods. However, in times of crop failures, agriculture cannot be expected to expand; because that would require capital, and capital especially will be scarce and in part be used up as a result of bad harvests: agricultural decline is the consequence. Russia's crop failure of 1891, for instance, caused a noticeable reduction of agricultural land.[415] But a reduced production would not put too severe a restriction on Russian exports if there was not a second more important aspect involved: The need for Russia's grain is today no longer as urgent in the West as it was 5 or 10 years ago. Its demand has decreased, which again is a result of climate variations. For, in the West rainfall has decreased over a number of years as well, and consequently higher yields were achieved. The year of 1893 was so dry that agriculture suffered damages in certain areas, hay production in particular. Nevertheless this is the exception; generally agriculture is expected to benefit from a dry period.

It is quite remarkable that the director of trade statistics in the Swiss Department of the Exterior, T. Geering, emphasised that Europe's current trade policy and in particular the grain taxes of Germany and France are in complete accordance with the theory of climate variability.[416] "Grain prices have fallen as low as never before. The demand (of the West) for imports is relatively small because production at home is sufficient at an unusually high degree and consequently it will be more likely that grain imports from Russia and North America to Central and Western Europe will decline. The tax policy of the continental Great Powers (i.e., France and Germany) should therefore, irrespective of other economical considerations, continue to get support from the prevailing climate as a factor in production over the next 10 years to come."

During the next wet period however which is expected to peak in about 20–25 years, a tremendous increase of grain imports from those countries with harvests subjected to continental climate variations and the price increases in the West may perhaps reverse trade policies in a free-trade direction as it happened in England in 1846 as a result of bad harvests.

[415] According to the official publication Résultats généraux de la récolte en Russie 1892 (St. Petersburg 1893), the agricultural land in the European part of Russia had diminished by 927,000 hectares in that year compared with its size in 1887, when the last estimates were made. This reduction is very extensive in areas hit by the crop failure of 1891, whereas the western and northwestern districts show increases.

[416] Geering

There is a peculiar kind of interaction between countries in which harvests are exposed to either the oceanic or continental type of climate variations: a relationship of mutual compensation regarding grain production. If one country has a good harvest, the other has a bad one and vice versa. This definitely must lead to prosperity as has been indicated in the figures for the seventies and eighties in our Tables III [page 215] and IV [page 216]. Long-term variations in grain prices will at first be subdued in the wet regions of Western Europe by the fact that price increases do no longer occur during wet periods. Prices have always been low in dry periods as a result of good harvests at home; they now will also be low during wet periods because of the rich crops of the big grain suppliers Russia and America as well as India. The costs of this development must be carried by the agricultural industry of the West which will have to go through critical times because each period of wet weather will produce poor harvests at home and an oversupply of cheap grain from countries enjoying good harvests at that time, as it happened in the eighties and in part still happens today. The less self-sufficient the West will be in satisfying its own demand for bread and the bigger the surplus produced by America and Russia is, forcing them to sell at any price, the closer the West will follow the pattern of price variations in these large grain producing countries. Most likely, this trend in England will be noticeable in the very near future, while Germany is still basically self-sufficient in bread supply.

Let us end here!

Up to now it was considered, as we emphasised earlier, that those factors beyond human power which influence harvest losses are mere chance occurrences, "to such an extent that they were completely ignored in the practical considerations concerning agricultural production and tax policies, and it was assumed instead that chance would take care of them".[417] The human mind was regarded as the sole factor in economical development. Trade policies, social conditions, the relentless fight for survival, these were the only factors taken into consideration when economical phenomena needed explaining. I am far from underestimating the enormous force of the human mind in this matter. But one factual influence of tremendous importance in the form of climate variations, that is the change between favourable and unfavourable weather conditions, plays an additional role that can no longer be overlooked. Climate variations influence crops and grain prices. I feel we have proven this here. That the entire economy and the course of history is affected is rather obvious. How much more important this influence is as opposed to those factors created by the human mind, I do not dare decide. It has so far not been completely dismissed while trade policy has repeatedly, if unintentionally, taken climate variability into account.

[417] Geering as mentioned before.

APPENDIX: TABLES AND SOURCES

The method of tabulation has been explained above on page 202. Extremes are marked in bold or with an asterisk. My essay *Climate Variations since 1700*[*] is initialised as "C. V." when cited.

Table I. Variations of Crop Yields in Prussia[1]

	Rainfall[2] (%)	Yield per Average Harvest[3] (%) Wheat	Rye	Barley	Oats	Yield per Hectare[4] (kg) Wheat	Rye
1846/50	2	7.4	9.6	4.0	–2.0	—	—
1851/55	3	–2.2	–3.8	2.6	1.8	—	—
1856/60	–9*	2.4*	6.6*	–9.8	–9.0	1385	1218
1861/65	–7	2.0	2.6	7.0*	8.2*	1399	1250*
1866/70	6	–2.0	–2.2	–2.6	–3.4	1410	1220
1871/75	–7	1.6	–1.0	–0.6	–1.8	1519*	1249
1876/80	**10**	–11.0	–13.8	–9.4	–8.4	1360	1066
1881/85	5	**–18.6**	**–23.2**	**–19.2**	**–21.6**	**1175**	**926**
1886/90	3	12.6	–21.6	–17.6	–11.6	1264	947
1891/93	–2	–5.0	–14	–17.3	–19.3	1367	1038
Dry[5]	–4.2	1.0	1.6		–1.5	1428	1234
Wet[6]	4.0	–11.8	–18.2		–15.2	1291	994

Addendum and notes to Table I:

1 Old State, i.e., the 8 old Provinces.

2 Average of 23 Measuring Stations in North and Central Germany in C.V., p. 158, updated. These unpublished results from observations of 1891–1893 were kindly given to me by the Königlich preußisches meteorologisches Institut.

3 Up to 1877, according to estimates from the annual harvest table that were taken prior to the actual harvest by estimating the crop in the fields. These estimates are known to be too low (compare also A. Kremp in Hildebrand's Jahrbuch N. F. IX, p. 358). Based on the data for 1859–76, Engel has defined the term "average harvest" (*Mittelernte*) by using the yields (in bushel per acre) from the *Erdrusch*table (Zeitschrift des preußischen statistischen Bureaus, 1878 p. 401). By setting the average of the estimates from the harvest table of 1859–76 equal to 100, I determined the average deviation by which these estimates are too low, namely the estimates of the grape harvest by 12%, rye by 16%, barley by 13%, and oats by 10%. I increased the estimates accordingly, then expressed the lustra averages in deviations from 100. Although these reductions distort the resulting figures somewhat with regard to climate variations, harvests vary by much more than those reduction percentages, therefore even non-reduced figures will still reflect their variations. In 1877 these estimates end. From 1878 on the amount of harvested grain per hectare (ha) is included in the Prussian statistics (results of the crop yields). These numbers were put in percentages of Engel's average value and the columns were continued until 1893. For rye, only winter rye was included because summer rye is hardly ever grown, but for wheat and barley the average yield from summer and winter crop combined was added. Some of the average values for the 'old' [Prussian] state had to be determined anew with East and West Prussia combined as one province as it used to be before.

[*] [See Chapter 4 of this collection].

4 Up to 1877 according to the *Erdrusch* tables, which up to 1866 are published by Meitzen (Boden und landwirtschaftliche Verhandlungen des preußischen Staates III), and until 1870 by Engel (Zeitschrift des preußischen statistischen Bureaus). For the years 1871–77 I obtained data from hand-written notes of the königlichen preußischen statistischen Bureaus. From 1878 on crop figures for grain are included in the Prussian statistics. The years from 1856 to 1858 are missing

5 1856–75.

6 1876–93.

Table II. Variations of Grain Prices 1701–1800

	England		France			Germany		Austria
	Rain-fall (%)[1]	Wheat Price[2]	Rain-fall (%)[3]	Wheat Price[4]	Date of Grape Harvest[5]	München Wheat Price[6]	Berlin Wheat Price[7]	Wheat Price[8]
1701/05	0	28.6	—	—	5.2	17.6	31*	—
1706/10	0	**42.1**	—	—	–1.0*	12.6*	33	—
1711/15	**11**	40.9	—	—	**5.4**	**18.1**	**42**	—
1716/20	–6	34.0	—	—	2.6	15.5	42	—
1721/25	–6*	31.0*	—	—	4.6	14.7	34	—
1726/30	**3**	**41.4**	—	—	–1.2*	15.9	32*	—
1731/35	–6	27.1	—	—	4.6	13.4*	—	—
1736/40	0	36.1	—	—	2.4	18.4	—	—
1741/45	–12*	29.0*	—	—	**6.8**	**24.3**	—	—
1746/50	–3	29.8	—	—	1.7	18.2	—	—
1751/55	6	33.6	—	—	5.8	14.9	—	—
1756/60	4	42.0	–0.7	11.1	0.5	19.9	—	—
1761/65	5	34.8	–3.3*	10.1*	–2.3*	18.8	—	—
1766/70	9	51.1	**6.5**	15.5	**7.8**	22.7	50	—
1771/75	**10**	**51.5**	0.3	**16.2**	2.3	**32.3**	**63**	—
1776/80	–6*	40.2*	–0.7	13.4	–1.3	17.6*	40*	—
1781/85	–2	48.7	–7.2	14.8	–4.0	18.5	48	—
1786/90	–2	47.2	0.5	17.2	0.1	22.3	54	2.9
1791/96	–2	53.6	—	—	–5.7*	23.9	56	3.0
1796/00	–3	73.4	—	—	0.7	30.5	64	3.3
Dry[9]	—	—		—	2.1	15.1	32	—
Wet[10]	4	37.2		—	4.2	16.1	40	—
Dry[11]	–6	32.5		—	1.9	15.9	32	—
Wet[12]	–1	34.9		—	4.8	19.1	—	—
Dry[13]	–8	29.4	–2	10.6	–0.9	19.3	—	—
Wet[14]	7	42.6	3	15.8	5.0	27.5	56	—
Dry[15]	–3	52.6	–4	14.1	–2.8	20.6	50	—

Addendum and notes to Table II:

1 Up to 1730 the averages are from several West European stations and later only from England. C.V., pages 188 and 190.

2 Up to 1770 prices in Shillings (with decimal fraction) per Winchester Quarter at Eton, average per year. Later for all of England—official average per year and Imperial Quarters. Tooke and Newmarch, *Geschichte und Bestimmung der Preise* [*History and Definition of Prices*], translated by Asher. Dresden ,1859, Vol. II, p. 512.

3 The Dates for Harvesting Grapes in Central France (C.V., p. 264); Minus (–) is too early, meaning warm and dry. Numbers stand for days.

4 Francs per hectolitre. Statistische Monatsschrift, 1877, p. 357.

5 In Southwest Germany and Switzerland. C.V., p. 225.
6 Mark per Bavarian bushel. Zeitschrift des preußischen statistischen Bureaus, 1886, p. 228.
7 Silver Penny per Berlin bushel, ibid., p. 225.
8 Austrian Guilders. Tafeln zur Statistik des Steuerwesens 1858, Wien
9 For Germany 1701–10.
10 For England 1701–15, Germany 1711–25.
11 England 1716–25, Germany 1726–40.
12 England 1726–40, Germany 1740–55.
13 England 1741–50, France and Germany 1756–65.
14 England 1751–75, France and Germany 1766–75.
15 England and France 1776–1800, Germany 1776–95.

Table III. Variations of Grain Prices in Western Europe 1801–85

	England		France		Belgium		København	
	Rain-fall (%)[1]	Wheat price[2]	Rain-fall (%)[3]	Wheat price[4]	Rain-fall (%)[5]	Wheat price[6]	Rain-fall (%)[7]	Rye price[8]
1801/05	–12*	80.2	0	21.8	—	—	**16**	8.6
1806/10	–3	87.9	**4**	18.1	—	—	12	8.3
1811/15	0	**94.3**	**1**	23.9	—	—	—12	—
1816/20	–2	80.8	–3	**25.3**	5	**25.6**	4	—
1821/25	**8**	57.3	–6	16.6*	1	13.2	–2	2.3*
1826/30	1	61.6	1	20.2	3	18.9	4	4.4
1831/35	–2*	52.7*	–8*	18.2*	–10*	17.4*	–4	4.0
1836/40	–2	**61.2**	1	19.9	–5	19.2	–10*	4.0
1841/45	3	54.8	**9**	19.6	**6**	19.9	8	4.7
1846/50	3	51.9	–1	19.9	–3	20.1	–3	4.6
1851/55	–3	55.9	3	**22.5**	1	**25.3**	1	**7.3**
1856/60	–1	53.3	4	21.8	–9*	22.8	–10*	5.4*
1861/65	–11*	47.5*	–8*	20.3*	–9	21.3*	1	5.9
1866/70	–4	54.6	**–6**	22.7	10	24.1	–3	7.0
1871/75	1	**54.7**	–1	**23.8**	–2	—	**12**	**9.5**
1876/80	**18**	47.5	**11**	22.4	**12**	—	–3	8.7
1881/85	6	40.1*	–1	19.5	10	—	0	7.4
Wet[9]	–8	84.0	—	22.3	—	—	—	—
Wet[10]	2	72.0	1	18.3	3	19.2	14	8.4
Dry[11]	–2	57.	–4	20.7	–8	18.3	–3	3.7
Wet[12]	3	53.4	3	22.3	1	21.8	2	5.5
Dry[13]	–5	52.8	–5	20.9	–9	22.0	–4	6.1
Wet[14]	8	47.4	5		10	24.1	3	8.5

Addendum and notes to Table III:

1 See note 1 in Table II.
2 In Shillings and decimal fractions per Imperial Quarter. Up to 1855 according to Tooke, later according to Francke in the Zeitschrift des preußischen statistischen Bureaus, 1887 p. 124.
3 Mean value of the averages of three groups each (without Mediterranean France) for France in C.V. on p. 166.
4 French Francs per hectolitre. Up to 1870 according to "tableaux des prix moyens mensuels et annuels de l'hectolitre de froment en France 1800–1870". Paris 1872. Later, according to Francke, op. cit.

5 Up to 1830 averages from Northern France and Holland, later from Holland and Belgium, C.V., p. 167.

6 Fr. Francs per hectolitre. Statistische Monatsschrift III, p. 396. Figures have been adjusted to a single unit.

7 Averages from København and Lund, C.V., p. 161.

8 Price in Rigsdaler Silver (rounded to one Decimal) per Tönde = 1.3912 (hl) hectolitres. Up to 1870. Statistische Monatsschrift III, p. 397. Later in Francke, Zeitschrift des königlichen preußischen statistischen Bureaus, 1887, p. 124. 1871 and 1872 are missing; for the lustrum 1871/75 which is not available, the years 1873–77 were used as substitute.

9 Refers to England in 1801–10.

10 Refers to England in 1811–30, France 1801–20, Belgium 1816–30, København 1801–10.

11 England 1831–40, France 1821–31, Belgium 1831–40, København 1821–40.

12 England 1841–50, France 1836–60, Belgium and København 1841–55.

13 England 1851–70, France 1861–75, Belgium 1856–65, København 1856–70.

14 England and København 1871–85, France 1876–85, Belgium 1866–85.

Table IV. Variations of Grain Prices in Central Europe 1801–1885

	Prussia			Southern Germany and Switzerland			Austria	
					München	Zürich		
Year	Rain-fall[1] (%)	Wheat Prices[2]	Rye Prices[3]	Rain-fall[4] (%)	Wheat Prices[5]	Wheat Price[6]	Rain-fall[7] (%)	Wheat Prices[8]
1801/05	—	—	—	5	44.9	—	—	**5.1**
1806/10	—	—	—	12	33.0	—	–3	4.4
1811/15	—	—	—	**12**	33.7	16.2	**9**	4.3
1816/20	–3	**206**	**152**	–8	**45.9**	**20.4**	1	4.7
1821/25	–10*	113*	76*	–15*	21.0*	10.5*	–6	2.8
1826/30	–3	131	98	–2	22.4	11.6	–3	2.7*
1831/35	–13*	134	103	–10	24.8	13.3	–14*	3.1
1836/40	–1	143	99	–1	22.6	12.3	–1	2.8
1841/45	**3**	154	114	**7**	29.3	14.3	2	2.9
1846/50	2	181	131	–1	33.0	15.4	9	4.1
1851/55	**3**	**214**	**177**	2	**41.3**	**16.3**	2	4.6
1856/60	–9*	209	154	–11	—	14.6	–3	5.2
1861/65	–7	188*	138*	–17*	—	14.0*	–14*	6.1
1866/70	6	220	172	3	—	15.8	2	**6.6**
1871/75	–7	**235**	**179**	–5	—	**17.8**	–3	6.6
1876/80	**10**	211	166	**12**	—	14.8	**22**	5.9
1881/85	5	190	160	2	—	13.5	10	5.2
Wet[9]	–3	206	152	5	39.4	18.3	2	4.6
Dry[10]	–7	130	94	–7	22.7	11.9	–6	2.8
Wet[11]	3	183	141	3	34.5	15.3	4	3.9
Dry[12]	–8	198	146	–14	—	14.3	–8	5.6
Wet[13]	4	214	169	3	—	15.5	8	6.1

Addendum and notes to Table IV:

1 Averages of the 23 stations in Northern and Central Germany; see notes 2 of Table I. Prior to 1831 averages of the Central European stations, C.V., p. 158.

2 Annual average for the Prussian State in Mark per 1000 kg. Francke in Zeitschrift des königlichen preußischen statistischen Bureaus, 1887, p. 121.

3 ibid.

4 Average according to C.V., p. 167, but from 1871 on without Mannheim, Karlsruhe, Zürich, and Aarau, because observations at these locations were no longer comparable with the older ones (comp. Schultheiß in Jahresbericht des Centralbureaus für Meteorologie und Hydrographie im Großherzogtum Baden, 1888 p. 57 a. 1890 p.75).

5 Mark per Bavarian bushel. See above note 6 of Table II

6 Fr. Francs per 50 kg. Statistische Monatsschrift, III p. 365. From 1871 on Zeitschrift für schweizerische Statistik 1883 and 1887.

7 Average for Bohemia and Austrian Alpine regions, C.V., p. 167.

8 Austrian Guilders per peck, Statistische Monatsschrift, III, p.365. From 1871 by Francke op. cit. Converted under the assumption that one peck [*Metze=3.44 litre*] equals 47 kg.

9 1801–20.

10 1821–40.

11 1841–55.

12 1856–65.

13 1866–85.

Table V. Variations of rye prices and rye export in Russia and yields in America

	Russia					United States	
			Rye Export in 100,000 Hectolitres				Wheat
Year	Rainfall[1] (%)	Rye Price Moscow[2]	Observ.	Calcul.	Diff.	Rainfall (%)	Bushel p. Acre[5]
1801/05	—	38					
1806/10	—	29					
1811/15	—	31					
1816/20	—	36					
1821/25	—	35					
1826/30	—	26					
1831/35	−18*	42*					
1836/40	2	39					
1841/45	8	39					
1846/50	−2	37	—	—	—	**15**	—
1851/55	6	38	21.4	13.1	8.3	−2	13.2
1856/60	−8	51	34.2	37.7	−3.5	2	12.5
1861/65	−11*	} 63*	30.2	62.3	−32.1	−5*	10.7*
1866/70	1		46.2	86.9	−40.7*	−1	12.1
1871/75	2	70	124.0	111.5	12.5	−5	13.7
1876/80	**12**	79	**193.4**	136.1	**57.3**	**10**	**15.3**
1881/85	8	90	141.0	160.7	−19.7	4	15.0
1886/90	2	—	188.4	185.3	3.1	—	—
Dry[6]	−18	42			—	—	
Wet[7]	4	38			8.3	0	12.8
Dry[8]	−6	57			−24.5	−4	12.2
Wet[9]	6	80			13.3	7	15.0

Addendum and notes to Table V:

1 Mean value of the 3-group averages: NW-, SW-, and SE-Russia as in C.V. on p. 167, completed up to 1890.

2 Kopecks per Pud. According to *Annuaire statistique de la Russie* [*Annual statistics of Russia*], 1890, p. 131. The years 1863–69 are missing.

3 Juraschek, *Übersichten der Weltwirtschaften 1885/89* [*Summaries on World Economics 1885–89*] p. 44. 1890 according to the Statistical Abstract for the principal and other foreigne countries in each year from 1881 to 1890/91.
4 Group B.S. Interior, East, C.V., p. 168.
5 In Ohio, C.V., p. 277.
6 For Russia 1831–35.
7 Russia 1871–55, Ohio 1851–60.
8 Russia 1856–70, Ohio 1861–75.
9 Since 1871 respectively 1876.

Chapter 7

Weather Prophets*

The desire to look into the future is deep-rooted in mankind. Hence the fact that fortune-telling is so enormously popular, having more followers even in our time than one would be inclined to assume. Those who consider themselves educated do of course no longer revert to palmistry nor do they consult the cards, and if so, only in secrecy: They are ashamed of an action incompatible with their level of education. There is only one category of fortune-telling that flourishes openly even today, which has its prophets as well as its believers even among the highly educated: it is the foretelling of the weather. Even respectable newspapers from time to time print predictions from weather prophets. Besides Falb's weather prognoses, Swiss newspapers carry weather prognoses by local prophets as well, such as by the high school teacher *C. Marti* in Nidau, the engineer *Gladbach* in Aarau, and *Jules Capré* in Chillon for the French-language newspapers. The public reads these prognoses, follows them up, and finds that in some cases the predicted weather actually occurs. Such correct forecasts win followers over and they gather around their prophets in considerable number. Is the belief of this group justified? We will try to answer this question.

Weather predictions have been around since the earliest documented times, their oldest versions being based on religious convictions.[418]

In antiquity, Greek and Roman gods controlled the weather in an absolutist manner, headed by Zeus, father of the Olympus. During a thunderstorm, several gods got into the action at the same time: Zeus through his thunderbolts, Aeolos sent out the winds under his command, and Iris stretched the rainbow across the sky. The ancient Germanic tribes did not think much differently, and even today we find similar views amongst the beliefs of many peoples. The North American native Indian hears the voice of the 'Great Spirit' in the

* *Wetterpropheten*, Jahresbericht der Berner Geographischen Gesellschaft, 1886.

418 Compare C. Lang, *Weather prophesies in old and modern times*, publ. in Sammler, Addition to the Augsburger Abendzeitung; 1889/90.

thunder; and the Caffre [member of the Bantu-tribe] will say when it thunders: "The old man is bowling[419]", in reference to someone living beyond the clouds.

The role of mediator between humans and gods fell to the priests; consequently they were the ones who had to consult the gods about future weather outlooks and to announce the answers. Apollo for instance had established some kind of information bureau in Delphi where all kinds of predictions, including weather forecasts, were handed out in exchange for money and kind words. In many cultures, priests were not only weather prophets but also weather makers, using symbolic acts, prayers, sacrifices to the gods, etc., in order to persuade or even to force the gods to send a certain weather system. If the friendly gods did not respond, the evil ones were called upon—in medieval times, even the devil himself. Members of the weaker sex in particular were often suspected of having influenced or bewitched the weather in cohorts with the devil to the detriment of their kinsmen; many proclaimed weather-witches were burned at the stake. The belief in witchcraft was seemingly justified as some voluntarily claimed to be witches and have carried out symbolic acts with the intention to exert witchcraft; in such cases their confession came from a bad conscience.

In Central and Western Europe, weather prophecies based on religious grounds are rarely published or believed in today. This is different with weather prophecies based on astrology. These, too, are ancient: Chaldean, Babylonians, and Egyptians attempted to interpret future events from the constellation of the stars. Greek natural scientists of the rank of Aristotle, Hypocrates, and Strabo engaged in this art. Rome paid homage to astrological superstition; even first-rate minds, such as Cicero, Virgil, or Tacitus were swayed by it. There were many astrological schools where the skills of fortune-telling were systematically acquired. This pseudo-science had its prime time in the Middle Ages.

Astrological superstition is based on the so-called geocentric ideology in which Earth is the centre of the universe. Anything beyond Earth is there only to serve this planet and has a certain purpose in relation to it. The constellation of fixed stars is seen as a clock's dial and the planets whose positions in relation to the fixed stars are constantly changing, as the hands of the clock pointing out future events, including the weather. As seven planets were known to exist, it was logical to connect each day of the week with one of them; but also each year had a specific planet that determined that year's weather. Saturn, the highest ranking planet, was supposed to be an enemy and destroyer of nature, malevolent, cold, and wet by nature; Jupiter, rather wet than dry and warm; Mars hot and rather dry than wet; Venus rather wet than dry—thereby warm; Mercury cold and dry; and the moon rather wet than cold and dry, at the same time windy. The sun, in contrast, was to be everyone's friend, not too cold and not too hot, yet dry. Because each weekday was controlled by a

[419] In the original words, 'plays'.

particular planet, the influence of the planet reigning throughout the entire year could be weaker or stronger, etc., on this day. The ruling planet of each year can be determined quite easily: you only have to divide the year by 7; the remaining number refers to the number of the planet that rules the year (1. Sun, 2. Venus, 3. Mercury, 4. Moon, 5. Saturn, 6. Jupiter, 7. Mars); e.g., the year 1902 divided by 7 leaves 5; the ruling planet of that year is Saturn. The year 1902, should therefore, under the reign of this "malevolent destroyer of nature", turn out to be cold and wet.

How deep-rooted the belief in these astrological prophecies was at the time is best exemplified by the impression Johann Stöffler's prognosis had on his contemporaries.[420] In his calculations of the planets' future constellations, this—by the way quite able—astronomer from Tübingen discovered in 1518 that, in February of 1524, the planets Saturn, Jupiter, and Mars would meet in the constellation of Pisces. As Saturn and Jupiter were both considered wet planets, and Mars to be misanthropic, and the sign of Pisces where the constellation was expected was a priori associated with water, Stöffler, in his letter to the emperor Charles V, did not hesitate to predict a flood of biblical proportion for February 1524. Though some scientists contradicted Stöffler, the general population was in great fear of the prophesied flood. Those who lived near the sea or next to a river tried to sell their property and moved to higher grounds. Others built arks to survive the flood as Noah had done. Many were driven mad with fear. Mayor Hendorf of Wittenberg , according to Luther, made extensive arrangements in the attic of his house for a survival of the flood, which included a substantial stock of beer—"in order to at least have plenty to drink should the deluge come." The dreaded February 1524 came: the weather in Europe was predominantly bright and nice, there was little rain—no trace of a flood. One would now assume that such a total failure would cure mankind of astrological superstition, but far from the truth. There was no shortage of excuses for the misdiagnosis. The monks declared that their prayers prevented the flood that otherwise inevitably would have occurred. Biblical scholars pointed out that God's promise to Noah was overlooked in the prophecy, namely that there would be no second flood after the biblical one, which He sealed with a rainbow; that is why the flood did not take place despite the constellation. Historians however did not at all relate the disastrous constellation to a meteoric deluge but to a political one—the peasants' uproar that broke out in 1524!

Astrology continued to thrive happily although its foundation was soon taken away by the Copernican world view that no longer considered earth the centre of the universe, but the sun. This reduced the earth to the rank of just another planet. Even such astronomers as Tycho de Brahe or Kepler could not escape the prevailing views of their time and included astrological prophecies

[420] Compare G. Hellmann, *Meteorologische Volksbücher*, Himmel und Erde, III, 1891.

in their own prognoses without personally believing in them. "Astrology is the foolish daughter of astronomy; but she provides for her mother," said Kepler. Science had to earn its bread and had to take it where it could find it.

Astrological superstition regarding weather developments has been codified in the hundred-year calendar first published by the abbot Martin Knauer. Even today, this popular book is being re-edited over and again finding faithful buyers.[421] The farming community in particular swears by its "Centennial...". A calendar that does not include the forecasts of the "Centennial" will not be bought. 'The Limping Messenger' ["*Hinkender Bote*"] of Neuenburg learned this the hard way: When during one year the astrological forecasts were omitted, the number of subscribers dropped so dramatically that the publisher was forced to reintroduce them in next year's edition, bowing to superstition. Even today all relevant calendars in Switzerland, e.g., 'The Swiss Farmer' ["*Der Schweizer Bauer*"], the various 'Limping Messengers', include the predictions of the hundred-year calendar, which, together with the astronomical data of each year, are edited by qualified Swiss scientists. Kepler's proverbial words are just as valid today! Even the calendar's producers seem to agree with the view point expressed by one of their editors, who signed on the front page of one of the calendar's editions in the 18th century with "*Tiehrhawnu*", which, read backwards, means "*Unwahrheit*"["untruth"].[422]

Recently, weather prophets have emerged who seem to rely on a more scientific basis turned up lately. Most of them give the moon a major role. *A priori,* it seems indeed likely that the moon has a major impact on the weather, as obviously everyone can see its effect on the ocean along the coasts in the spectacular phenomenon of the tides. It is mainly the moon's gravity that makes the ocean level rise and fall twice within 24 hours. A similar, though weaker, impact is exerted by the sun. If these two effects are combined, as is the case during full moon and new moon, the tidal movements are particularly strong. The ocean is in a state of tremendous agitation. It was natural to assume that the moon had a similar tidal effect on the earth's atmosphere and consequently on the weather. However, a close scrutiny of air pressure observations showed no noticeable atmospheric tidal movements. Nonetheless a number of weather prophets seized on this idea and developed systems of weather prophesies based on the moon's tidal effect; as Overzier did 20–15 years ago, also Baron Friesenhof, and Gustav Jäger, who is more known as a would-be apostle than a weather prophet, and also Rudolf Falb.[423]

Falb is the most popular of all, partly perhaps because he invented a slogan for his weather predictions: his "critical days" are known all over the world. He

[421] I have for example: Dr Martin Knauer's centennial calendar for the 19th and 20th century. Bern, J. Heuberger's Publishers, 1883.

[422] According to C. Lang.

[423] When this lecture was held, Falb was still alive.

calls those days 'critical' during which the tidal factors' effect is particularly strong. Just as the ocean's tidal movement is strongest under the influence of a new moon and a full moon, Falb applies this phenomenon to the earth's atmosphere. His 'critical days' are always days when there is a full moon or a new moon. Furthermore, just as these tides during full moon or new moon are particularly strong whenever the moon, or even sun and moon at the same time, are close to earth, the atmospheric 'tides' are supposedly of equal intensity during such days. In this manner, Falb differentiates between 'critical days' of the first, second, and third order.

Faithful believers in Falb's predictions are to be found among both city and country dwellers; his "weather calendar and index of critical days" that he publishes every six months is available everywhere. Newspapers pay large sums for the copyright to print his prognoses. In fact, in many cases these prognoses prove right; the weather turns out to be just as Falb had predicted it to be during the 'critical days'. Chance hits like these are exploited by Falb and dazzle the public. That, nonetheless, his prognoses are worthless and their fulfilment says nothing in his favour shall be outlined briefly.

Let us first ask which kind of weather conditions Falb considers typical for the critical days. He writes in his weather calendar of 1901 I: "1. An increase in barometric minima and low pressure, cyclonic winds and increased precipitation in general. 2. Thunderstorms in winter or at a time of day when they rarely occur (nights, mornings). 3. Snowfall in the upper mountains in the summer or in regions where it seldom occurs. 4. Thunderstorm, accompanied by a snow storm, at the same location. 5. The first thunderstorms in spring and the first snow fall in autumn. 6. In high altitudes influx of warm southern air masses saturated with moisture which is indicated either by a sudden thaw or a deep blue sky and an unusually clear atmosphere, and which then engages in a struggle with opposed northern air masses characterised by Cirrus clouds or clouds in general with a tendency to stripe formations," etc., "rainbows, local showers and frequent changes between rain and sunshine, the so-called April weather, seem typical conditions." As we can see, it is a whole bouquet of many different types of weather conditions, and all of them are typical for Falb's "critical days"!

The critical days are not only in effect on that particular day itself, but may have an impact two days earlier or even occasionally two or three days later. Falb considers a total of five to six days to be under the influence of one critical day. As, according to Falb, each year has 24–25 critical days, more than a third of the entire year is affected. Falb adds: "That does not mean that the predicted events have to occur on each of the days described, but that as far as the moon is a factor, weather conditions tend to be unbalanced on these days." In other words, the fact that the predicted weather does not materialise cannot be used against Falb's theory, while on the other hand, correct predictions serve as proof of his method.

It must be pointed out however, that correct predictions in themselves are just as useless in proving Falb's theory as incorrect predictions are in disproving it. The frequency of correct predictions is completely meaningless; it blends no one but the amateur.

This becomes immediately evident if we ask the following question: Is the occurrence of an event on a critical day in fact the necessary outcome of this critical day? Obviously this would only then be the case if this event does not occur on any other days or at least less frequently than on critical days. An explicit example will make this clear. Let us suppose someone prophesied that the sun will rise on Thursdays. He then observes every Thursday and finds out that the sun does indeed rise each and every Thursday. He then proclaims: I am right—Thursday is the sun's critical day for rising. Here again we have a prognosis and a whole lot of positive returns. Yet, this result is completely meaningless because the sun rises on all other days as well, not just on Thursdays. This is the type of mistake Falb makes. To verify his theory, he first of all needs to identify the frequency of the events which he deems characteristic for critical days for all the other days, and can only then conclude that they occur more often on critical days and affected neighbouring days than on any other day. Yet Falb always refused such an approach.

J.M. Pernter, currently director of the Austrian Meteorological Observation-Network, has carried out this inquiry and published it in 1892.[424] He, for example, examined the frequency of low pressure systems or barometric minima in Europe and found out that, during the three years of observations from 1888–90, each critical day had an average low pressure of 1.67, and on each of the neighbouring 4 days—according to Falb affected by the critical days—this figure was also 1.67, as well as on each of the so-called non-critical days. Therefore, in Europe, low pressure systems are just as frequent on critical days as on any other day. The same applied to storms whose average occurrence was 1.05 on each of the 5 days affected by the critical days, and 1.04 on any other day. The average number of stations observing rain or snow was 19.4 on a critical day and the same on other days. The total rainfall measured per day was 132 mm and 138 mm respectively, the number of floods 0.08 and 0.08. As we can see, the two values are identical everywhere or almost identical, i.e., those occurrences that Falb defined as characteristic for the so-called critical days are just as frequent on any other day. This outcome is nothing short of devastating for Falb. It shows that his prognoses have no major value, just like the one in our example forecasting a sunrise every Thursday. A year does not have 24 or 25 nor 5 × 24 or 5 × 25 critical days but 365 and in a leap year 366!

Let us now refer in detail to our Swiss weather prophets.

[424] Himmel und Erde, IV.

Jules Capré in Chillon, the weather prophet of western Switzerland, bases his weather predictions on the movements of the moon. They are published annually in the "Almanac des chemins de fer du Jura-Simplon".

Based on the moon's position, Capré predicts the occurrence and duration of low and high pressure systems for a particular day or a number of days in certain places of Europe and the resulting future weather conditions for these regions linking them to cyclones and anti-cyclones—just as modern meteorologists do. Capré, describing the misfortunes of a weather prophet in his humorously written preamble to the 1901 prognoses, openly admits that his 1900 predictions were wrong. He blames this failure on the fact that, as an amateur in the area of astronomy, he was unable to correctly project the positions of the moon. After having closed this knowledge gap and corrected the moon's positions he is now confident to be able to achieve better prognoses in the future. A comparison between prognosis and actual occurrence reveals again, just as in Falb's prognoses, that they are frequently correct, but often they are not. Low pressure, rainfall, etc., occur just as often on the predicted days as they do on days not predicted.

The weather prophet of Aarau, engineer Gladbach, keeps his method in the dark. Although his *Wetterprognosen*[425], in which he forecasts the weather for a number of months, contain the "theoretical reasoning and practical training for observing the barometer with regard to weather predictions", impressing the amateur with differential equations and research diagrams, Gladbach's method remains nonetheless unclear. He claims that his diagrams of "Europe's Cloud Belt", which seem to be the basis of his prognoses yet remain unexplained, are graphic constructions of the air pressure conditions of previous years. He thinks it "inopportune" to explain how this is done because of "the danger that an unauthorised reproduction might be attempted", "discrediting his work". Gladbach's brief report during a session of the Swiss Society of Natural Science in Zofingen in August 1901 again failed to give insight into his method. Only that much became clear that his work relies on the gravity of the planets and of the moon.

High school teacher C. Marti, the weather prophet of Nidau, whose prognoses frequently appear in German-Swiss papers, also kept his method in the dark for a long time; he only revealed it in November 1900 during a session of the Society of Natural Science in Bern, where he handed out printed excerpts, including a.o. th. an invitation to test his method.[426] Whereas Falb and Capré base their prognoses on the gravitational and tidal powers, respectively, of the moon and the sun, Marti refers to a mysterious, completely unknown power. He assumes that apart from "local constants" and the annual change of the

[425] Aarau published by the author.

[426] An explanatory account was published after this lecture in the essays of the Society of Natural Science of Osnabrück.

sun's position weather is affected by the so-called "fast weather factors". According to his theory, these factors are caused by the planets: If two planets surrounded by a sufficient layer of atmosphere, one on the near side and one on the far side of the planetary system, align, that is when rotating around the sun they come into such a position that a straight line joining the two extends to the sun, a "disturbance" occurs in that spot on the sun's surface that is facing the planets. Marti does not disclose the nature of this disturbance. The sun rotates around its axis, that is once every 26 1/2 days. The agitated spot rotates with it; when it faces the earth, it is presumed to cause a disturbance in the earth's atmosphere. This disturbance is repeated with a further rotation of the sun when the spot faces the earth for a second time, a third time, in some cases even a fourth time. Each effective conjunction causes rain or thunderstorms in the summer, and rainfall or storms in the winter. The most effective conjunctions are presumed to be those of Mercury and Saturn and Mercury and Uranus, followed by those of Venus and Jupiter as well as Venus and an asteroid. Marti has tried to verify his method by examining excerpts from meteorological yearbooks. Unfortunately his effort suffers from the same methodological error as Falb's approach: He recounts those cases where in fact storms did occur on his critical days. This rate of success proves nothing as we have already seen. In order to justify his method, Marti would have to prove that rainstorms, etc., actually occurred more often on those days identified as critical than on any other day. Marti has so far failed to submit evidence to this effect. Based on an index of the critical days provided to me by Marti, I have examined the frequency of rainstorms as defined by Marti, using the weather reports of the Swiss Meteorological Centre and allowing for the delay in the years 1882–86 and 1894–98 which Marti claimed to have discovered. On any given day of the entire time span, the average frequency was 0.25, i.e., on one out of four days it was stormy somewhere in Western Europe. The storm frequency rate on those days which according to Marti were affected by the conjunction of Mercury/Saturn was also 0.25, of Mercury/Uranus again 0.25, of Venus/Jupiter 0.22, of Venus/Juno again 0.25, and of the double conjunction of two pairs of planets again 0.22. Again the numbers are quite the same and by chance on critical days even smaller. Storms occur just as often on days identified 'as critical' by Marti as they do on non-critical days. Therefore even Marti's critical days are of no use. This should not be surprising since his entire theory of a mystical "disturbance" on the sun's surface reflecting onto the earth is without any physical foundation. If I nevertheless took on the task of checking his prognoses with great scrutiny, it was done in respect for the energy of a man who, although in total vain, has invested into his method an enormous amount of computing.

Therefore, neither Marti's, nor Falb's, Capré's, Gladbach's prophecies stand up to scientific scrutiny. Weather predictions, or rather, weather prognoses simply cannot be established by ignoring the results of the meteorological science based entirely on physical evidence and of the calculation of probabilities.

What does modern meteorology tell us?

The earth's atmosphere as a whole may be compared to a huge machine. Its heating is provided by solar energy, largely in the tropics; heat loss is highest in the polar regions. The permanent temperature differences developing between equatorial and polar regions cause enormous air streams. If we open the door of a warm room in the winter, we notice—for example with the aid of a burning candle—that on top, warm air flows out of the room into the open, while at the bottom the cold air moves into the room. In the same way the warm tropical air rises and moves towards higher latitudes, while down below cooler air moves down from higher latitudes towards the equator. This process is affected by the rotation of the earth around its axis in a manner that we will not elaborate on in this context.

Streams of air moving side-by-side from different directions or at different speeds tend to produce cyclonic systems , and frequently to such an extent that we can recognize the entire convection process in those progressing cyclones. Any rapid river exhibits this phenomenon: there are descending eddies seen as a small funnel-shaped indentation in the surface and ascending eddies marked by gyrating waters; all of these move downstream with the general flow. The same process takes place in the atmosphere, particularly in medium and high latitude regions. As a result of those large prevalent air streams, cyclonic systems develop at low altitudes, covering an extraordinary large area that quite often may reach 1500–2000 kilometres and more in diameter. Sometimes these are ascending cyclones distinguished by low air pressure and are, therefore, called barometric minima or depressed systems; at other times these are descending cyclones characterised by high air pressure. The cyclones are surrounded by winds blowing from different directions that at low pressure, for instance, consist of the following: west winds at the south side, south winds at the east side, east winds at the north side, and north winds at the west side. As the weather is primarily determined by winds, we find the cyclones surrounded by different weather systems. These cyclonic movements proceed as they are carried along by the large prevalent air stream in middle and high latitudes mostly from west to east in successive short intervals of a few days. Their position is constantly changing in relation to a geographical location and so is that location's weather.

Careful observation of the skies allows the forecast of imminent changes by interpreting certain meteorological signs. In our age 'of the telegraph', these observations include all of Europe. The Meteorological Centre in Zürich, for instance, is notified during the morning via telegram from different parts of Europe about the particular weather situation at each weather station on the same day as of 7.00 or 8.00 a.m. Data about air pressure, temperature, wind direction, cloud formation, rainfall are reported in this way. Based on these data that morning's weather map is drawn up for Europe, including those cyclonic systems previously described. By comparing this weather chart with those of previous days and relying on previous experience, it is then attempted to establish what the cyclones' potential position will be the next day. On the basis of their anticipated course a conclusion is reached for next day's weather and is published as a prognosis. We are of course today not yet in the position to determine with mathematical precision where a vortex, which today we discover somewhere over the ocean near Ireland, will reach us tomorrow. But within a certain margin it is possible and so are forecasts of the weather. Out of 100 prognoses, about 80 are correct.

Despite the fact that they can only be made for one day in advance, these prognoses are of practical importance. Proof of this are the gale warnings which are issued with success at some coastlines as navigational guidelines, and above all the very precise signal service of the United States, where these forecasts are particularly important for agriculture.

In Zürich, the prognosis issued by the Swiss Meteorological Centre appears each afternoon together with the weather chart under the title "Weather Report."[427] The public in Zürich also contacts the Centre directly by telephone and will ask, for example, on a Saturday for Sunday's prognosis. As long as the personnel are not too occupied, these questions will be answered. It may even happen that a member of the fairer sex inquires if she should wear a light or a dark dress for the planned Sunday excursion and the witty officer replies: "Wear the one that suits you best!"

But among the greater public, the number of followers of weather prophets the likes of Falb, Marti, and others is still high because the public is hypnotised by their rate of success. In general, people are not critically inclined. When education is lacking in this area, observations are done uncritically: successful predictions stay in mind, failed ones are forgotten. "But he got it right before," is the objection. That this "getting it right" is not any different than playing a lottery with many winnings, is overlooked. Additionally, the public prefers long-term weather forecasts over prognoses of one day in advance. Weather prophets cater to this preference by announcing their prognoses months in advance. Even if they are totally useless, they still find followers. About this subject as about many other areas of superstition, it can be said: *Mundus vult*

[427] Besides, the annual subscription of the weather reports through the Swiss mail service is only 12 SFr.

decipi: the world wishes to be deceived. Weather prophets will not die out for a long time to come.

Chapter 8

An Inquiry About the 35-Year-Period Climatic Variations*

When in 1889 I finished my research about climatic variations,[428] I had to rely on meteorological observations going no further than 1880 and 1885, respectively. From the pattern of the rainfall fluctuations I nevertheless concluded that there had been a maximum of rainfall around 1880 over global land masses, as well as a minimum around 1860 and a maximum around 1850. Later in 1895, I described the rainfall variations for Prussia up to the year 1893 and for the European part of Russia up to 1890 in my article about the influence of climatic variations on harvests and grain prices in Europe.[429] In comparison to the years around 1880 rainfall was slightly less but still above average. After several years have gone by it may now be of interest to find out whether or not during the past decade rainfall did indeed decrease all over the globe's land masses. Professor Woeikof has in fact dealt with this question and counters my findings in his article "*Die Seespiegelschwankungen zwischen Aralsee und Baraba und die Brücknersche Hypothese*" [*The Lake Level Fluctuations between Lake Aral and the Baraba* [steppe in Sibiria] *and Brückner's Hypothesis*] published in the September issue of this journal (1901, p.199).[430] Being occupied with the final stages of another work, I am unable at present to start a thorough investigation for all parts of the globe. It is also advisable to at least wait for the results from the observations of 1900 in order to complete one full lustrum. Yet, I would like to refer briefly to Professor Woeikow's article now and, above all, draw attention to a large region where the decrease in precipitation since 1880 was proven.

* *Zur Frage der 35jährigen Klimaschwankungen.* Petermann's Mittheilungen, 1902.

428 *Klimaschwankungen seit 1700* [*Climatic Variations since 1700*], Wien 1890; see Chapter 4 of this collection.

429 Hettner's Geographische Zeitschrift, Volume I, p. 39.

430 Dr. A. Petermanns Mittheilungen, Gotha. Justus Perthes.

8.1 WATER LEVEL FLUCTUATIONS IN THE KIRGHIZ STEPPE AND FLUCTUATIONS OF RAINFALL IN RUSSIA SINCE 1860.

According to the observations by Berg, Ignatow, and Tangniliew reported by Woeikof, Lake Aral and other lakes within its vicinity had reached their lowest level around 1880 and have been rising ever since. A water level decrease of Lake Aral was also observed over a longer period of time immediately prior to 1880, yet no details are available as to the beginning of that decrease. Professor Woeikof states that this pattern contradicts the climatic variations I claimed to have discovered. In reply, I would like to point out that, based on the observations available to me in 1889, I had explicitly declared the Kirghiz steppe, where these lakes are located, an exceptional area in which the pattern of rainfall variations is reversed, i.e., maxima are reached at times when the majority of the other land masses—3/4 to 4/5 of it—shows minima.[431] I concluded from the situation at Lake Alakol that "the exceptional region of the Kirghiz Steppe has to be extended into the corner area between Tarbagatay and Alatau."[432] If the lake levels in this region were lowest around 1880 and began to rise thereafter, it only confirms the exceptional situation of the Kirghiz Steppe, as I stated in 1890.

I do not agree with Professor Woeikof's claim, that the variations of these lake levels "coincide closely with the rainfall fluctuations at Barnaul. " What is characteristic at Barnaul is the low precipitation during the years 1859–69, which is entirely consistent with the dry period I had identified for most of the continental areas. At Barnaul rainfall increases sharply thereafter, while the lake levels, among those Lake Aral's level in particular, decrease and do not reach their lowest levels until around 1880. Consequently, the minimum lake levels occur 15 years after the minimum rainfall at Barnaul, and exactly at a time when Barnaul is experiencing a very wet period. The lakes' fluctuations are much more in accord with the rainfall fluctuations at the station Irgiz located 200 kilometres further north of Lake Aral in the centre of the steppe. The following lustra averages (mm) for Irgiz with the exception of the missing years 1861, 1862, 1884, and 1885, show this. I have included the related averages for Barnaul.

[431] *Klimaschwankungen seit 1700* [*Climatic Variations*], p. 170, Table p. 168 f.
[432] op. cit., page 176.

	1861/65	66/70	71/75	76/80	81/85	86/90	91/95	96/99
Kirghiz Steppe	Lake Levels Decrease			Lowest Level			Increase	
Irgiz (mm)	193	200	144	148	153	183	197	—
Barnaul[433]	150	173	258	349	338	380	380	375

Since 1861, precipitation and the lakes' water levels show nearly opposite trends from Barnaul's precipitation. The highest lake levels are measured at the end of the 90s, a time when Barnaul's precipitation does not rise further.

Professor Woeikof concludes further that the rainfall at Barnaul does not correspond with the 35-year climatic variations, and he would suggest a 55-year period instead. He draws this conclusion from the fact that the centres of the wet periods at Barnaul supposedly do not coincide with the centres of the wet periods as I have identified them. I am unable to agree with him in this. There is no reason to assume that the centre of the first wet period at Barnaul was reached before the meteorological observations were started, as Professor Woeikof does. The first observation year of 1838 was dry (251 mm), 1839 was very wet (449 mm), as was 1842 (448 mm). The maximum occurs at the beginning of the 40s according to available observations; we do not know what happened prior to that. As to the second wet period, I would like to point out that the observations at Barnaul are not homogenous because the rainfall measuring device up to 1882 was placed at 3.1 m above ground level, thereafter at 1.0 m, and some years even at 2.1 m. This lowering of the device changed the measurements of precipitation in winter quite substantially, as can be read from the following comparison:

Time Span	Height of the Rainfall Indicator (m)	Ratio of Summer (May–Sept., 5 mos.) and Winter (Oct.–April, 7 mos.) Precipitation
1838–82	3.1	100 : 49
1883–89	1.0	100 : 98
1890–97	2.1	100 : 83
1898–99	1.0	100 : 116

Whether the increase is due to snowdrifts or whether earlier measurements indicated too little snow as a result of the wind's impact remains unclear. It is obvious though that the measurements of the winter precipitation after 1882 are much too high in comparison to earlier measurements, in some years even around 100 mm and more.

By smoothing these values[434] and by determining a 62-year mean (289.8 mm), we arrived at the following periods whereby each 5-year average refers

[433] See further down about the adjustment to the winter precipitation that was done because of the lowering of the rainfall measuring device at Barnaul after 1882.

to the centre year of that time span. For comparative purposes I have included the climatic variations of wet and dry time span which I had established in 1890.[435]

Precipitation at Barnaul		Climatic Variations as Established in 1890	
Above the Mean	1840–49	Wet	1841–55
Below the Mean	1850–73	Dry	1856–70
Above the Mean	1874–97	Wet	1871–85

There is a perfect compliance with the wet and dry periods of the climatic variations as far as those have been traced for the majority of countries, i.e., up to 1885. However, at Barnaul even after 1885, precipitation remains high. Whether this is a single event or is repeated in other regions is now the question. According to the following tabulation (Table 8.1) the latter seems in fact to be the case for Russia, at least in part.

Table 8.1. Deviations of the ten-year averages of precipitation (mm) from the mean in the river regions of the European part of Russia

	Upper Wolga and Oka	Central Wolga and Kama	Lower Wolga	Upper Dnjepr	Lower Dnjepr	Don	Mean of River Regions
Mean	513	462	354	533	435	450	458
1861–70	1	–28	–14	5	–32	–15	–14
1862–71	–2	–33	–23	8	–37	–13	–17
1863–72	7	–32	–9	18	–28	–20	–11
1864–73	19	–12	10	25	–19	–4	3
1865–74	23	1	21	11	–18	–4	6
1866–75	19	16	36	5	–19	2	10
1867–76	27	22	37	15	–4	11	18
1868–77	16	16	35	10	5	23	18
1869–78	29	16	31	18	15	27	23
1870–79	26	23	37	22	**24**	42	**29**
1871–80	**29**	**29**	38	6	18	40	27
1872–81	25	25	**48**	–6	15	51	26
1873–82	11	19	41	–5	18	**58**	24
1874–83	7	3	30	3	19	54	19
1875–84	1	–2	30	12	20	58	20
1876–85	4	–11	18	7	19	46	14
1877–86	–1	–14	20	–3	23	44	12
1878–87	5	–3	14	–1	14	33	10
1879–88	11	–2	14	–10	13	24	8
1880–89	8	1	11	–13	4	10	4
1881–90	–11	3	1	–9	–1	–1	–3
1882–91	–14	3	–8	–2	–3	–22	–8

[*continued on next page*]

[434] Adjustment factor of the winter precipitation 1883–89 and 1898–99, 49:98 = 0.5; 1890–97, 49:83 = 0.6. Without this adjustment the average would be 317 mm, and the years 1840–42 as well as 1876–97 would show above average values.

[435] See also in *Klimaschwankungen* [*Climatic Variations*], p.192.

[continued from previous page]

	Upper Wolga and Oka	Central Wolga and Kama	Lower Wolga	Upper Dnjepr	Lower Dnjepr	Don	Mean of River Regions
Mean	513	462	354	533	435	450	458
1883–92	−12	7	−14	−4	0	−31	−9
1884–93	−6	18	−10	2	6	−28	−3
1885–94	6	25	−19	13	11	−30	1
1886–95	8	24	−17	23	19	−22	6
1887–96	14	23	−16	32	19	−21	8
1888–97	−4	5	−24	26	15	−30	−2
1889–98	−9	−4	−30	25	11	−32	−7

The table shows the deviations of the averages of successive ten-year periods[436] in relation to the mean of all years and is based on the figures for the total annual precipitation provided by E. Heintze in his essay about the deviation of precipitation from the mean in the river regions of European Russia during the time span of 1861–98.[437]

The raw rows of data sufficed to convince E. Heintze that his findings about the fluctuations of precipitation, in general, come pretty close to my own results. The smoothed rows show this even more clearly: With one exception (region of the upper Dnjepr) all river regions show a pronounced maximum of precipitation in the decades of 1870–79, 1871–80, 1872–81, or 1873–82, which was preceded by a minimum in the 60s. Precipitation begins to decrease considerably in the 80s, showing a sharp and well-pronounced decline at the lower Wolga and the Don respectively; it is less noticeable at the central Wolga and lower Dnjepr, where precipitation remains fairly high up to 1889/98 even though it is lower than in the 70s. Thus only the central sections of the Wolga river and the Dnjepr region show a similar pattern of rainfall as at Barnaul. In contrast, the other parts of Russia show a considerable decrease of rainfall after 1885, as well as East Siberia's Nertschinsk, and Nikolajewsk at the Amur. The smoothed figures for the rainfall observations in these latter places are included in Table 8.5 below, as well as those of most of Central Europe represented here by the two cities of Bruxelles and Bremen.

436 The multi-year averages were re-determined for each single year according to the instructions by Heintze.

437 Meteorologische Zeitschrift, 1901, p. 219.

8.2 DECREASING RAINFALL IN THE UNITED STATES SINCE THE MIDDLE OF THE 80S.

One area that is known to suffer a general decline in precipitation since 1888 is that of the United States.

Alfred J. Henry, departmental director with the U.S. Weather Bureau, discovered while examining rainfall patterns in the U.S.,[438] that it was too low from 1887–96 in almost all of the country. Based on his data, I put together the following table, which includes the deviation of precipitation, that is the deviation from the multi-year means in each of the different parts of the United States for the decade of 1887–96. Henry provides the deviations for each year, which limited space does not allow here. To show the relative extent of the deviations, they are also expressed in percentages of the multi-year averages (1870–88) which Hann[439] provided for the individual regions. These percentages are, of course, estimates because Hann's data refer, at least in part, to somewhat different stations than those used by Henry. Nonetheless these percentages highlight the lack of rain in the arid regions.

Only the New England states and the central Atlantic states, the southern part of the eastern slope of the Rocky Mountains, the Pacific coast, and California show extensive or normal rainfall. The rest of the regions show too little rainfall. This lack of rain is particularly evident in the Gulf states and the upper Mississippi valley; in the Gulf states, for instance, rain decreased by 105 mm and 104 mm respectively or 7 and 9 percent compared to the multi-year average. A large decrease is also evident in the percentages from the upper Great Lake region, the Missouri valley, the central part of the Rocky Mountains' eastern slope as well as on their southern plateau. Henry emphasises correctly that, as far as the United States are concerned, too little rain in one region is not compensated by too much rain in another. In fact, Table 9.3 shows that in only one year (1896) out of the ten were there more regions with too much rain than with too little, while in four years three quarters of the regions suffered too little rain all at the same time.

[438] Report of the U.S. Weather Bureau for 1896/97, p. 328, Washington 1897.
[439] Compare *Klimatologie* [*Climatology*], 2. Edition, Vol. III, p. 289

Table 8.2. Deviations of rainfall 1887–96 in the different regions of the United States

Regions (Districts)	Number of Stations	Deviations from Normal (mm)	(%)	Number of Years with: Too Little Rainfall	Too Much Rainfall
New England	9	+1	0	5	5
Central Atlantic States[440]	11	+12	+1	5	5
Southern Atlantic States	11	–66	–5	7	3
Eastern Gulf States	5	–105	–7	9	1
Western Gulf States	8	–104	–9	8	2
Ohio Valley and Tennessee	11	–68	–6	7	3
Lower [Great] Lakes Region	8	–38	–4	7	3
Upper [Great] Lakes Region	10	–68	–8	9	1
Outer North West	4	–18	–4	8	2
Upper Mississippi Valley	13	–83	–9	7	3
Missouri Valley	10	–52	–7	7	3
Mountain Range—East/North	6	–2	–1	5	4
Mountain Range—East/Central	7	–43	–8	7	3
Mountain Range—East/South	4	+6	+1	4	6
Mountain Range—Northern Plateau	4	–21	–5	6	3
Mountain Range—Central Plateau	5	–5	–2	5	5
Mountain Range—Southern Plateau	9	–23	–8	5	5
Northern Pacific—Coast	8	+41	+4	5	5
California	12	+21	+2	4	6

Table 8.3. Number of Regions with

	Too Little Rainfall	Too Much Rainfall		Too Little Rainfall	Too Much Rainfall
1887	15	4	1892	11	7
1888	10	8	1893	12	7
1889	14	5	1894	15	4
1890	10	9	1895	15	4
1891	10	9	1896	8	11

The above figures are deviations from the multi-year mean; the ten years from 1887–96 were too dry compared to the standard value. More extreme is the difference to the immediately preceding decade, which on the average was too wet. This is shown in the next table which, based on Henry's data, includes the ten-year averages of 1877–86 and the deviations of the averages of 1887–96 from those means by the [U.S.] Weather Bureau. I also added the mean values of 1897–99 determined on the basis of the Weather Bureau's report.

440 This category is almost identical to the category "Southern Atlantic States" in my book because I did not include stations located south of Washington [D.C.].

During the decade of 1887–96, all stations, with the sole exception of New Bedford, Mass. and Augusta, Ga., had less rain than during the previous decade, and in some places even substantially less. Towards the end of the century the aridity has increased even more, so that in 1897/99 the deviations are considerably higher still. As the deviations of 1877–86 are higher than the above-mentioned standard deviations, it must be concluded that the decade of 1877–86 was too wet.

The tabulated figures clearly show a maximum rainfall in 1877–86 followed by a considerable decrease that has been observed until 1899. Only the New England states and the neighbouring central Atlantic states do not fit this pattern. Instead these states show an increase in rainfall and must therefore be considered exceptional regions among the rest of the U.S., as I had discovered earlier for the time around 1880.

Table 8.4.

Station	Mean	Deviations from the Mean 1877–86			
	1877–86	(mm)		(%)	
	(mm)	1887–96	1897–99	1887–96	1897–99
New Bedford, Mass.	1133	+127	+206	+10	+18
Providence, Rh. I.	1290	−17	+68	−1	+5
Lenoir, N.C.	1305	−66	+31	−5	+2
Hatteras, N.C.	1865	−398	−435	−21	−24
Wilmington, N.C.	1494	−293	−467	−20	−31
Charleston, S.C.	1443	−113	−235		−16
Augusta, Ga..	1182	+23	+42	+2	+4
Savannah, Ga.	1293	−55	+31	−4	+2
Jacksonville, Fla.	1465	−217	−237	−15	−16
Mobile, Ala.	1692	−200	−208	−12	−12
Montgomery, Ala.	1322	−55	−157	−4	−12
Vicksburg, Miss.	1576	−390	−313	−25	−20
Memphis, Tenn.	1442	−203	−311	−14	−22
New Orleans, La.	1558	−179	−512	−11	−33
Shreveport, La.	1378	−351	−518	−26	−38
Galveston, Tex.	1331	−316	−374	−24	−28

8.3 RAINFALL FLUCTUATIONS FROM 1830 TO 1900 IN THE UNITED STATES, AS WELL AS AT SOME STATIONS IN CENTRAL EUROPE AND EAST SIBERIA.

Henry, who appears to be unaware of my own research about climate variations and, hence, seems to be unbiased in his observations, raises the question of the rainfall pattern in former times. In an effort to demonstrate this, he includes in his table the observations from three stations in New England (Boston, New Bedford, and Providence), three in the upper Ohio valley (Marietta, Portsmouth, and Cincinnati) and four in the central Mississippi valley (Muscatine, Monticello, Marengo, and Peoria). Henry smoothed these numbers according to the following formula:

$$(a + 4b + 6c + 4d + e) / 16 = c'$$

For each of the three geographical areas, he then determines a mean value. These figures, which he also expresses in graphical curves, clearly reflect climatic variations in the Ohio and Mississippi valleys, while the New England values deviate.

Still, there remain certain secondary distortions because the above formula takes only three years into account and factors "*a*" as well as "*e*" have very little weight. I have made further adjustments by combining five years each of Henry's smoothed figures into one mean; regrettably I was unable to go back to the original observational data for each of these years because Henry does not provide them. Fully adjusted, the formula reads:

$$(a + 5b + 11c + 15d + 16e + 15f + 11g + 5h + i) / 80 = e'$$

In accordance with his report, I have supplemented the data sequences up to the year 1899, enabling a smoothing until the year 1897.

I am adding the data sequences for Bruxelles and Bremen which have been smoothed by forming 5-year averages, i.e., for 1832 the lustrum 1830–34, for 1833 the lustrum 1831–35, etc., and which are the only stations in Central Europe for which observations of the year 1900 are presently[441] available to me, as well as the figures for the ore smelter of Nertschinsk and Nikolajewsk at the Amur in Siberia. The adjustment was done on figures rounded to full centimetres. Missing monthly measurements by the stations Nertschinsk and Nikolajewsk were interpolated beforehand according to those months' precipitation in the four neighbouring years. If an entire year was missing, four- instead of five-year averages were used. All numbers are in millimetres.

Aside from shorter oscillations that Henry concentrates on, there are sharply defined rainfall variations evident in the upper Ohio and the central Mississippi valley. Aridity prevails in the middle of the 1830s; rainfall then increases and reaches a maximum toward the end of the 1840s, decreases again and stays about average for some time arriving at a minimum at the beginning of the 1870s; a rapid increase to another maximum in the early 1880s is then followed by a considerable decrease towards the end of the century. This pattern is in complete accordance with the periods of climatic variations as I have established them earlier. Only the minimum of 1871/72 occurs quite late at a time when, in most areas, rainfall is already slightly

[441] December 1901.

Table 8.5.

	New England (Non Appl. Region)	Upper Ohio Valley	Central Missis-sippi Valley	Bru-xelles	Bre-men	Nertschin-sky Smelter Plant (Siberia)	Nikola-jewsk at the Amur (Siberia)
Mean Value	1163	1049	894	725	690	413	454
1830	38	—	—	—	—	—	—
1831	20	—	—	—	—	—	—
1832	–23	5	—	—	–24	—	—
1833	–84	–41	—	—	–68*	—	—
1834	–145	–89	—	–45	18	—	—
1835	–190	–99	—	–33	62	—	—
1836	–208*	107*	—	–75*	90	—	—
1837	–201	–84	—	–9	88	—	—
1838	–175	–76	—	–5	74	—	—
1839	–142	–48*	—	–15	58	—	—
1840	–114	–36	—	–37	28	—	—
1841	–99	–8	—	3	52	–17	—
1842	–102	46	—	7	70	41	—
1843	–122	41	—	39	108	77	—
1844	–147	66	—	9	54	109	—
1845	–160	97	—	5	52	159	—
1846	–163	124	—	5	56	122	—
1847	–152	155	—	–17	56	75	—
1848	–119	168	—	–9	64	35	—
1849	–81	145	—	17	108	13	—
1850	–53	99	—	75	130	–15	—
1851	–28	61	—	47	68	–35	—
1852	–18	25	—	55	60	–23	—
1853	-28	–5	272	21	52	–53	—
1854	–43	–23	150	27	24	–9	—
1855	–41	–18	66	–61	–48	–17	—
1856	-30	8	48	–91*	–80	–33	—
1857	-15	30	64	–87	–80	–57	—
1858	0	41	74	–59	–50	–63	—
1859	23	51	76	–63	–38	–103	—
1860	46	43	66	–21	34	-117*	—
1861	58	8	41	–3	88	–101	—
1862	58	–15	18	–53	18	–103	–124*
1863	53	–5	10	–91*	–82	–87	–116
1864	51	18	–5	–87	–104	–67	–116
1865	53	30	–25	–59	–88	–51	–82
1866	71	48	–23	–35	–92	–41	–72
1867	99	43	–5	39	–42	–43	–102
1868	122	5	–10	41	–20	13	–106
1869	130	–53	–23	9	–48	–7	–118
1870	124	–91	–33	7	–90	26	–102
1871	104	–109*	–53	13	–136	7	
1872	86	–107	-71*	–25	–154*	13	–112
1873	81	–81	-51	–25	–136	–5	
1874	84	–46	-8	11	–104	21	46
1875	91	–30	33	19	–74	–27	47
1876	99	–33	58	101	–16	–17	101

[continued on next page]

[Continued from previous page]

	New England (Non Appl. Region)	Upper Ohio Valley	Central Missis-sippi Valley	Bru-xelles	Bre-men	Nertschin-sky Smelter Plant (Siberia)	Nikola-jewsk at the Amur (Siberia)
Mean Value	1163	1049	894	725	690	413	454
1877	94	–30	78	115	36	15	94
1878	76	–15	91	155	88	31	130
1879	61	18	97	147	98	65	36
1880	36	74	97	139	120	119	38
1881	10	119	102	67	112	135	26
1882	8*	127	112	69	136	115	14
1883	18	97	94	37	94	59	11
1884	33	51	53	31	82	7	6
1885	38	28	10	–33	112	–27	28
1886	104	–28	-25	–1	–6	–55	28
1887	142	–18	-56	9	–50	–55	47
1888	168	15	-71	35	–66	–25	50
1889	170	38	-69	19	–40	–43	56
1890	142	38	-56	19	–24	–37	18
1891	97	13	-53	–33	–24	–7	2
1892	46	–41	-66	–33	–6	–19	–10
1893	0	–99	-81	–53	8	–3	–18
1894	-18	–130*	-102	–45	–30	33	–6
1895	-7	–109	-107	–23	34	23	–16
1896	18	–76	-109*	–33	–16	7	–30
1897	36	–30	-104	–49	–58	–23	–80
1898	—	—	—	–67*	–74*	—	—

above average. Nonetheless, the maximum of 1882 complies again with my findings.

The periodicity of a complete variation reaching from the minimum of 1836 to the minimum of 1871 is 35 years, and from the maximum of 1848 to the maximum of 1882, 34 years.

The New England states deviate completely from this pattern showing an arid period from the early 1830s to the early 1850s, a maximum around 1869 and a secondary minimum around 1882. The New England states are exceptional regions, as explained earlier in my book. But the most striking feature of the New England data is the increase in rainfall that manifests itself as early as 1816 and interferes with the observed variations. This increase is evident in all of the available long-term data rows. It has never been fully explained. Could it be symptomatic of a long-term climate period, such as the 160-year periodicity that Sieger[442] has shown to be probable for Scandinavia?

Bremen and Bruxelles, representing the Northwest corner of Central Europe, show similar fluctuations: a minimum in 1833 and 1836

[442] Zeitschrift der Gesellschaft für Erdkunde, Berlin 1893, p. 444.

respectively, a maximum in 1850, in Bremen a minimum in 1872, and in Bruxelles in 1856–63, a maximum in Bremen in 1882, and in Bruxelles 1882. Bremen shows the same delay in the occurrence of the minimum of the 60s as was observed in the central parts of North America, while in Bruxelles the maximum (1878) occurs slightly earlier. In general, the trend of the curves is the same.

That is also the case for Nertschinsk in East Siberia and Nikolajewsk at the Amur. At Nertschinsk the first maximum (1845) occurs some years earlier than in the interior of the United States, at Nikolajewsk as in the European part of Russia it is the second maximum (1878).[443]

The interior of the United States, Bremen, Bruxelles, the European part of Russia, Nertschinsk, and Nikolajewsk at the Amur, all these locations show an obvious 35-year periodicity of climatic variations, while there is no trace of a 55-year period. The epochs vary in some cases slightly—they occur earlier or later. The early or late occurrences will however be evened out with the next epoch. The time frames of the wet and the dry periods show similar irregularities. I am adding a table about the dry and wet periods at this point.

	dry	wet	dry	wet	dry
Ohio-Valley	1833–41	1842–52	1853–78	1879–91	1892–97
Mississippi-valley		18? –63	1864–74	1875–85	1886–97
Bruxelles	18? –40	1841–54	1855–66	1867–90	1891–98
Bremen	18? –33	1834–54	1855–76	1877–85	1886–98
Europ.Russia	—	—	18? –67	1868–84	1885–?
Nertschinsk	—	1842–49	1850–67	1868–84	1885–97
Nikolajewsk a. A.	—	—	18? –73	1874–91	1892–97
Centr. Germ. border	18? –38	1840–54	1855–71	1872–87	1888–?
Brückner 1890.	1826–40	1841–55	1856–70	1871–85	

If we look at the average years in which the periods ended we almost get the same results, i.e., with deviations of not more than two years, as I had established in 1890.

The discrepancies in the individual regions should not surprise us: we are dealing with meteorological periods and not with mathematical ones. Just as the daily temperatures vary from one day to the next and differ in their extremes and length of time, just as their maximum and minimum shifts according to cloud cover, wind, etc., in short according to weather conditions, so do climatic variations. Only the use of a large amount of data, as I have done in my book, can help to eliminate the local and temporary effects

[443] Op. cit.

of random occurrences.[444] It should be strictly differentiated between these temporary irregularities of unknown causes and the lasting exceptional discrepancies; these refer to regions where the pattern of the oscillations is more or less reversed. An inquiry into these exceptional regions in particular might shed some light on the mechanism of climatic variations.

Let us take a look at the amplitude of these variations expressed in percentages of the multi-year averages for the different rows of data and compare it with the figures which I arrived at earlier[445] using a different method (through lustra averages) and dealing largely with a totally different subject matter.

	Average Minimum	Average Maximum	Variations	My Results of 1890
Lower Ohio area	–11%	+14%	25%	} 19 %
Central Mississippi-region	–10	+13	23	
Bruxelles	–11	+16	27	} 20 %
Bremen	–16	+20	36	
Nertschinsk	–28	+36	64	} 47 %
Nikolajewsk	–28	+29	57	

According to the new figures, the oscillation is considerably larger, especially in those data rows provided by one station only. This cannot be otherwise: a row of data that, for each single year, includes a five-year mean must show a larger oscillation than the same row based on lustra averages alone. These variations of the rainfall are in fact slightly larger than I had identified them in 1890 with the help of lustra averages.

Let me summarise briefly what I described above:

1. Professor Woeikof's findings about the variations of Lake Aral and its surrounding lakes confirm my conclusion of 1890 that the Kirghiz Steppe is an exceptional case with regard to climatic variations.
2. The rainfall fluctuations at Barnaul do not correspond with the fluctuations at these lakes but show instead a 35-year periodicity of alternating wet and dry climate periods. Noticeable is the long duration of the wet period at the end of the century.
3. In contrast, East Siberia and the selected Central European stations in most parts of European Russia show a considerable decline in rainfall towards the end of the century.
4. Significant is the decline in precipitation in the United States since the middle of the 80s. Regions exempt from this phenomenon are

[444] Please compare *Klimaschwankungen* [*Climatic Variations*], pp. 175 and 192, regarding these irregularities.

[445] We are unable to supply data for the European part of Russia because the absolute minimum of 1860 is not included in our table.

predominantly the New England states, as well as the central Atlantic states, which is in full accordance with my earlier assumptions.

5. The 35-year periodicity of precipitation is also clearly evident when the data are smoothed as suggested by P. Schreiber[446] by determining progressive group averages. The resulting figures are identical with my own findings which I gained earlier through the shorter method of lustra averages; except that the fluctuation of the rainfall is slightly larger than in my study.
6. Occasionally the timing of the epochs is irregular; climatic variations are after all a meteorological phenomenon and not a mathematical one.
7. None of the observed rows show any trace of a 55-year fluctuation.

[446] Excerpts of the Königlich Sächsisches Meteorologisches Institut Leipzig 1896, Issue 1, p. 46. (Also in *Civilingenieur*, Volume 42, Issue 1 and 3.)

Chapter 9

About Climate Variability*

The common interest in the question of changes in the climate of smaller or larger regions can be explained by the fact that all life on earth, especially the plant life of a region, depends on the climate to a large degree. Temperature and rain primarily have a decisive impact on plant life, each playing a slightly different role. Because each plant requires a certain amount of warm temperature in order to exist, which may differ from species to species, and since the summer temperatures decrease closer to the polar regions, it is quite obvious that temperature is a particularly significant factor in determining the polar line of plant growth. For example, an imaginary line connecting all those spots where the temperature rises to more than 10° during three months in the summer roughly coincides with the polar tree line. The amount of precipitation, on the other hand, often determines plant life in its growth from West to East. Due to the fact that in the Old World rainfall decreases from the Atlantic Ocean in an easterly direction, plants requiring a lot of moisture are limited to western regions and those which thrive only when there is little moisture, to the East.

If weather conditions would remain the same year after year, the growth line of a plant species would be indicated world-wide, and all other conditions being equal, by the same isotherm, i.e., the line connecting places with the same average temperature, or the same isothyte, i.e. with the same annual amount of precipitation. Of course this is not the case, because any extremes of the two factors are much more important for the distribution of plants than average temperatures and average precipitation recorded over many years of observations. Over several years, a forest-tree in a certain location may have received a sufficient amount of moisture to grow on, yet if in the following year the required minimum amount is not reached, the tree and its seedlings will perish. This is a common fact. In a series of good years a plant will be able to extend its growth area; yet the first bad year

* *Über Klimaschwankungen*, Mitteilungen der Deutschen Landwirtschaftsgesellschaft, 1909.

which fails to supply the warm temperatures and the moisture needed for its survival, will force it back to its original habitat. In the long run, plants are only able to survive in areas where even in bad years the required minimum temperature and moisture is provided. Obviously the weather pattern of bad years determines the boundaries of the plant's growth area. This primarily refers to the natural habitat of the plant.

Wherever humans interfere, conditions are partially altered. Often it can be an advantage to grow a plant that thrives only in good years and fails to produce or perishes in bad years. The yield of one good year may compensate for many bad years. For example, in the Middle Ages and later, wine was grown in many regions of Northern Germany and Northern France, where nowadays viniculture is out of the question, even though some years did not produce at all. Because of high freight costs, it was more advantageous to put up with poor crops than to import grapes from the south. Later on as infrastructure improved and transportation became cheaper, and also the taste became more refined, wine-growing moved southward and vineyards in the North were re-cultivated with a product more suitable for that climate.

A similar development is taking place today in the area of grain production.

Temperature conditions determine how far north the different types of grain are grown. Temperature restricts wheat growing in Europe primarily to regions south of the 60th or 61st parallel of latitude, while allowing the growth of the unexacting grain types, such as barley and oats, almost as far north as the Northern Cape and the White Sea. Nonetheless, in the main agricultural regions of Europe temperature has only a limited local impact and its influence on the outcome of the crop is therefore secondary. Only close to the polar boundary of the grain's growth area, e.g., in Scandinavia and Northern Russia, crops are often damaged by frost. Too much heat does not damage grain as long as enough water is available, e.g., wheat grows well in the wet tropical part of India's subcontinent!

Moisture is a much more important factor in growing wheat. Without water, there would be no agriculture, and with too much water, none either. This rule comes to mind when we look at the expansion of agriculture around the globe or rather when we examine the causes of crop failures. Droughts and poor harvests go hand in hand in all those regions with little rainfall. Where rainfall is plenty, poor harvests are mainly evoked by years of heavy rainfalls. Of course, it is not the amount of rainwater *per se* that determines this, but the ratio between rainwater and evaporated water. The same quantity of water that inevitably drowns the crop in a cold climate may just be sufficient enough for the irrigation of grain fields in a hot climate. In contrast, a low amount of rainfall causing droughts in a hot country due to heavy evaporation might often be entirely sufficient in a cool climate. For Europe's wet coastlines of the North Atlantic Ocean and Central Europe, the effect is the exact opposite to the dry interior of the continent. Southern Russia on the one hand and Great Britain and Ireland on the other are extremes in this regard. In the majority of cases, droughts are the causes for crop failures in southern Russia, whereas the many crop failures that England endured at the end of the 1830s, during the 1840s, and at the beginning of the 1850s, as well as the 1870s and 1880s, occurred in years with excessive rainfalls.

Southern Europe and most of the Tropics, at least as far as they are producing grain, experience similar conditions as in Southern Russia. The years of famine in the East Indies coincide with dry years.

Central Europe, and the German Empire in particular, is halfway between these two extremes. Dry years are often good years for a number of agricultural activities, especially wine and fruit growing, but also for grain, specifically in the West, whereas they are bad years for dairy farming or cattle raising. Although occasionally dry weather does damage crops, crop failures due to too much precipitation are much more common. A detailed inquiry into the quite complicated causes of crop failures is still outstanding.

Under the climate factors' determining influence on crop failures, a radical shift in agricultural production has taken place over the last decades starting around the middle of the past century. Regions where crop failure used to be a frequent occurrence because of too much rain have mainly abolished grain cultivation and have almost exclusively turned to the cultivation of grasslands and of fodder-plants instead, both of which are perfect for a wet climate and a prerequisite for keeping and breeding cattle. This can be said of Ireland and England, but also of Denmark and of larger regions of the German Empire, in particular the northwestern regions (e.g., Holstein, Friesland, and Western Prussia). Due to the modernised transportation system in these regions, the supply of grain from other regions became easier and allowed agricultural production to adjust much better to the climatic conditions. This is also the case for Switzerland.

In view of the close link between plant life and weather and climate, their influence on the outcome of harvests is widely undisputed. If for that matter the climate was constant and the weather changed from year to year according to random law, then, on the average of a long series of years, the harvested crops would, on the whole, have to be constant as well. Is the climate actually constant?

Recent studies have shown that this is not the case; in fact climate has many fluctuations that vary around a middle range. I am not referring to the 11-year weather cycle that has often been claimed yet never been verified completely, which disappears entirely from time to time and is caused by the

11-year cycle of sunspot activity, but rather to the far more significant 35-year climate variations, the probability of which I was able to declare in 1888 and which I could verify for larger parts of the globe in 1890.[447] Since then I have continued with my research and summarise as follows.

The climate variations involve multi-year changes in temperature, air pressure, and rainfall that take place simultaneously all over the globe. The temperature is the one factor in this process upon which all others depend. Temperature variations are common to almost all regions on earth. As shown in Table 9.1, they all experience cold and warm periods at the same time. Figures for the three continents Europe, Asia, and North America indicate the deviations by 1/100 degree Celsius of the five-year (lustrum) averages from the multi-year (30–50) average. The minus symbol (–) stands for a lustrum that was too cold by the indicated degree and the positive figures (+) for lustra which were too warm.

Table 9.1. Temperature Variations of the Three Continents of the Northern Hemisphere 1806–1885. (1/100° C)

Five Year Lustrum	Europe	Asia	North America	Globally
1806/10	–05	—	–16	–18
1811/15	–34*	—	–38*	–46
1816/20	–05	—	–37	–35
1821/25	**+46**	**+54**	+29	**+56**
1826/30	–03	+17	**+93**	+14
1831/35	+14	–01	–18	+03
1836/40	–54*	–31*	–63*	–39
1841/45	–08	+09	–17	+00
1846/50	+04	+02	+17	–08
1851/55	+03	+29	+16	**+11**
1856/60	–08	–05	–28	+06
1861/65	+11	–07	**+29**	–06
1866/70	**+26**	**+31**	–07	**+11**
1871/75	–02	+25	—	+04
1876/80	–09	–05	—	–07
1881/85	–08	—	—	–08

It is evident from these figures that Europe, Asia and North America experienced a relatively cold period from 1806 to 1820, followed from 1821

[447] Compare Ed. Brückner, *Klimaschwankungen seit 1700* [*Climatic variations since 1700; Chapter 4 of this anthology*] Wien 1890. — *Der Einfluß der Klimaschwankungen auf die Ernteerträge und Getreidepreise in Europa [Influence of climate variability on harvest and grain prices in Europe*; Chapter 6] Geographische Zeitschrift, I, 1895.— *Zur Frage der 35jährigen Klimaschwankungen* [*An inquiry about the 35-year period of climate variations;* Chapter 8], Petermann's Mittheilungen 1902, issue 7.—*Schwankungen des Niederschlags im Deutschen Reiche [Variations of precipitation in the German Empire*], Zeitschrift für Gletscherkunde, für Eiszeitforschung und Geschichte des Klimas. 1 (1906).

to 1835 by a fairly warm one, and then alternating between cool (1836–45), warm (1846–75), and cool (1876–85) periods. In all continents the maxima of the cold periods, marked by an asterisk (*), fall consistently into the periods 1811–15, 1836–40, and (to a lesser extent) 1876–85. The maxima of the warm periods, highlighted in **bold**, occur with the same consistency in Europe and Asia in 1821–25 and 1866–70, while in North America they shift by five years (1826–30 and 1861–65). Space does not permit to include the data for the Southern Hemisphere; they however show a similar picture. I have not yet followed up on global temperature variations after 1885.

Temperatures vary on the average by almost 1° C. That is quite substantial. For it means nothing else than that during the five years of 1836–40 the average annual temperature of Berlin, for instance, was a full degree lower than during the five years of 1821–25, as if that city from one period to the other had moved further north by 3 degrees latitude. During the last 30 years—since approximately 1875—the temperature fluctuations have become less distinct and, therefore, appear less conspicuous at individual stations.

The distribution of air pressure is affected by temperature variations. In Europe, as we will later find out in detail, the transfer of moist air from the ocean to the continent seems more difficult during warm periods, yet easier during cold periods. This in turn must affect the rainfall of the region. And indeed, rainfall fluctuations are much more pronounced and can be traced much better than temperature fluctuations. Mind you, the rainfall measurements of one single station fail at first to show a pattern of regularity from one year to the next; rain is, after all, an extremely unsettled meteorological phenomenon. One or two major thundershowers coming down in one particular spot can significantly affect that location's annual amount of precipitation. It is advisable to combine the measurements of a number of stations for a somewhat larger area. This should be done in such a manner, that for each station the difference between the amount of precipitation during the observed year and the multi-year average of rainfall is expressed in percentages indicating whether too much (+) or too little (-) precipitation occurred. In addition the average deviation is then determined for all the stations combined. In the table below this has been done for the large river regions of the German Empire.

Table 9.2. Variation of Precipitation for the River Regions of the German Empire: Deviation of the Amount of Precipitation in Percentage of the 50-year Mean Value

Year	Weichsel-Oder	Elbe	Lower Rhein	Upper Rhein	Mean
1851	+8	+6	+2	+8	+6
1852	+3	+17	+24	+5	+12
1853	+5	± 0	−2	−3	±0
1854	+17	+9	+13	+5	+11
1855	+12	−3	−5	−2	±0

Year	Weichsel-Oder	Elbe	Lower Rhein	Upper Rhein	Mean
1856	−3	−8	+10	+12	+3
1857	−31	−31	−32	−34	−32
1858	−19	−13	−20	−21	−18
1859	−12	−5	+13	−3	−2
1860	−8	+15	+17	+20	+11
1861	+3	+4	−3	−7	−1
1862	−15	+8	+3	+7	+1
1863	−13	−2	−8	−11	−9
1864	−8	−16	−25	−28	−19
1865	−12	−25	−12	−27	−19
1866	−2	+9	+21	+16	+11
1867	+23	+14	+17	+16	+18
1868	−9	+4	−4	+4	−1
1869	+4	+2	± 0	−3	+1
1870	−7	+1	−2	+2	−2
1871	−9	−5	−4	−13	−8
1872	+9	−4	+13	+3	+5
1873	−11	−15	−15	−8	−12
1874	−21	−21	−15	−12	−17
1875	−1	−2	+15	+11	+6
1876	−2	+8	−1	+5	+3
1877	+7	+14	+25	+15	+15
1878	+1	+2	+14	+26	+11
1879	+9	+14	+7	+8	+10
1880	+14	+22	+17	+20	+18
1881	−10	−1	−4	−10	−6
1882	+18	+26	+34	+36	+28
1883	+5	−8	−5	−6	−3
1884	±0	+4	−12	−12	−5
1885	+8	−6	−1	+7	+2
1886	−13	−7	−1	+4	−4
1887	−6	−17	−14	+17	−13
1888	+16	+8	+13	+4	+9
1889	+9	+5	−5	±0	+2
1890	+10	−3	−2	±0	+1
1891	+14	+6	±0	−1	+5
1892	−13	−18	−14	−7	−13
1893	+4	−6	−11	−13	−6
1894	+5	+13	±0	−3	+4
1895	+1	−2	−6	+2	−1
1896	−3	−1	−4	+10	±0
1897	+5	+3	−8	−4	−1
1898	+8	+2	−10	−4	−1
1899	+11	+3	−6	−3	+1
1900	−1	+4	−4	+5	+1

Here again quite a few irregularities are still evident. Years with too much precipitation (+) and with too little precipitation (–) alternate repeatedly. However, some general trends are noticeable. The period of 1856–74 includes a number of very dry years which is indicated by an accumulation of minus symbols; all years from 1877 to 1880 are too wet, and plus symbols continue to outnumber the minus symbols slightly beyond that date. Still, even these figures leave the impression that a relatively drier period in the middle of the 70s was followed by a somewhat wetter period. This trend becomes even clearer if we try to adjust the irregularities mathematically. This can be achieved in different ways. In 1890, I chose the method of combining five years each as shown in the following table for the entire area of the German Empire. Figures indicate in percentages of the multi-year average by how much the precipitation was too high or too low in the respective five-year period.

Table 9.3. Variations of Rainfall in the German Empire, Expressed in Lustra Averages

1816–20	–1	1861–65	–9
1821–25	–1	1866–70	+3
1826–30	–6*	1871–75	–6
1831–35	–6	1876–80	+9
1836–40	+2	1881–85	+3
1841–45	+3	1886–90	+1
1846–50	–1	1891–95	–2
1851–55	+8	1896–00	+2
1856–60	–7	1901–05	—

The lustra show an obvious pattern, 1816–35 are years with low precipitation, 1836–55 with a lot of rain, 1856–75 with little rain, and from 1876–90 again with a lot of rain. Lately, I have been using a different formula that I consider to be much better. I determine the averages of ten-year periods (compare Table 9.2), e.g., 1851–60, 1852–61, 1853–62 etc. The resulting figures depict a full decade and show the development of the precipitation from year to year in averages of decades. Table 9.4 shows these 10-year averages next to the year which marks the centre of each time span for which the mean value was established.

Table 9.4. Variations of Precipitation in River Regions of the German Empire (in Percent). Smoothed by 10-year Mean Values

Centre Year of the 10-Year Average	Weichsel-Oder	Elbe	Lower Rhein	Upper Rhein	Mean
1855/56	–3	–1	+2	–1	–1
1856/57	–3	–2	+1	–3	–2
1857/58	–5	–2	–1	–3	–3
1858/59	–7	–3	–1	–3	–4
1859/60	–9	–5	–5	–7	–7
1860/61	–12*	–7*	–6*	–9*	–9*
1861/62	–12	–6	–5	–9	–8

Centre Year of the 10-Year Average	Weichsel-Oder	Elbe	Lower Rhein	Upper Rhein	Mean
1862/63	–6	–1	±0	–4	–3
1863/64	–5	+1	+2	–1	–1
1864/65	–4	+1	+1	–1	–1
1865/66	–4	±0	–1	–3	–2
1866/67	–5	–1	–1	–4	–3
1867/68	–2	–2	±0	–4	–2
1868/69	–2	–3	–1	–4	–3
1869/70	–3	–4	±0	–2	–2
1870/71	–2	–2	+3	+2	±0
1871/72	–2	–2	±0	+1	–1
1872/73	–4	–2	–1	±0	–1
1873/74	–3	–2	+3	+3	±0
1874/75	–2	–1	+4	+4	+1
1875/76	±0	+1	+6	+6	+3
1876/77	±0	+2	+6	+6	+3
1877/78	+1	+5	+8	+9	+6
1878/79	+2	+5	+9	+9	+7
1879/80	+4	+8	+9	+9	+8
1880/81	+5	+8	+7	+9	+7
1881/82	+4	+6	+7	+9	+7
1882/83	+3	+3	+4	+6	+4
1883/84	+4	+3	+3	+3	+4
1884/85	+4	–3	+2	+3	+3
1885/86	+4	±0	±0	+1	+1
1886/87	+6	+1	+1	+2	+2
1887/88	+3	–1	–4	–3	–2
1888/89	+3	–3	–5	–4	–2
1889/90	+3	–3	–4	–3	–1
1890/91	+3	–2	–4	–3	–2
1891/92	+4	–2	–4	–3	–1
1892/93	+5	±0	–4	–1	±0
1893/94	+4	±0	–6	–2	–1
1894/95	+4	±0	–6	–2	–1
1895/96	+3	±0	–6	–2	–1

It becomes quite apparent now that all river regions of the German Empire experienced a period of low precipitation from 1855–69, those in the East up to 1874. This was followed by a period of high precipitation up to 1886, when rainfall decreased again. The decrease was much more pronounced in the Rhein region than in the Weichsel and Oder regions where precipitation continued to be too high up to 1900. I have not yet examined in detail the precipitation variations beyond 1900.

Table 9.5. Variations of Precipitation in Central Germany, Asia, and North America

Year	Central Northern Germany by P. Schulz (% adjusted)			Asia by Brückner (cm adjusted)		North America by Brückner (cm adjusted)	
	Summer	Fall	Year	Nertschinsk	Nikolajew (Amur)	Upper Ohio Valley	Central Mississippi Valley
1831/32	–4	–4	–2	—	—	± 0	—
1832/33	–2	–5	–1	—	—	–4	—
1833/34	–3	–10	–4	—	—	–9	—
1834/35	–2	–15*	–2	—	—	–10	—
1835/36	–10	–12	–6	—	—	–11*	—
1836/37	–7	–14	–5	—	—	–8	—
1837/38	–10	–10	–6	—	—	–8	—
1838/39	–10	–9	–6*	—	—	–5	—
1839/40	–18*	–2	–6	—	—	–4	—
1840/41	–9	–1	–1	–2	—	–1	—
1841/42	–7	–5	±0	+4	—	+5	—
1842/43	–6	–6	–3	+8	—	+4	—
1843/44	–8	+6	–1	+11	—	+7	—
1844/45	–11	+8	–2	**+16**	—	+10	—
1845/46	–4	+5	±0	+12	—	+12	—
1846/47	–8	+12	+2	+8	—	+16	—
1847/48	–2	+10	**+5**	+4	—	**+17**	—
1848/49	–2	**+12**	+4	+1	—	+14	—
1849/50	+2	+10	**+5**	–2	—	+10	—
1850/51	+2	+8	+2	–4	—	+6	—
1851/52	+4	+8	+2	–2	—	+2	—
1852/53	+2	+2	+2	–5	—	±0	**+27**
1853/54	+7	–8	±0	–1	—	–2	+15
1854/55	+8	–9	±0	–2	—	–2	+7
1855/56	+10	–13	±0	–3		+1	+5
1856/57	+12	–18	–2	–6		+3	+6
1857/58	+17	–22	–1	–6		+4	+7
1858/59	+14	–21	–2	–10		+5	+8
1859/60	+11	–23	–5	–12*		+4	+7
1860/61	+7	–24	–8	–10		+1	+4
1861/62	+2	–25*	–9*	–10	–12*	–2	+2
1862/63	+1	–20	–5	–9	–12	±0	+1
1863/64	–4	–16	–3	–7	–12	+2	±0
1864/65	–4	–10	–3	–5	–8	+3	–2
1865/66	–4	–8	–4	–4	–7	+5	–2
1866/67	–4	–13	–5	–4	–10	+4	±0
1867/68	–10	–6	–6	+1	–11	±0	–1
1868/69	–9	–6	–7	–1	–12	–5	–2
1869/70	–13	–8	–7	+3	–10	–9	–3
1870/71	–12	+1	–4	+1		–11*	–5
1871/72	–15*	+5	–4	+1	}–11	–11	–7*
1872/73	–12	+5	–5	±0		–8	–5

Year	Central Northern Germany by P. Schulz (% adjusted)			Asia		North America	
				by Brückner (cm adjusted)			
	Summer	Fall	Year	Nertschinsk	Nikolajew (Amur)	Upper Ohio Valley	Central Mississippi Valley
1873/74	–8	+2	–5	+2	+5	–5	–1
1874/75	–4	±0	–3	–3	+5	–3	+3
1875/76	–5	+4	–2	–2	+10	–3	+6
1876/77	–6	+8	–1	+2	+9	–3	+8
1877/78	+4	+10	+3	+3	+13	–2	+9
1878/79	+6	+14	+4	+6	+4	+2	+10
1879/80	+9	**+19**	**+8**	+12	+4	+7	+10
1880/81	+9	+16	+6	**+14**	+3	+12	**+11**
1881/82	**+13**	+12	+7	+12	+1	**+13**	+11
1882/83	+10	+11	+6	+6	+1	+10	+9
1883/84	+10	+16	+6	+1	+1	+5	+5
1884/85	+8	+17	+6	–3	+3	+3	+1
1885/86	+9	+14	+5	–6	+3	–3	–2
1886/87	+9	+12	+6	–6	+5	–2	–6
1887/88	–1	+2	±0	–2	+5	+2	–7
1888/89	–4	+4	±0	–4	+6	+4	–7
1889/90	±0	+5	±0	–4	+2	+4	–6
1890/91	–1	+4	+2	–1	±0	+1	–5
1891/92	+1	+6	+2	–2	–1	–4	–7
1892/93	+2	+7	+2	±0	–2	–10	–8
1893/94	+1	+2	+2	+3	–1	–13*	–10
1894/95	±0	±0	+2	+2	–2	–11	–11
1895/96	–1	–1	+2	+1	–3	–8	–11*
1896/97	–5	+5	+2	–2	–8	–3	–10

Upon my recommendation, Dr. Paul Schulz in his thesis[448] determined the average precipitation of each meteorological season exactly as I had done for the annual precipitation. His figures for central Northern Germany are included in Table 9.5 according to year, summer, and fall; they have been adjusted by forming ten-year averages and are expressed in percentages of the multi-year average.

It became evident that the figures for summer and fall as well as the annual averages indicate an arid period in the 1830s and up to the beginning of the 40s, then a wet period in the 1840s and the first half of the 1850s which, however, has a late start in the summer season, followed by a dry period in the 1860s up into the first half of the 1870s, and then again a wet

[448] *Klimaschwankungen im mittleren Norddeutschland und ihr Einfluß auf die Ernteerträge* [*Variability of the Climate in central Northern Germany and its Influence on the Harvest*]. Inaugural-Dissertation, Halle a. p., 1907.

period which according to Schulz' figures does not fade out until towards the end of the century. In contrast, winter and spring (omitted here) show irregularities.

In order to demonstrate the simultaneous occurrence of these variations in most parts of the global land masses, I have added to Table 9.5 the results for the upper Ohio- and central Mississippi-valley in North America, as well as for Nertschinsk (smelter plant) and Nikolajewsk on the Amur in Asia.[449] It is obvious how completely coincidental the variations are, as shown in the following table for each continent (annual averages, percentages).

Table 9.6. Variability of Precipitation for Each Continent

Five-year Span	Europe	Asia	North America	Central and South America	Australia	Globally (Land Masses)
1806/10	**+3**	—	—	—	—	+3
1811/15	±0	−7	—	—	—	−3
1816/20	−1	**+22**	−4	—	—	**+6**
1821/25	−2	±0	+4	—	—	+1
1826/30	−1	+14	−4	—	—	+3
1831/35	−10*	−7	−12*	−6	—	−8*
1836/40	−1	−14*	−7	−19*	—	−5
1841/45	**+4**	+9	−3	−11	−11	+1
1846/50	+1	**+13**	**+16**	+8	**+17**	+3
1851/55	**+4**	+2	+3	**+16**	+16	+1
1856/60	−4	−9	−8	+5	+2	−4
1861/65	−10*	−13*	−10*	−11	−6*	−5*
1866/70	±0	−9	+7	−12*	+10	−1
1871/75	±0	+8	−1	−9	**+14**	+2
1876/80	**+10**	+20	+4	+7	−1	**+7**
1881/85	+6	**+23**	**+13**	**+10**	−19	+6

It is evident that the dry period of the 30s and of the 60s occurred in all the continents. So does the wet period of 1850, as well as the one around 1880, which peaks in Australia five years earlier than in the other continents. Based on the averages for the global land masses, I was able to determine more precisely that the time from 1826 to 1840 was dry, 1840 to 1854 wet, 1855 to 1871 dry, 1872 to 1887 wet, and since 1888, fairly dry again. In this tabulation the general pattern emerged (compare Table 9.1 and 9.6) that the cool and the wet periods coincided, and so did the warm and the dry periods.

The variations of precipitation are quite pronounced, especially so in the largely continental regions. In East England they are 16%, in Northern Germany 20%, in Southern Russia 24%, in Southeast Russia 35%, in West

[449] These numbers from my essay of 1902 are not expressed in percentage of the long-term average but state rainfall height in cm. The adjustment of a ten-year mean has not been made either but through a five year mean. The second figure for the year of each row show a five-year average in relation to precipitation figures.

Siberia 86%, in East Siberia 47% of the multi-year average. They are also significant in the interior of the United States of America (36%).

It is noteworthy that the islands in the middle of the ocean, such as Iceland and Ireland, but also parts of the oceanic coast lines show the opposite pattern, i.e., in those periods when the average precipitation is low for the global land masses, it is high for the ocean regions, and vice versa. This is easily explained if air pressure is taken into account. High temperatures accompany the dry periods of the continental areas, which means that in middle and higher latitudes the air pressure over oceanic areas is slightly lower than during the wet and cool periods of the continental areas. If, however, the air pressure over the ocean is relatively lower than on land during warm periods, which has been established in particular for the region of the North Atlantic Ocean on the one hand and Central Europe on the other, the transfer of wet ocean air to the continent is more difficult, whereas it is easier when the air pressure over the ocean is not quite as low, as is the case during cool periods.

Meteorological observations have been around for slightly more than a hundred years. Climate variations, however, can be traced back much further by other methods. It has been shown, for instance, that the precipitation fluctuations of the 19th century are quite poignantly reflected by the water level fluctuations of rivers. Moreover, the temperature variations accompanying the changes in precipitation affect the timing of the grape harvest in the wine regions of Europe as well as the freezing and melting of the rivers in Russia and Siberia. Regularly kept records about the harvesting of grapes in France go back to the year 1391 and those about the melting and freezing of Russia's rivers to the year 1556. Based on these data, I was able to trace the climate variations back over five centuries. In addition it was possible to establish the average time period of a climate variation; it is about 35 years, i.e., the length of time between the peaks of one wet and cool period and the next is about 35 years on the average. Occasionally, however, this time period varies in length so that periods may last 40 or even 45, and others 30 or only 25 years. This irregularity is not surprising: meteorological cycles never occur with mathematical precision. For example, the length of time which lies between the ten hottest days of one year and those of the next is not always the same depending on whether these hottest days fall into the month of June, July or August.

My research results on climate variability have been confirmed by a number of scientists. J. Hann was the first meteorologist among my contemporaries to do so; based on extremely thorough research methods involving a number of stations in the region of the Alps, he was able to provide strong evidence of climate variations for the alpine countries. Other scientists, like B. Kremser, are more reluctant. I would like to stress that in long-term mean values, as demonstrated above using ten-year averages, the climate variations are clearly evident for almost all continents. However, if individual years alone are considered, the random irregularities which the precipitation is subject to from year to year, partially obscure the large variations.

Climate variations affect human life in many ways. I have commented on this in detail in previous essays. Using long-term averages I have demonstrated in 1895 that agricultural crops in particular are influenced by climate variations, which Dr. Paul Schulz confirmed in his dissertation in 1907. This is shown in the following table. The figures up to 1890 quoted from my article in the Geographical Journal, Volume I are average values for the eight old Prussian provinces; those from 1891 onward (in brackets) are taken from Dr. Schulz' dissertation and refer to the provinces Brandenburg and Saxony. Yields are expressed in percentages of an average crop and their deviations from 100. For example, the figure −19 indicates that on the average the wheat crop of 1881–85 was 19% below average. The three bottom rows combine the years according to wet and dry periods.

Table 9.7. Variability of Crop Yields in Prussia

Five-year Period	Precipitation	Wheat	Rye	Barley	Oats
1846/50	−1	+7	+10	+4	−2
1851/55	+8	−2	−4	+3	+2
1856/60	−7	+2	+7	−10	−9
1861/65	−9	+2	+3	+7	+8
1866/70	+3	−2	−2	−3	−3
1871/75	−6	+2	−1	−1	−2
1876/80	+9	−11	−14	−9	−8
1881/85	+3	−19	−23	−19	−22
1886/90	+1	−13	−22	−18	−12
1891/95	−2	(+6)	(−15)	(−3)	(−2)
1896/1900	+2	(+28)	(−12)	(+16)	(+27)
1856/75	−5	+1	+2	−2	−2
1876/90	+4	−14	−20	−15	−14
1891/1900	±0	(+17)	(−16)	(+6)	(+13)

During the two fairly arid decades 1856 to 1875 the wheat, rye, barley, and oats yields in Prussia were relatively good; in contrast the wet 15 years 1876 to 1890 had poor yields—according to assessed average crop. Particularly the wheat, barley and oats yields increased under dryer weather conditions.

Also grain prices reflect the impact of climate variability. These prices can be traced back further than data about the outcome of crops. For Germany, e.g., according to records of wheat prices for München, it clearly indicates that the wet periods of the 18th Century and in the first half of the 19th Century show high prices and the dry periods low prices. The same goes for England and France. Crops in these regions suffer more likely because of too much rain than of too much dryness. The reverse is the case in Russia. Here arid years are characterised by high prices, wet periods by low prices. However, the impact of climatic variability on wheat prices in European regions near the ocean is less noticeable since the middle of the past century. The outcome of the home crop is no longer the main guideline for pricing but rather the one of the world's major grain producers: Russia and the United States.

In Russia and the United States good harvests occur during wet periods; on the other hand, poor harvests occur in the oceanic regions like England, especially in Western Europe, and in most of Germany. Consequently, during wet periods in oceanic regions an increased demand for imports is created, whereas increased export opportunity of continental [inner European] grain suppliers occurs as well. Wet periods stimulate an international wheat trade. As a result, this triggers price reductions in regions with already bad harvests, unless prevented by protective duties. The result is a severe distress in agriculture because in addition to poor harvests, prices are low. This situation becomes different during arid periods. In oceanic regions, harvests are better than before, and continental grain suppliers suffer losses because of dryness. Therefore, demand of the first is lower and export capability of the latter has also diminished. An arid period will therefore coincide with a period of diminished international grain trade, and pricing in oceanic regions will respond more positively to their own harvest conditions.

Such an impact of climate variability may not fully affect market conditions. Political and economic events naturally also play a major role. Nonetheless, our reports show that the influence of climate variability on grain prices cannot be cancelled out completely by human interference.

Finally, I would like to raise the question if weather predictions for individual future years are possibly based on climate variability. I would like to point out, and as undoubtedly seen in the above, this is not feasible. From the past climate variability curve range we can only conclude that a certain future long-term period (10 to 15 years) could be wetter or dryer on the average. Within a wet period more wet years than dry years will occur, but dry years will not be lacking completely as shown in Table 9.2. On the other hand, wet years will also occasionally occur during arid periods. Based on climate variability for single years, it is not possible to predict weather conditions for agriculture under such circumstances. If we look at the climate variability pattern of the past 200 years, the conclusion that this will also occur accordingly in future cannot be fully ignored. Around 1880, land masses on earth experienced a wet period. Precipitation has more or less decreased ever since. Based on the average length of climate variability, the centred maximum of the next arid period should be around 1900. However, an exact prediction of the situation will only be possible when precipitation observations of additional ten years are available. Still, I believe to be able to predict that we are approaching a wet period again.

Chapter 10

Climate Variability and Mass Migration*

We may look upon today's economy as a huge mechanism. Multiple factors interlink like wheels of a clockwork. They all have an impact on the end result and through probability studies alone can the significance of one or the other factor be isolated. It is obvious and quite evident that the natural environment on earth affects the development of human life and culture. Yet one generally tends to attribute changes in the economic conditions of large or small regions to man-made causes. Nature's factors are assumed to be constant, and only because man's acquired knowledge changes over time is their effect different at different times. Overall this is quite correct—but not always. In some instances natural phenomena can be contributing factors of considerable significance in economic processes. Their significance is masked by the fact that the mechanism of economic life is so very complicated. The effect of these natural causes is enhanced by political and social factors. Permit me to concentrate on such combined influences at this point.

One of the most striking aspects in the history of the XIX Century is the mass immigration from Europe to the New World. From 1805 to 1911, no less than 28 1/2 million people left old Europe and emigrated to the United States of America. The rapid settlement in vast regions of the United States was the immediate consequence of this emigration from Europe. The boundary of the "oecumene", as F. Ratzel called the populated land, advanced further west each year. The massive number of people mobilised during this time is certainly not smaller but considerably larger than the number of people who mass migrated in the early Middle Ages. Emigration to America is the largest mass migration of all times. Economic conditions in some parts of Europe had suffered because of overpopulation and were the reason for emigration, while especially the sparsely populated western part of the United States, where as a result wages were high, was a strong incentive for immigration. Though these undoubtedly are the key elements setting this

* *Klimaschwankungen und Völkerwanderungen*, Vortrag Kaiserliche Akademie der Wissenschaften, Wien 1912.

modern mass migration in motion, there are a number of aspects indicating that other than human factors were involved, sometimes precipitating migration, sometimes slowing it down.

Economic conditions in the United States as in major parts of Europe are predominantly dependent on harvest conditions. Years of poor harvests, particularly when suffered in succession, stimulate emigration from the affected European regions. If at the same time in another area harvests are good and the economy is prospering, these factors are an additional incentive to leave the country. Regarding the turnout of crops, there is in fact this polarity between Western and Central Europe on the one hand and the United States on the other.

There would be no agriculture without water, and then again no agriculture with too much water! This dependency is obvious when one looks at the extent of agricultural land around the globe and examines the causes of poor harvests. In the regions with little rainfall, droughts and poor harvests go hand in hand; in regions with excessive rainfall, poor harvests occur mainly during wet years. It is of course not the absolute amount of rain that makes the difference, but the ratio between rain and evaporation. The same amount of precipitation that drowns the crop in a cold climate may barely be sufficient for growing grain in a hot climate. In contrast, low rainfall that in a hot country may cause drought due to high evaporation is frequently entirely sufficient in a cool climate. In Europe, conditions in the wet countries near the North Atlantic Ocean, as for instance Norway, Denmark, Ireland, and Great Britain in particular, but also Sweden and Central Europe are the exact opposite to those in the dry interior of the continent. Southern Russia on the one hand and Great Britain on the other are the extremes in this regard. Drought is almost always the cause of crop failure in Southern Russia, while in England the numerous crop failures at the end of the thirties, during the forties, at the beginning of the fifties, and again in the seventies and eighties occur in excessively wet years. Southern Europe and most of the tropics, at least as far as they are grain-growing countries, have similar conditions as those prevailing in Southern Russia. Years of famine in India coincide with dry years. The same pattern is clearly evident in the United States of North America where crops increase and decrease with the amount of precipitation. As a result we may conclude that the simultaneous occurrence of a number of wet years in Western and Central Europe and in the United States of North America must affect the crops in these areas in a completely different way. Western and Central Europe will suffer from poor harvests, regions of the United States will have good ones. These are the conditions which encourage emigration from central and western Europe to the United States. The situation is different, when the two regions simultaneously go through a number of dry years. In that case, poor harvests make it less attractive to immigrate to the United States while good harvests at home persuade people to stay.

If wet and dry years would change purely by chance in place and time, poor or good harvests could not succeed one another over a number of years; they would show the same random pattern. That is in fact not the case. My own research all around the globe has shown that wet and dry years often occur in groups, so that climatic conditions are not completely constant. The climate varies instead, oscillating around a middle range. My results have been confirmed by various parties, above all by Julius Hann.

Climate variability consists of multi-year fluctuations of temperature, air pressure and rainfall, all of which occur simultaneously all around the globe. Temperature is the one element on which all the others depend. Temperature variations are common to nearly all of the countries on earth. All of them go through cold periods and warm periods at the same time. For example, the five-year periods of 1806–20 have on average been too cold all over the globe, those of 1821–35 too warm, 1836–50 again too cold, 1851–75 too warm, 1876–90 again slightly too cold.

Temperature variations have an impact on the distribution of atmospheric air pressure. In Europe as well as in North America the transfer of moist oceanic air from ocean to land seems more difficult during warm periods, and easier during cold periods. Consequently, on both continents warm periods tend to be, at the same time, dry periods, whereas cold periods show a higher amount of rainfall. The cold periods around 1815, 1850 and 1880 for instance were wet periods in both Europe and North America, while the warm periods around 1830 and 1860 show very little precipitation. Rainfall has slightly decreased since 1890; this is clearly evident in the United States of America but also in Central and Western Europe.

Over the past two centuries, the centres of cold and, for the land masses, wet periods fall into the years 1705, 1740, 1775, 1815, 1850, and 1880; centres of warm and, for the land masses, dry periods into the years 1720, 1760, 1790, 1830, and 1860. Hence, one climate variation takes an average of approximately 35 years from one maximum to the other. The centre of the latest warm and dry period should occur around 1900. This however is a preliminary estimate only because the climatological data from all over the world have not yet been interpreted in this regard.

This climate variability clearly affects crops even if in a different sense for different areas. Some examples may demonstrate this more clearly.

The variability of crop yields is very pronounced in Prussia. The top curve in the following graphic (Figure 10.1) shows the rainfall fluctuations in periods of 5 years each.[450]

[450] Smoothed according to the formula $(a+2b+c):4$ = smoothed value of b.

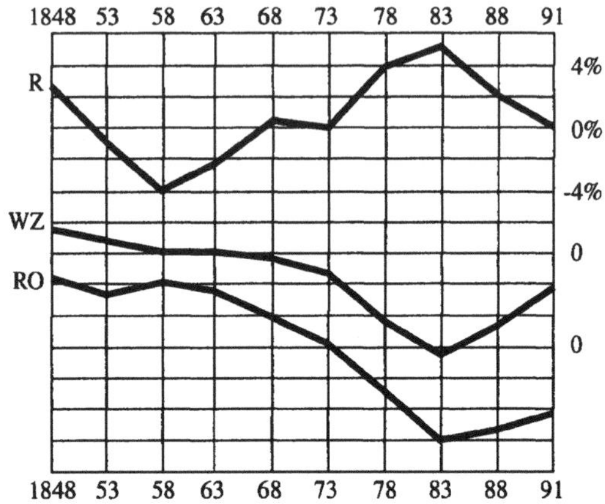

Figure 10.1. **Variability of rainfall in relation to the grain crop in Prussia.** The grain crop (*WZ* = wheat crop, *RO* = rye crop) is in percentages of an average crop, i.e., in deviation from a multi-year mean (1 increment = 5% deviation); rainfall (*R*) is also in deviations (%) from the multi-year mean (1 increment = 4%).

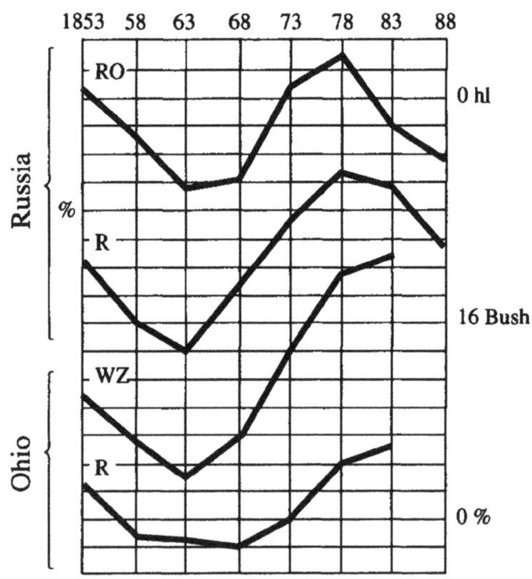

Figure 10.2. **Variability of rainfall and of the increase in rye export from Russia and of wheat crop in Ohio.** One scale mark equals 2% for the rainfall (*R*), 1.5 million hectolitres for the increase in the export of rye from Russia (*RO*) and 0.8 bushel per acre for the wheat crop in Ohio (*WZ*). The variation in the increase of the rye export from Russia was determined as follows: A straight line was drawn, according to the method of least squares, through the recorded quantities exported from 1851 to 1890, and for each five-year period, the differences between this figure and those actually observed were then depicted.

Around 1860 rainfall was low, around 1880 it reached a maximum. Because in Prussia harvests suffer more frequently from too much rain than from too little, yields were high during the dry period around 1860, and much lower during the wet period in the early eighties. The curves of the crop yields[451] are the exact mirror image of the rainfall curve. The same pattern applies to the crop situation in the entire area of Western Europe and also to Great Britain in particular. The pattern is reversed, however, in the United States of North America and in Russia, as demonstrated in Figure 10.2. Again, in the bottom curve for Ohio we recognize a minimum of rain-

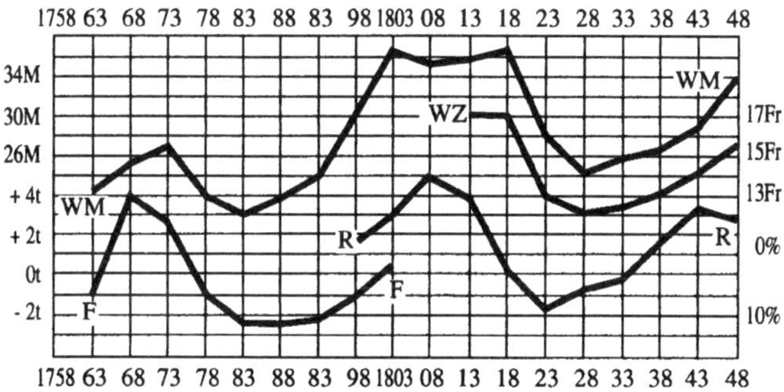

Figure 10.5. **Variability of rainfall and wheat prices in Southern Germany and Switzerland.** *WM* = wheat price, in Marks, of a Bavarian bushel of wheat in München (1 scale mark = 2 Marks): *WZ* = price, in Swiss Francs, of 50 kilograms of wheat in Zürich (1 = 1 SF); *R* = rainfall in southern Germany (1 = 3% deviation from the multi-year mean); *E* = date of the vine harvest in southern Germany and Switzerland which varies proportionately with the rainfall, being late in moist and early in dry years (1 = 1 day; the multi-year mean is marked *Ot*).

fall around the year 1860 and a maximum at the beginning of the eighties, as was the case in Prussia. But since the U.S. crops are mainly affected by droughts, the dry period around 1860 results in poor harvests, the wet period at the beginning of the eighties in good ones. For Russia harvest statistics dating further back in time are missing. I have used the increase in rye export from Russia as an indicator of harvest conditions. The bottom curve again reflects variability of precipitation with a minimum around 1860 and a maximum around 1880. The obvious parallels between the two curves is so effective that the crop yields fall and rise with the rain.

Harvest statistics do not go back very far. But at a time when there was no world trade in grain, the outcome of a country's harvest also determined grain prices at home. To a certain extent the grain price is, therefore, an indicator for the quality of the crop yield in a region. In times of poor harvests grain is expensive, in times of good harvest it is cheap. Figure 10.3 shows the close correlation between wheat prices and harvest conditions and also the rainfall in southern Germany and Switzerland. Again slightly

[451] Smoothed as before.

smoothed five-year averages are used as a basis. Rainfall shows pronounced maxima around 1770, 1810, and 1850, minima around 1785, 1825, and 1860. Wheat prices in München and Zürich fluctuate accordingly, as can be read from the curves. This obvious correlation is missing from the middle of the last century on because Switzerland started to import grain from Russia and the Balkan states. Grain prices in Zürich were no longer determined by harvest conditions at home, but by the conditions of the harvest in those continental countries.

England has data about rainfall and wheat prices dating back to the beginning of the 18th Century. Figure 11.4 illustrates this.[452] During the wet periods around 1713 and 1768 up to 1773 grain prices were high and harvests poor; in the dry period in between, however, the price was low. The similarities of the curves comes as a surprise, yet it disappears completely towards the end of the 18th Century. At this time political events, such as the implementation of the continental blockade are responsible. Later the world price for grain evolves, a criterion even for England's grain prices, which is gradually abolishing grain duties.

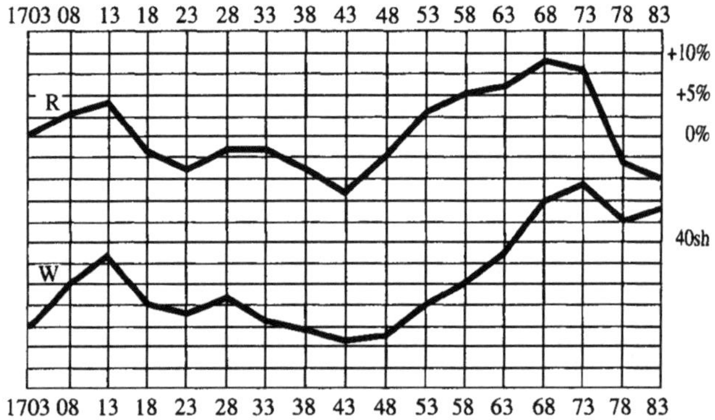

Figure 10.6. **Variability of rainfall and wheat prices in England.** Rainfall (*R*) is indicated in deviations from the multi-year mean (percentages) (1 increment = 2.5%), the annual average wheat price (*W*) in Shillings per Imperial Quarter (1 increment = 2 sh.).

In summary, we can state that as a result of the wet years around 1815, 1850 and 1880, harvests in the oceanic regions of Europe, including central Europe, were poor, whereas in the United States and Russia they were good. The situation was reversed during the dry periods around 1830, 1860, and at the end of the century. It is important to note that we are not dealing with single occurrences of good or poor harvests but rather with a number of harvests of the same quality grouping around those years.

[452] Again smoothed as before.

That it is mainly the farming population which suffers from poor harvests is obvious. On the other hand, it is a known fact that the large number of emigrants to the United States came from the farming population of Central and Western Europe. Therefore, the question arises whether perhaps climatic variability by way of varying crop yields is also reflected in the emigration figures to the United States.

In fact U.S. immigration statistics are very detailed and date back a long time. The following graph (Figure 10.5) gives an overview. The top curve shows the rainfall rate in the United States, the second one the rainfall in Western Europe. Rainfall maxima occur around 1850 and 1880, the minima around 1860. The next curve shows the U.S. immigration figures.

The first thing that catches the eye is the enormous increase in the number of immigrants. Yet it is not a steady increase. Periods of rapid increase alternate with times when immigration figures remain the same or even decline. In the years 1821–35 immigration was relatively low, after which it began to increase rapidly. Numbers continued to grow until the middle of the fifties. It coincides with the weather period which brought a lot of precipitation to both Europe and in the United States, resulting in poor harvests in Western Europe and in good ones in America. From approximately 1855 until the middle of the sixties immigration figures fall by more than half. This coincides with the arid period leading to good harvests in oceanic Europe and bad harvests in the United States. The American Civil War was certainly a contributing factor at the beginning of the sixties, but the flow of immigrants had already started to ebb 6 years earlier. As a consequence of the wet period peaking at the beginning of the 80s, immigration rapidly increases peaking at more than 400,000 people per year between 1880 and 1893 and reaching a maximum of 780,000 people in 1882. This is followed by a period of low immigration from 1894–1900, which then however is followed again by a period of rapid increase, leading to an unprecedented high of 1,285,000 immigrants in 1907. This last rapid increase coincides with an arid weather period and appears to contradict our claims. This, however, is not at all the case as we shall see when we take a closer look at the demographic factor of the immigration movement.

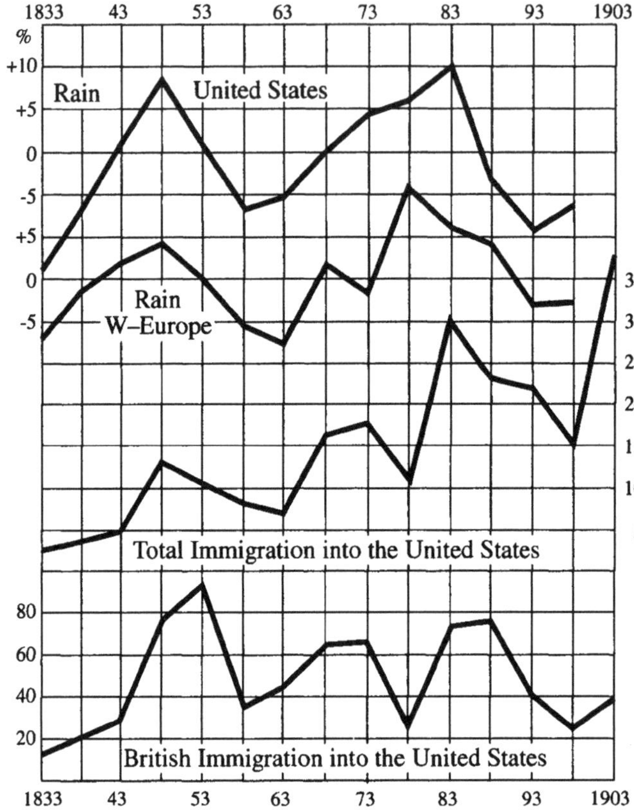

Figure 10.7. **Variability of rainfall in the United States and Western Europe and the total number of immigrants to the U.S. and from Britain.** The curves are based on five-year totals (non-smoothed). The number of immigrants is given in 10,000s, rainfall in deviations (percentages) from the multi-year mean.

Until the end of the past century the majority of emigrants came from Great Britain and Ireland as well as the German Empire, that is from the oceanic regions of Europe. It is a striking fact that emigration from these oceanic regions increases during wet periods and decreases considerably during dry periods. It is high in the wet period around 1850 and 1880, and considerably lower in the dry period around 1860. If, for example, the data about emigrants from the German Empire to America and about the precipitation in the German Empire are presented in a graph the resulting curves show a striking similarity (compare Figure 10.6). Even during the secondary maximum of precipitation in the second half of the sixties emigration does increase rapidly. After a slight delay the secondary minimum of rainfall during the first half of the seventies is followed by a minimum number of emigrants. The number of extremely wet years in the late seventies and early eighties which are particularly pronounced in the German Empire, resulted in the highest number of emigrants to America which this country had ever reached. Several times the emigration curve is lagging behind the rainfall curve by five years. This is not surprising, because of the logical delay between cause and effect.

For Great Britain the situation is similar (Figure 10.5, bottom curve). Again emigration to the United States increases and decreases with the rainfall. And again, an occasional delay of a few years can be observed.

A closer look at each year reveals that emigration to America from Great Britain reached a maximum in 1851, from Germany in 1854, and a second time in 1888 and 1881 respectively. During the past century the largest number of immigrants to the United States was reached in the years 1854 and 1881. All these years fall into wet periods.

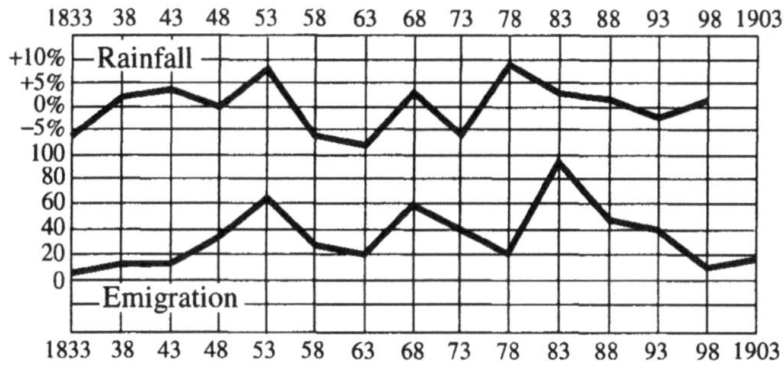

Figure 10.8. **Variations of rainfall and emigration from the German Empire to the United States.** The curves are based on five-year totals (non-smoothed). Rainfall is the deviation (percentages) from the multi-year mean. Emigration to the United States is in 10,000s.

As we have seen, starting in 1900 the number of immigrants to the United States rises sharply. Emigration, however, from oceanic Europe does not reflect this development. The rapid increase is instead caused by a factor that heretofore had played only a minor role: massive emigration from Russia, Galicia, Hungary, and Italy. This conforms again with the climate variations. The same dry period that produced good harvests in the oceanic regions of Europe caused poor harvests in Russia. In addition, the sad state of Russian politics, the persecution of Jews, the Japanese war, and the revolution encouraged emigration. In contrast, emigration from Germany to the United States remained very low during that entire dry period beginning in 1893 and barely reached half of the numbers that had emigrated in the eighties and at the beginning of the nineties. No doubt, the enormous industrial boom had something to do with it, keeping large numbers of the

German labour force in the country which otherwise might have emigrated. Nonetheless, the arid period and its good harvests should not be overlooked.

In the same way as the impact of climate variations manifests itself in the massive emigration from Europe to the United States, it also does so in the regional settlement of the United States.

If we compare the census data of several years regarding the distribution of the population, we find a prevailing pattern. The majority of the population is slowly moving westward. This trend led to the rapid settlement of the far West of the United States, especially in the seventies and eighties of the past centuries. Population in the states West of the Mississippi increased by 79% from 1870 to 1880, then from 1880 to 1890 by 71%. These are average values; in some of the states this increase was many times higher. In contrast, from 1890 to 1900 the population increased by not more than 32%. These western states are obviously going through a decline at that time which the Atlantic and Central States are not experiencing.

The reasons for this decline lies, no doubt, in the fact that the Western States in the short period from 1870 to 1890 had increased their population to capacity through immigration. What has remained unnoticed and, in my opinion, is certainly not a coincidence, is the fact that this settlement rush in the far Western United States occurred during a period of high precipitation. Statistics about precipitation in the United States show a series of predominantly wet years in the interior of the United States at the beginning of the seventies until the mid eighties. This was followed by a very severe dry period in the nineties. I am inclined to attribute the slowdown of the population growth in the Western States from 1890 to 1900 at least in part to that arid period.

Moreover, with the beginning of the series of dry years, people in the Western States were forced to abandon those regions that had been cultivated during the previous wet period. A sufficient water supply was available for agricultural use at that time, which now during the dry period of the nineties of the 19th Century and at the beginning of the 20th Century was no longer available. Density of population charts as part of the census data shows that from 1890–1900 the settled area of the western states shrank considerably. On the basis that areas with at least two people per Imperial square mile are defined as settlements, New Mexico has a decrease of around 61% from 1890 to 1900, Nevada of 86%, Idaho of 58%, Kansas of 12%, etc. North Dakota is an exception. The total populated area of all 13 western states combined went down by 15%, or about 25,000 sq. km from 1890 to 1900. The area that people have abandoned is about four-fifths of the size of Austria.

What causes this population backflow from those affected Western regions? Again, we cannot blame a single event but rather a complex system of several events. Since the settlement declined mainly in already sparsely populated areas, a reduction in the agriculturally used land had to follow. What causes such a depletion? Reasons vary from state to state. In states where farming is closely connected to the mining industry, serving the demands of the local mining communities, a reduction in mining naturally results in a reduction in farming. That is indeed the case in Nevada and perhaps in some parts of Colorado. The situation is different in states where farming plays a major role, such as Kansas, Wyoming and Nebraska. Here the decline in farming can undoubtedly be attributed to the dry period that is evident since the second half of the eighties and particularly since the beginning of the nineties. Abandoned farms are proof of this decline in these states. Yet the overall population has not decreased. People moved from the affected farmland to more fertile areas or into the cities where industry and trade were booming. The number of small country towns, which had mainly developed as centres for these farming communities, has now gone down considerably. If the foremost agricultural state of North Dakota continues to show strong population increases from 1890 to 1900 and the size of South Dakota's population does at least stay the same, it can be attributed to the fact that these states were settled relatively late and were still able to accept additional settlers at the onset of the dry period. In Nebraska and Kansas as well as in all the other states with similar conditions vast areas had been converted into farmland during the wet period, which in the following arid period did not produce. These are examples of the correlation between climate variations and the varying sizes of agriculturally used land. Utah is another example. This state experienced a rapid increase in population during the seventies. The size of the farmland grew enormously while, at the same time, the water level of the Great Salt Lake rose. The Mormons attributed the rising lake level to the increase of agricultural land in the area and assumed this was the cause of higher precipitation. But the arid period starting around the middle of the eighties caused the lake's water level to fall again and proved that the fluctuations of rainfall were unrelated to man's cultivation efforts. Instead they occur in connection with the global phenomenon of climate variability.

Let us repeat! In the oceanic regions of Europe, poor harvests as a consequence of wet periods led to a large wave of emigration. At the same time, these same wet periods enable the cultivation of vast areas for agriculture in the continental regions of the United States. At this point the dry period sets in. Emigration becomes less attractive in Europe's oceanic regions and in the continental states of America people are driven back from large areas of previously cultivable land. Mass migration, which we see happening as a result of climate variations, is indeed an enormous phenomenon. Though climate change is not the only cause for mobilising this stream of people, its strong influence is apparent in the ebb and flow of the migration movement. Political and economic crises affecting industry, but which also to some degree depend on harvest conditions, to some degree cannot disguise this influence. This is always clearly manifested in the long-term averages.

The enormous phenomenon of mass emigration from the Old World to the New World is generally not called a mass migration, although the number of people mobilised in the course of this event was much larger than those in history's many mass migrations from Central Asia to Europe or China. No doubt, the process of this modern mass migration is more complex and its causes are more difficult to isolate. But we have the great advantage that we can trace it statistically and that in our inquiry into the influence of climate variations we can rely on statistics about the amount of precipitation as well as about harvest yields and grain prices. Such data is lacking for an inquiry into the causes of historical mass migrations. For instance, we know nothing about the conditions in the home countries of those hordes of riders from inner Asia, and are unable to identify the cause for their migration. We can only assume that again climatic variations played a role. Aurel Stein has introduced us to the remnants of buildings in the middle of the sands of the Tarim basin where nowadays the desert makes human settlements impossible. Only a small part of his extensive research material found in these desert cities has so far been published, but already the outline of those regions' history of settlement is coming to light. It has been found that a number of settlements from the beginning of our time have disappeared by the end of the 3rd Century. It cannot be coincidental that only a few decades later a massive number of migrating people reached Central and Western Europe. This does not mean that those who reached Europe were the occupants of the settlements now in ruins. Rather the demise of these settlements seems to be indicative of deteriorating climatic conditions, which made the steppe of central Asia uninhabitable.

Climate deterioration in Asia also preceded the invasion of the Mongols. We have evidence of this climate change: According to buildings along its shoreline whose age could be determined, we know that the Caspian Sea had reached its lowest water level in the 12th Century, which was unprecedented and was never again reached thereafter. A low water level of this magnitude could only have been triggered by a long arid period. This aridity cannot have been a local occurrence but must have affected the entire Volga region as well as large regions of central Asia. I am inclined to see this drought as the incentive for the invasion of Europe, India and China by the Mongolian hordes. As a consequence, many aspects support the theory that even the mass migrations of the past were caused by strong climate variations. Huntington even speaks of the pulse of Asia. In the rhythmic intervals of a pulse beat, Asia, from time to time and in accordance with the fluctuations of the climate, releases waves of people into the peripheral regions of the Old World. They surge threateningly against the borders of Europe's old cultures; some disappear in these cultures, others are rejected. It should not be overlooked that political conditions play a role here as well which, however, are again linked to economic conditions. Undoubtedly, the migratory movements of an underdeveloped nation are much more strongly affected by climate change than those of a highly developed nation, reflecting the rhythmic variations of the climate in the ebb and flow of their migrating people.[453]

[453] Compare the following publications:

- Brückner, Ed., *Klimaschwankungen seit 1700* [*Climate change since 1700*], Wien, 1890. [Partly reprinted in Chapter 4 of this anthology]
- *Über die praktische Bedeutung der Klimaschwankungen* [*The practical implications of climate change*], Compte rendu du Vme Congr. inter. des Sc. geogr. Bern, 1892, p. 618 ff.
- *Rußlands Zukunft als Getreidelieferant* [*Russia's future as grain exporter*], Insert of the Münchener Allgemeine Zeitung of Nov 19, 1894.
- *Der Einfluß der Klimaschwankungen auf die Ernteerträge und Getreidepreise in Europa* [*Influence of Climate Variability on Harvest and Grain Prices in Europe*] Geographische Zeitschrift , I, 1895, p. 39 ff. [reprinted in Chapter 6]
- *Zur Frage der 35jährigen Klimaschwankungen* [*An Inquiry About the 35-Year-Period Climatic Variations*], Petermann's Mittheilungen, 1902, p. 173 ff. [reprinted in Chapter 8]
- *Klimaschwankungen und Völkerwanderungen im XIX. Jahrhundert.* [*Climate variations and mass migration in the XIX century*] Internationale Wochenschrift für Wissenschaft, Kunst und Technik, 1910, p. March.

Chapter 11

The Settlement of the United States as Controlled by Climate and Climate Oscillations*

In the two months of our Transcontinental Excursion, arranged in so hospitable a manner by the American Geographical Society, each of the European members visiting America for the first time has without doubt obtained impressions of the greatest value and so numerous that a long time would be necessary to work them out. I think one of the most important impressions was the observation of the enormous differences in the climatic conditions of the various parts of the United States. We started from the humid east, where woods and meadows cover the ground and an abundant agriculture is possible. A ride of one night brought us from Duluth with its great woods to the borders of the prairies in the neighborhood of Fargo. Nevertheless rich crops of wheat are harvested here. As the rainfall diminishes to the west the grain fields gradually disappear, and in the "bad lands" of the Little Missouri we were in a half-desert region, where vegetation requires a regular water supply, in so far as it is found only along the rivers. Much of the western region is of this character and may be called half-desert. Regions that are fully desert are rare. We have seen such regions only in the Great Basin on the bottom of the extinct Lakes Lahontan and Bonneville. Fortunately only the plains and the lower mountains show these features. The higher mountains, on the other hand, are able to condense the vapor of the air and therefore enjoy a greater rainfall.

Nevertheless the half-desert regions today are to some extent inhabited and now yield good crops. Man has by his skill and by his work in many cases made out of a desert a paradise. In the neighborhood of Coulee City on the Columbia Plateau we saw the admirable results of dry farming. By an ingenious choice of the sequence of grain and by a not less ingenious use of the natural water supply furnished by rain, crops are produced, not very rich, but giving a good return to the farmer who practices extensive agriculture.

* Presented on October 18, 1912, at the scientific meetings held after the return of the Transcontinental Excursion to New York.

Much greater success attends his efforts where it is possible to utilize, in the streams that flow down from them, the water fallen as rain in the mountains, for the irrigation of the arid and semi-arid plains. Where once was a desert we now see rich orchards. During the Excursion we saw extensive oases of this nature, which had developed in a few years in the Yakima valley, Washington, at the Dalles on the Columbia River, at Salt Lake City, at Grand Junction, Colorado, and at Phoenix, Arizona. Indeed fruit trees find here the best conditions that can be imagined. In humid regions the water supply by rainfall decreases the temperature of the air by diminishing the amount of radiation from the sun because of clouds. In the irrigated regions of the West that is not so: the fruit trees receive their water from beneath without an interruption of the radiation of the sun.

Wonderful results are obtained by the co-operation of governmental and private work, the government studying through its Geological Survey and its Reclamation Service the available water supply and making it possible of utilization by building canals, reservoirs, etc. Today the water available for irrigation is not yet exhausted. In the Yakima valley the government will provide the water supply for 34,000 acres beyond the area now under cultivation. In the neighborhood of Phoenix, 160,000 acres are under irrigation today. But the area might be increased by the water supply already available to 230,000 acres, of which 190,000 can be irrigated directly by surface supply, regulated by the Roosevelt Dam, and 40,000 by pumping ground water.

Surely today an increase in population by using to a greater extent the water available for irrigation is possible in some regions, but not indefinitely. The available water is controlled by climate, and, therefore, there is a limit beyond which man cannot go. But still more. This limit would be constant only if rainfall and the other meteorological conditions that control the water supply, or—in other words—only if climate were constant. Now we have proofs that climate is not constant but that there are climatic oscillations of importance that affect to a marked degree rainfall and temperature.

Some years ago I showed that such oscillations of climate are observable over the whole world. They consist, on the continents, of an alternation of relatively warm and dry with relatively cool and humid periods. Taking only the last century, we have series of humid and cold years about 1815, 1850, and 1885, series of warm and dry years about 1830, 1865, and 1900.

In an extensive paper published in 1890,[454] I have worked out the meteorological observations of about 800 stations, distributed over the whole

[454] The German titles are as follows: Ed. Brückner: *Klimaschwankungen seit 1700 nebst Beobachtungen über die Klimaschwankungen der Diluvialzeit* [*Climate change since 1700*

world. I proved that climatic oscillations are simultaneous on the whole earth, oscillations of temperature being the initial cause. The oscillations of temperature cause oscillations of the distribution of air pressure. During a warmer period, the pressure is distributed in such a manner that the overflow of air from the Ocean to the continents diminishes, while during a cooler period, on the contrary, it increases. Therefore during a cooler period the continents receive more humid air and more rain than during a warmer period. Coolness and humidity coincide on the continents, while the oceans and also some of the coastal regions of the continents receive less rainfall during the cool periods.

It is not possible to reproduce here the detailed tables, given in the book mentioned above, that illustrate climatic oscillations up to 1885. I will only give some examples of the oscillations of rainfall.

They are shown by the curves of Figure 1. They give the rainfall of various meteorological stations that represent different regions of the world: Brussels represents western Europe; Bremen central Europe; Nertschinsk, the central part of eastern Siberia; Nikolaevsk on the Amur, the Pacific coast of Siberia; and Madras, India. European Russia is represented by the average of a great number of stations in the basin of the Don River. The United States is represented by three curves, one calculated from the observations of a great number of stations in the upper Ohio Valley, the second calculated in the same manner for the central part of the Mississippi Valley, and the third for New England. I do not give the data for each year, because there are many minor irregularities due to local thunder storms, local heavy rains, etc. But for each year I use the average of the ten years of which it represents the

---

etc.; Chapter 4 of this anthology], Wien 1890. 324 pp.; see also the following papers treating on climatic oscillations:

Ed. Brückner: *Über die Bedeutung der Klimaschwankungen für das praktische Leben* [*The practical implications of climate change*], Compte rendu du Vme Congr. intern. des Sc. geogr., Bern, 1892, pp. 616–623

– *Rußlands Zukunft als Getreidelieferant* [*Russia's future as grain exporter*], Beilage zur Münchener Allgemeinen Zeitung Nov. 19, 1894.

– *Der Einfluss der Klimaschwankungen auf die Ernteerträge und Getreidepreise in Europa* [*Influence of Climate Variability on Harvest and Grain Prices in Europe*], Geographische Zeitschrift, Vol. I, 1895, pp. 39–51, 100–108.

– *Zur Frage der 35jährigen Klimaschwankungen* [*An Inquiry About the 35-Year-Period Climatic Variations;* reprinted in Chapter 8], Petermann's Mittheilungen, 1902, pp. 173-178.

– *Klimaschwankungen und Völkerwanderungen im XIX. Jahrhundert* [*Climate variations and mass migration in the XIX century*], Internationale Wochenschrift für Wissenschaft, Kunst und Technik, 1901, March 5.

– *Klimaschwankungen und Völkerwanderungen.* [*Climate variations and mass migration;* reprinted as Chapter 10], Vortrag gehalten in der feierlichen Sitzung der K. Akademie der Wissenschaften am 31. Mai 1912. Wien, 1912, 24 pp.

center, viz.: for the year 1835, the average of the years 1831–40; for 1836, the average of 1832–41; for 1837, the average of 1833–42, etc.

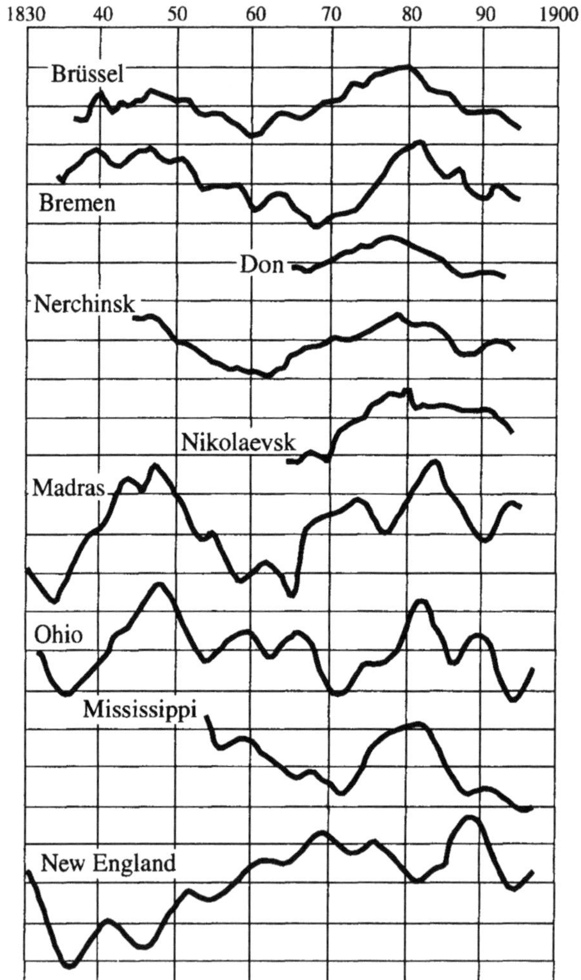

Figure 11.1. Oscillation of Rainfall in Various Regions of the World. Horizontal divisions = 100 mm rainfall

It will readily be seen that in all parts of the world represented in the diagrams, with the single exception of New England, there is a maximum of rainfall about 1845–50, a minimum about 1860–70, another maximum about 1880, followed by a decrease of rain until the end of the last century. The oscillations are rather great, the difference between the maximum and the minimum amounting, in the most continental regions of the earth, to 50 percent and more. Only the curve representing the fluctuation of rainfall in New England has another rhythm, having maxima in 1869 and in 1889; but we must keep in mind that the climate here is quite oceanic, and therefore the oscillations of rainfall follow the oscillations of rainfall on the Ocean, which are the converse of the oscillations on the continents, as stated above.

This raises the question: Are these oscillations of climate felt in the history of the settlement of the United States?

One of the most characteristic features in the distribution of the population in the United States is the displacement of the center of population in a westerly direction. The censuses of the different decades show this very clearly. It is caused by the rapid settlement of the Far West, which took place principally in the 70s and 80s of the last century. The following table shows this. It concerns the states west of the first tier of states adjoining the Mississippi on the west. The increase in population is given in percentage of the population at the beginning of each decade. For example, in Montana the increase of population from 1870 to 1880 was 91% of the population in 1870; from 1880 to 1890, 238% of the population in 1880, etc.

Table 11.1. Increase of Population (%) in the United States from 1870 to 1900

	1870–80	1880–90	1890–1900
Montana	91	238	70
Idaho	117	118	83
Wyoming	128	192	48
Nevada	46	–26	–11
Utah	66	44	31
Colorado	388	112	31
Arizona	319	47	39
New Mexico	30	29	22
North Dakota	1435	395	67
South Dakota	734	235	15
Nebraska	267	134	0
Kansas	173	43	3
Texas	94	40	36
North Atlantic Division	18	20	21
South Atlantic Division	30	17	18
North Central Division	34	29	18
South Central Division	39	23	26
Western Division	79	71	32

The table demonstrates very clearly the rapid settlement of the Far West and the enormous growth of the population from 1870 to 1880 and also from 1880 to 1890. But afterwards the growth became slow.

In the table are also given the corresponding data for the great divisions of the United States used in the census reports. It is very easy to see that the enormous growth from 1870–90 is confined to the western states and that the retardation of the growth since 1890 is also marked only in the western states. This phenomenon is without doubt due in large measure to the fact that the population in the western states had grown by immigration from 1870–90 so much as to fill the region completely or nearly completely. But one point seems to me to have been overlooked: the rapid settlement of the Far West from 1870–90 coincided with the period of great humidity. The rainfall in this period was relatively great; the years in which it exceeded the average were especially numerous, more so than before. Since 1890 there has come a dry period. I suppose that the desiccation of the climate since 1890 must also be taken into account if we wish to explain the small increase of the population since 1890, the drier weather resulting in poor crops.

Again, because of the dry period that began in 1890 the population of the Far West was obliged to abandon large areas that were settled in the preceding wet period. In the wet period there was a sufficient water supply for the farmer—not so at the end of the last and at the beginning of our century, with its smaller rainfall. The diminution of the inhabited area is very clearly to be seen on the maps of density of population published in the United States census reports. I have measured upon the map giving the density of population for 1890 the area that had a population of at least 2 per square mile. This measurement I repeated upon the map for 1900. The following table gives the comparison of the results. The states with an increase in the inhabited area are designated plus; those with a decrease, minus. Because the inhabited region is of different size in the different states I have also calculated the decrease in the inhabited area in percentage of the total area of each state.

Most of the states show a very conspicuous decrease in the area with a population of 2 or more per square mile or, in other words, in the inhabited area. If we take all thirteen states together, we find a total decrease in the inhabited area of 242,800 square kilometers (100,000 square miles in round numbers), or of 15% of the inhabited area of 1890. That is a great deal and is very significant.

If we seek the reason for this remarkable phenomenon of the ebb of the population of the exposed parts of the Far West, we must bear in mind that it was especially the thinly populated regions that suffered the decrease. There is therefore no doubt that the recession of the population is accompanied by a decrease in the area used for agricultural purposes.

Table 11.2. Variation in the inhabited area in the Western States from 1890 to 1900

	Inhabited Area in Sq. Kilometers 1890	Variation 1890–1900 Sq. km	Per Cent
Montana	130,000	−15,000	−12
Idaho	101,900	−59,200	−58
Wyoming	61,300	−9,600	−16
Nevada	32,400	−27,900	−86
Utah	60,400	−2,100	−3
Colorado	199,100	−68,800	−35
Arizona	34,700	−5,800	−17
New Mexico	114,200	−69,800	−61
North Dakota	50,800	+38,500	+76
South Dakota	113,500	+1,100	+1
Nebraska	153,400	−6,100	−4
Kansas	186,000	−22,100	−12
Texas	368,800	+14,200	+4
Total	1,638,700	−242,800	−15

But what is the reason for the retreat of agriculture? Evidently the reason is somewhat different in the different states. In states like Nevada or a part of Colorado, where agriculture depends closely upon mining because it is especially for the use of the mining districts that grain is here cultivated—and that with great difficulty—the decrease in mining activities has also caused a decrease in agriculture. But in states where agriculture is of prime importance, such as Kansas, Wyoming, and Nebraska, I am inclined to think that the decrease in agriculture is due to the dry period that began in 1890. In these states, it is reported, one may often see abandoned farms as a sign of the retreat of agriculture. Nevertheless the population of the states as a whole has not diminished; it is only concentrated in the more favorable regions and in the towns, while the drier regions have been abandoned. One exception to the rule is seen in Dakota. Here no sign of a retreat of the population is apparent; on the contrary, here the inhabited area has increased from 1890 to 1900. To be sure, in 1890 this region was very far from being fully populated. In Kansas and Nebraska, on the contrary, wide areas had been put under the plough in the wet period which, in the following dry period, failed to yield crops. This is a very clear instance of who climatic oscillations control settlement near its border toward the desert.

Now we are in the beginning of a new humid period. The rainfall in the United States has been increasing during the past few years. About 1920 we may expect a maximum of humidity. For there can be no doubt that the oscillations of climate will continue, since they have been followed back to Europe over 700 years, each oscillation having from maximum to maximum a duration of about 35 years. Therefore it must not be forgotten that the irrigation works which are now under construction have better conditions than in the middle of a dry period. In the next dry period the rainfall will be less; therefore the flow of water and the water available for irrigation will also be diminished. For irrigation projects of long duration it is necessary to keep this in mind. They should be adapted to the water supply of dry periods. Therefore the climatic conditions of the dry periods will control the areas of permanent settlement. Outside of these there always will be a region fit for settlement in the wet period, but uninhabitable in the dry. Doubtless, in many districts that are now being irrigated, the limit imposed by the water supply of the dry period has not yet been reached. In other districts it may be already. Here the next dry period will first show that the settlements are controlled in a very high degree by the oscillations of climate.

In the case of Great Salt Lake, the oscillations of climate might influence human activity in other respects. The level of Great Salt Lake is not constant. As the lake has no outlet and the water brought in by streams and by rainfall is entirely absorbed by evaporation, it is highly dependent upon climatic oscillations. Since the close of the dry period just after the middle of the last century, the lake rose more than 12 feet up to about 1880, with a maximum during the wet period. When the dry period came, it fell to a low level. But now it is rising again. During our excursion, Mr. Hood, Chief Engineer of the Southern Pacific Company, gave us a table showing that the level of the lake was low at the beginning of the present century, but that since 1905 or 1906 it has risen no less than 6 feet. There can be no doubt that these oscillations of the lake are due to the oscillations of climate. I am aware that in general some other causes have Now the Union Pacific Railroad has built an embankment across the lake to shorten the road. The top of this embankment with the rails is today only a few feet above the level of the lake; I estimated from the window of our train about 5 feet, no more. If the future rise of the lake from the dry to the wet period will be as great as it has been before the maximum of about 1880, we must expect that in the middle of the next wet period, which we may expect about 1920, the level of the lake will be 6 feet higher than today and therefore will just cover the tracks.

But in another way also the settlement of the United States is controlled by climatic oscillations. At the beginning of this paper I showed by a series of curves (Fig. 1) that the oscillations of rainfall in Europe and in the United States coincide and that the two continents have simultaneous wet and dry periods. Now, the weather influences crops, but in a very different manner. In the United States the wet years are the good years, because in the cereal region with its high summer temperature, the rainfall is in general not very abundant. On the contrary, in the western part of Europe, including Germany, the dry years are good years since here in the cool climate the crops suffer because of too much humidity. The following Figures 2 and 3 may elucidate the relation between rainfall and crops in the western part of Europe. Above and beneath the diagrams are given the years, each year representing the middle of a lustrum, viz. 1848 the lustrum 1846–50, 1853, the lustrum 1851–55, etc. The curves marked R give the fluctuation of rainfall in percentage of the average. In Figure 2, WZ means crop of wheat expressed in percentage of an average crop; RO means crop of rye.

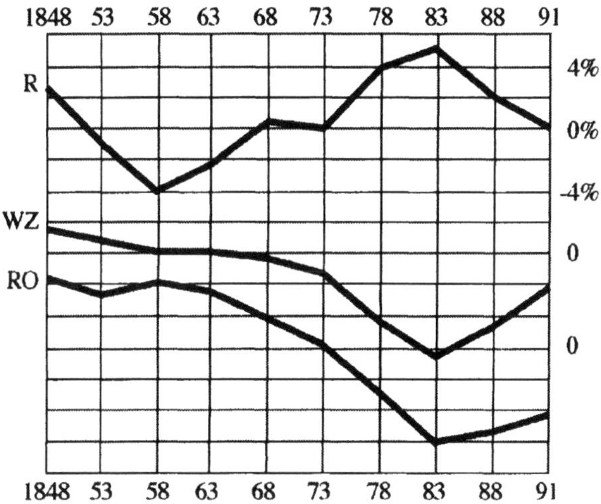

Figure 11.3: **Relation of Rainfall (R) to Crops (WZ and RO) in Prussia**. Grain production and rainfall are indicated in percentages of the variation from their respective means, which are based on a long series of observations. One division of the former is equivalent to 5% variation; of the latter to 4%.

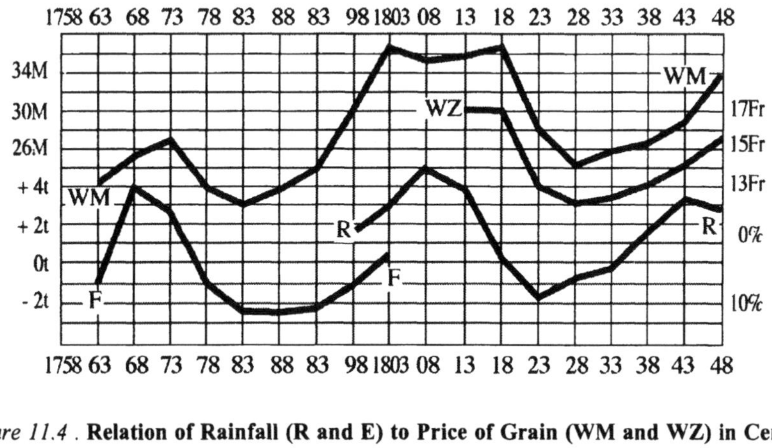

Figure 11.4 . **Relation of Rainfall (R and E) to Price of Grain (WM and WZ) in Central Europe**. WM = price, in marks, of a Bavarian bushel of wheat in München (1 division = 2 marks); WZ = price, in francs, of 50 kilograms of wheat in Zürich; R = rainfall in South Germany (1 division = 3% variation from the mean of many years); E = date of the vintage in

South Germany and Switzerland, which varies proportionally with the rainfall, being late in moist and early in dry years (1 division = 1 day; the mean of many years is marked 0 t).

Figure 2 shows that, when the rainfall increases in Prussia, the crops decrease and vice versa. The crops were good in the dry years about 1860, but very bad in the wet years about 1880; in the following drier period they increased again.

As the crop statistics do not go back very far, I have taken the price of grain as a criterion for the quality of the crop; for, before an international commerce in grain existed, the price of grain in a country was always a function of the crop of that country. When the crop was bad, the price was high and vice versa. In Figure 3 the curve WM shows the variation of the price of wheat at München (marks per Bushel); WZ, of the price of wheat at Zürich (francs per Bushel). E means the date of the vintage, which is a good measure of the rainfall, the grapes being harvested earlier in the dry years, as I have shown in the paper mentioned above.

The figure shows at a glance that in the wet periods about 1770, 1810, and 1845 grain was dear; in the dry periods about 1785 and 1830, cheap.

In the continental climates of Europe the relation between rainfall and crops is the same as in the United States. In Russia the crop of rye increases and decreases with the rainfall as in Ohio (see Fig. 4).

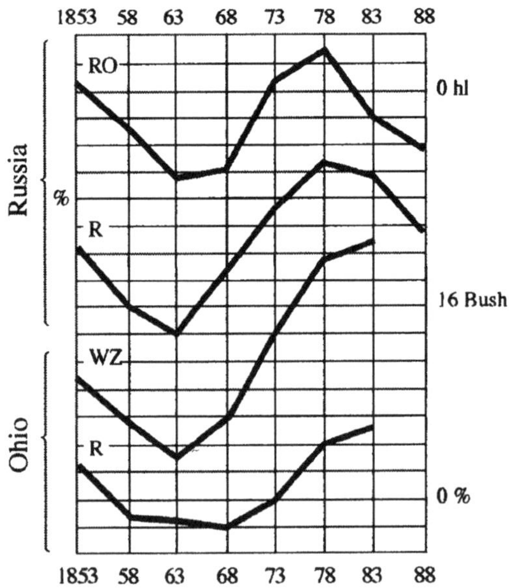

Figure 11.5. **Relation of Rainfall (R) to Crops (RO and WZ) in Russia and in Ohio.** One division is equivalent to 2% for the rainfall (R); to 1,500,000 hectoliters for the increase in the export of rye from Russia (RO); and to 0.8 bushel per Acre of the wheat crop in Ohio (WZ). (The variation in the increase of the rye export from Russia was determined as follows: A

straight line was drawn, according to the method of least squares, through the recorded quantities exported from 1851 to 1890, and the differences for every five years between these values and those actually observed were then plotted.)

These conditions are of great importance for the emigration from Europe and the immigration into the United States. The accompanying figure (Fig. 5) gives a diagram showing the oscillation of rainfall in the United States and in western Europe, and under it a diagram showing the immigration into the United States, all curves being constructed by using five-year averages. In general the immigration into the United States increased from the beginning to the end of the curve. But the increase was not regular. In the wet period about 1850, very well-pronounced in western Europe and in the United States, the immigration into the United States increased; in the following dry period it decreased. Five years after the 1880 maximum of rainfall in western Europe the immigration also had a maximum; it decreased materially in the next dry period. Only after 1900 it increased in an extraordinary manner.

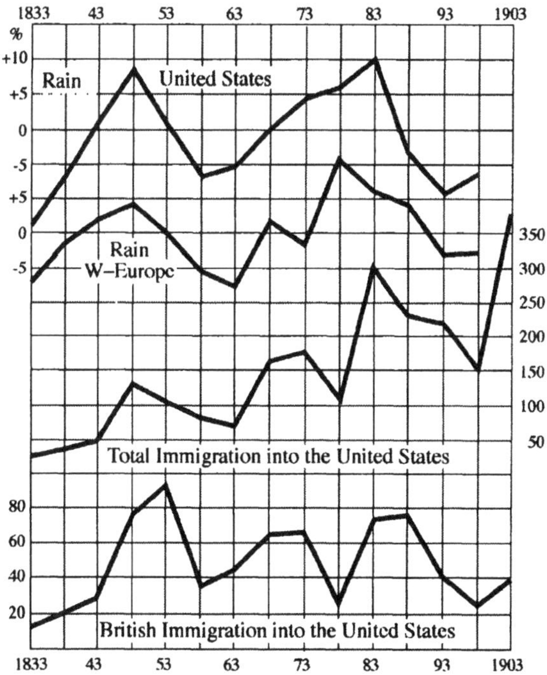

Figure 11.6. **The Fluctuation of Rainfall in the United States and Western Europe** as Compared, Respectively, with the Total Immigration and British Immigration to the United States. The curves are based on five-year totals and are not adjusted. The number of

immigrants is given in ten thousands, rainfall in deviations (percentages) from the mean of many years.

The parallelism between rainfall and the emigration from Europe is to be seen much better when we compare only the curve of rainfall for western Europe with British emigration (Fig. 5), or the curve of rainfall for Germany and the curve of emigration from Germany (Fig. 6). The parallelism of these two sets of curves is striking. The reason for the parallelism is clear: the greater part of the emigrants coming from Europe to the United States are agricultural. The rainy period causes bad crops in western Europe, including Germany, and therefore gives an impulse to emigration. In the same period the greater humidity is associated with good crops in the United States. This fact is communicated by correspondence to the relatives of the immigrants remaining in Europe, and this furnishes an additional incentive to immigration. As greater rainfall and the bad crops resulting therefrom are the cause of the increase of emigration out of Europe, it is not astonishing that sometimes the fluctuations of the curve of emigration are five years behind the fluctuations of rainfall.

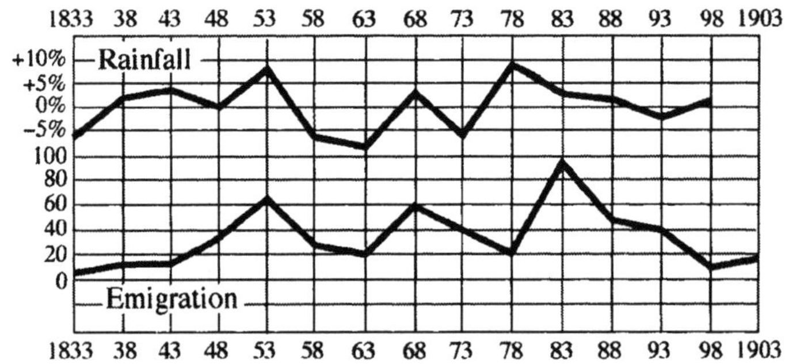

Figure 11.7. **Fluctuation of Rainfall in Germany and of German Emigration to the United States.** The curves are based on five-year totals and are not smoothed. Rainfall is indicated in percentage of variation from the mean of many years; the number of immigrants into the United States, in ten thousands.

Since 1900, the coincidence between rainfall and immigration into the United States ceases; since then immigration has reached numbers never attained before. That seems to be an exception to our rule. But if we examine the composition of this immigration, we find that the immigration from Great Britain and Germany is now very small, but that Russia and the eastern parts of Austria and Hungary are sending enormous numbers of emigrants. Here, where the summer is hot and the rainfall small, as in the United States, wet years are good years, dry years are bad years. Therefore in the dry period around 1900 the impulse to immigration has been great. To be sure, especially in Russia, there are also political considerations to be kept in mind. They might be of more importance in this connection than bad crops.

While we have seen that there is a correlation between the oscillations of climate and immigration into the United States, I am nevertheless far from overlooking other causes of emigration from Europe. The great density of population in Europe and the extended room available for settlement in the United States are constantly at work to induce emigration from Europe, and political causes are not lacking. But these forces, which are constantly at work, cannot veil the influence of climatic oscillations. The stream of immigrants to the United States ebbs and flows with the oscillations of climate, which give it a rhythmical impulse. And not only is immigration to the United States controlled by climatic oscillations, but also the settlement of the Far West, as we have seen.

List of Publications of Eduard Brückner

CLIMATE

1. *Über die Methode der Zählung der Regentage und deren Einfluß auf die resultierende Periode der Regenhäufigkeit.* Meteorologische Zeitschrift, Volume 3, 1887.
2. *Notre climat subit-il des changements?* Archives des sciences physiques et naturelles Sept.–Okt., 1888.
3. *Die meteorologische Station auf dem Säntis.* Meteorologische Zeitschrift, Volume 5, 1888.
4. ***In wie weit ist das heutige Klima konstant?***[455] Verhandlungen des VIII. Deutschen Geographentages, 1889.
5. ***Klimaschwankungen seit 1700 nebst Bemerkungen über die Klimaschwankungen der Diluvialzeit.***[456] Pencks Geographische Abhandlungen, Volume 4, 1890.
6. *Verdunstung einer Schneedecke.* Meteorologische Zeitschrift, Volume 7, 1890.
7. *Das Klima der Eiszeit.* Verhandlungen der 73. Jahresversammlung der Schweizer Naturforschenden Gesellschaft, Davos, 1891.
8. *Über die Bedeutung der Klimaschwankungen für das praktische Leben.* Compte rendu du V. Congrès International des Science Géographie, 1891.

[455] Chapter 3 in this book
[456] Excerpts in Chapter 4

9. *Materialien zur Verfolgung mehrjähriger oder säkulärer Perioden der Witterung*. Meteorologische Zeitschrift, Volume 9, 1892.
10. ***Über den Einfluß der Schneedecke auf das Klima der Alpen***.[457] Zeitschrift des Deutschen und Österreichischen Alpen-Vereins, 1893.
11. *Dürren in Ostasien*. Meteorologische Zeitschrift, Volume 11, 1894.
12. *Das Klima von Odessa*. Meteorologische Zeitschrift, Volume 11, 1894.
13. *Meteorologische Stationen in den Französischen Alpen*. Meteorologische Zeitschrift, Volume 12, 1895.
14. ***Der Einfluß der Klimaschwankungen auf die Ernteerträge und Getreidepreise in Europa***.[458] Geographische Zeitschrift, Volume 1, 1895.
15. *Über die Herkunft des Regens*. Verhandlungen des VII. Internationalen Geographenkongresses in Berlin 1899.
16. *Über die Herkunft des Regens*. Geographische Zeitschrift, Volume 6, 1900.
17. ***Zur Frage der 35-jährigen Klimaschwankungen***.[459] Petermann's Mittheilungen, 1902.
18. ***Wetterpropheten***.[460] 20. Jahresbericht der Geographischen Gesellschaft Bern, 1903/04.
19. *Höhengrenzen in der Schweiz*, Naturwissenschaftliche Wochenschrift, N. F., Volume 4 (Volume 20) No. 52, 1905.
20. *Die Bilanz des Kreislaufs des Wassers auf der Erde*. Geographische Zeitschrift, Volume 11, 1905. (In the same year also published in Russian.)
21. *Meer und Regen*. Naturwissenschaftliche Wochenschrift, N.F., Volume 4, No. 26, 1905.
22. *Schwankungen des Niederschlages im Deutschen Reich 1816–1900*. Meteorologische Zeitschrift, Volume 23, 1906.
23. *Schwankungen des Niederschlages im Deutschen Reich*. Zeitschrift für Gletscherkunde, Volume 1, 1906/07.
24. *Niederschlag, Abfluß und Verdunstung auf den Landflächen der Erde*. Meteorologische Zeitschrift, Volume 25, 1908.
25. *Schnee in der algerischen Sahara*. Zeitschrift für Gletscherkunde, Volume 11, 1908.
26. ***Über Klimaschwankungen***.[461] Mitteilungen der Deutschen Landwirtschaftsgesellschaft, Volume 24, 1909.
27. *Klimaschwankungen und Völkerwanderungen im XIX. Jahrhundert*. Internationale Wochenschrift für Wissenschaft, Kunst und Technik, 1910.

[457] Chapter 5
[458] Chapter 6 in this collection
[459] Chapter 8
[460] Chapter 7
[461] Chapter 9 in this collection

28. *Prozentischer Anteil des Schnees am gesamten Niederschlag in verschiedenen Höhen der Schweiz.* Zeitschrift für Gletscherkunde, Volume 5, 1910/11.
29. *Über die Klimaschwankungen der Quartärzeit und ihre Ursachen.* Compte rendu XI. Congrès géol. international Stockholm 1910 (1912).
30. Ergebnisse der Schneemessungen in den Schweizer Hochalpen. Zeitschrift für Gletscherkunde, Volume 5, 1910/11.
31. ***Klimaschwankungen und Völkerwanderungen.***[462] Vortrag in der feierlichen Sitzung der kaiserlichen Akademie der Wissenschaften, Wien 1912.
32. *Moorbildungen und postglaziale Klimaschwankungen am Nordsaum der Ostalpen.* Zeitschrift für Gletscherkunde, Volume 7, 1912/13.
33. With Albrecht Penck: *Über die Verschiebung der Klimagürtel in der Quartärzeit.* Zeitschrift für Gletscherkunde, Volume 9, 1914/15.
34. ***The settlements of the U.S. as controlled by climate and climatic oscillations.***[463] Memorial-Volumen of the transcontinental excursion of 1912 of the American Geographical Society of New York 1915.
35. *Klimaschwankungen 1813-1912 in Vorderindien.* Festschrift "Albrecht Penck", 1918.
36. *Änderungen der geographischen Breiten und des Klimas in geologischer Zeit.* Zeitschrift für Gletscherkunde, Volume 12, 1921/22.
37. *Verdunstung und Kondensation an Schnee und Eis im Gebirge. Zeitschrift für Gletscherkunde,* Volume 12, 1921/22.
38. *Die Schneedecke an der Straße zum Großen St. Bernhard in Wallis.* Zeitschrift für Gletscherkunde, Volume 12, 1921/22.
39. With W. Köppen und A. Wegener: *Die Klimate der geologischen Vorzeit.* Zeitschrift für Gletscherkunde, Volume 14, 1925/26.
40. *Forschungen über das diluviale Klima in Mexiko.* Zeitschrift für Gletscherkunde, Volume 15, 1926/27.

GLACIERS

1. *Die Hohen Tauern und ihre Eisbedeckung. Eine orometrische Studie.* Zeitschrift des Deutschen und Österreichischen Alpen-Vereins 1886.
2. *Die Höhe der Schneelinie und ihre Bestimmung.* Meteorologische Zeitschrift, Volume 4, 1887,
3. *E. Richters Untersuchungen über die Schwankungen der Alpengletscher.* Volume 1891.

[462] Chapter 10
[463] Chapter 11

4. *Der Gletscherabbruch an der Altels im Berner Oberland vom 11. Sept. 1895.* Himmel und Erde, Volume 8, 1896.
5. Mit S. Finsterwalder: *Protokoll der 3. Internat. Gletscherkonferenz in Maloja 6. bis 9. Sept. 1905.* P. M., Volume 51, 1905.
6. *Die Höhe der Firnlinie am Hüfigletscher und die Methode der Bestimmung der Höhe der Firnlinie im allgemeinen.* Vierteljahrsschrift der Naturforschenden Gesellschaft Zürich 1906, Volume 51.
7. *Zur Einführung. Zeitschrift für Gletscherkunde*, Volume 1, 1906/07.
8. Mit E. Muret: *Les variations periodiques des glaciers. XII. Rapport, 1906.* Zeitschrift für Gletscherkunde, Volume 2, 1907/08
9. *Die Schmelzformen des Firns im tropischen und subtropischen Hochgebirge.* (Nieve Penitente). Zeitschrift für Gletscherkunde, Volume 2, 1908.
10. *Schneeschmelzkegel in den Alpen und Nieve Penitente.* Zeitschrift für Gletscherkunde, Volume 2, 1908.
11. Mit E. Muret: *Les variations periodiques des glaciers. XIII. Rapport, 1907.* Zeitschrift für Gletscherkunde, Volume 3, 1908/09.
12. Mit E. Muret: *Les variations periodiques des glaciers. XIV. Rapport, 1908.* Zeitschrift für Gletscherkunde, Volume 4, 1909/10.
13. *Gletscherschwankungen in der Schweiz 1800-1900.* Zeitschrift für Gletscherkunde, Volume 4, 1909/10.
14. *Größe der Ablation am Rhonegletscher.* Zeitschrift für Gletscherkunde, Volume 5, 1910/11.
15. *Bericht der internationalen Gletscherkommission für die Jahre 1907–1910.* Zeitschrift für Gletscherkunde, Volume 5, 1910/11.
16. Mit E. Muret: *Les variations périodiques des glaciers. XV. Rapport, 1909.* Zeitschrift für Gletscherkunde, Volume 5, 1910/11.
17. *Die Firnhaube des Titlisgipfels in den Glarner Alpen, Schweiz.* Zeitschrift für Gletscherkunde, Volume 5, 1910/11.
18. *Bemerkungen zu der Abhandlung des Herrn Lamansky über das Absterben der Gletscher.* Zeitschrift für Gletscherkunde, Volume 8, 1913/14.
19. *Bolletino del Comitato Glaciologico Italiano.* Zeitschrift für Gletscherkunde, Volume 9, 1914/15.
20. *Beobachtungen über die Größenänderungen der Gletscher der Ostalpen in den Kriegsjahren 1914 und 1915.* Zeitschrift für Gletscherkunde, Volume 9, 1914/15.
21. *Veränderungen im Stande der Gletscher der österreichischen Alpen nach den Beobachtungen der Jahre 1914, 1915, und 1916.* Zeitschrift für Gletscherkunde, Volume 10, 1916/17.
22. *Eigenartige stengelige Anordnung des Eises in einem Glaszylinder.* Zeitschrift für Gletscherkunde, Volume 10, 1916/17.

23. *Das Vorrücken der Gletscher in den Ostalpen.* Mitteilungen des Deutschen und Österreichischen Apenvereins, 1917.
24. *Die Förderung der Wissenschaft von den Alpen durch den deutschen und österreichischen Alpenverein in den letzten 25 Jahren.* Zeitschrift des Deutschen und Österreichischen Alpen-Vereins 1919.
25. With Sikosak,: *Im Meer grönländischer Fjorde entstandene Firnfelder und Gletscher.* Zeitschrift für Gletscherkunde, Volume 11, 1918/20.
26. With J. Bowman: *Über Schneerosion und Entstehung der Kare.* Zeitschrift für Gletscherkunde, Volume 12, 1921.
27. *Vorstoß der Schweizer Gletscher.* Zeitschrift für Gletscherkunde, Volume 12, I921.
28. *Die meteorologischen Ursachen des Gletschervorstoßes in den Schweizer Alpen.* Zeitschrift für Gletscherkunde, Volume 12, 1921.
29. *Vermessung des Lysgletschers auf der Südseite des Monte Rosa.* Zeitschrift für Gletscherkunde, Volume 13, 1923/24.
30. *Die Vergletscherung der Kette Peters des Großen im Pamirgebiet.* Zeitschrift für Gletscherkunde, Volume 13, 1923/24.
31. *Fortbewegung des "Hotel des Neuchâtelois" auf dem Unteraargletscher im Berner Oberland 1842–1922.* Zeitschrift für Gletscherkunde, Volume 14, 1925/26.
32. *Verzeichnis der rezenten Gletscher Italiens.* Zeitschrift für Gletscherkunde, Volume 15, 1926/27.
33. *Dickenmessungen von Gletschern mittels seismischer Methoden.* Zeitschrift für Gletscherkunde, Volume 15, 1926/27.
34. Mit O. Lütschg: *Über Gletscher, Niederschlag und Abfluß im Mattmarkgebiet (Monte Rosagruppe).* Zeitschrift für Gletscherkunde, Volume 15, 1926/27.

GLACIAL AGES

1. *Die Eiszeit in den Deutschen Alpen, nach E. Penck.* Kosmos, I. Volume, 1884.
2. *Über die Vergletscherung Ostsibiriens.* N. Jahrbuch für Mineralogie, etc., Volume I, 1885.
3. *Die Eiszeit am Nordabhang der Alpen.* Der Naturforscher, Volume 19, 1886.
4. *Die Vergletscherung des Salzachgebietes nebst Beobachtungen über die Eiszeit in der Schweiz.* Pencks Geographische Abhandlungen, Volume 1, H. I, 1886.
5. *Die Eiszeit in den Alpen.* Mitteilungen Geographischen Gesellschaft Hamburg, 1887/88.

6. *Eiszeit-Studien in den südöstlichen Alpen.* 10. Jahresbericht der Geographischen Gesellschaft Bern, 1891.
7. *Le Système glaciaire des Alpes.* Guide publié à l'occassion du Congrès Géologique International 6me Session a Zürich en 1894. Bull. Soc. des scienc. natur. de Neuchâtel, T. 22, 1893/94.
8. *Die Eiszeiten in den Alpen.* Verhandlungen Gesellschaft Deutscher Naturforscher und Ärzte, 1904. Allg. Teil.
9. *Die Eiszeiten in den Alpen.* Geographische Zeitschrift, Volume 10, 1904.
10. *Die Eiszeiten in den Alpen und die "Einheitlichkeit" der Eiszeit.* Geographische Zeitschrift, Volume 11, 1905.
11. *Alpen und Eiszeit.* Mitteilungen der Geographischen Gesellschaft Hamburg, Volume 21, 1906
12. Mit A. Penck: *Die Alpen im Eiszeitalter.* 2. Volume, 2. Buch, 2. Kap.: *Linth- Reuß-, Aare- und Rhonegletscher auf Schweizer Boden*, 1903/04; 3. Volume, 3. Buch: *Eiszeit in den Südalpen, Brenta-, Piave- und Isonzogletscher,* 1908; 4. Buch: *Savegletscher.* Work completed in 1909.
13. *Palaeozoische Eiszeitspuren in der Kapkolonie.* Zeitschrift für Gletscherkunde, Volume 5, 1911.
14. *Über die Klimaschwankungen der Quartärzeit und ihre Ursachen.* Compte rendu XI. Congres geol. international Stockholm 1910 (1912).
15. *Interglaziale Torflager in den nördlichen Ostalpen.* Zeitschrift für Gletscherkunde., Volume 7, 1912/13.
16. *Zur Frage der Verschiebung der Eisscheide in Skandinavien.* Zeitschrift für Gletscherkunde, Volume 8, 1913/14.
17. *Lagerungsverhältnisse und Alter der Höttinger Breccie bei Innsbruck.* Zeitschrift für Gletscherkunde, Volume 10, 1916/17.
18. *Bemerkungen zum Aufsatz von Prof. Deecke über die tiefliegenden glazialen Reste in Südwestdeutschland und über die Lößtratigraphie Süddeutschlands.* Zeitschrift für Gletscherkunde, Volume 11, 1918/20.
19. With G. v. Zahn: *Über die angeblichen Moränen im Thüringer Wald.* Zeitschrift für Gletscherkunde, Volume 12, 1921/22.
20. *Geochronologische Untersuchungen über die Dauer der Postglazialzeit in Schweden, in Finnland und in Nordamerika.* Zeitschrift für Gletscherkunde, Volume 12, 1921/22.
21. *Die glaziale Entwicklungsgeschichte Nordwestskandinaviens.* Zeitschrift für Gletscherkunde, Volume 12, 1921/22, *13,* 1923/24.
22. *Albrecht Pencks neue Untersuchungen über die Eiszeit in den nördlichen Alpen.* Zeitschrift für Gletscherkunde, Volume 13, 1923/24.
23. *Die Salpausselkä-Randbildungen des Inlandeises in Finnland.* Zeitschrift für Gletscherkunde, Volume 13, 1923/24.
24. *Ein Institut für Eiszeitforschung.* Zeitschrift für Gletscherkunde, Volume 13, 1923/24.

25. *Geochronologische, geomorphologische und pflanzengeographische Untersuchungen im Bereich des alten Ragundasees in Jämtland, Schweden.* Zeitschrift für Gletscherkunde, Volume 14, 1925/26.
26. *Geochronologische Untersuchungen über den Rückzug der letzten Vergletscherung in den Neu-Englandstaaten.* Zeitschrift für Gletscherkunde, Volume 14, 1925/26.
27. *Die äußerste Grenze der letzten Vergletscherung in Nordwestdeutschland nach Gipp.* Zeitschrift für Gletscherkunde, Volume 14, 1925/26.
28. With H. F. Osborn: *Über die Gliederung des Quartärs und ihre Beziehung zur Prähistorie.* Zeitschrift für Gletscherkunde, Volume 14, 1925/26.
29. *Die großen Endmoränenzüge Norddeutschlands.* Zeitschrift für Gletscherkunde, Volume 15, 1926/27.
30. With Otto Ampferer: *Über geologische Methoden zur Erforschung des Eiszeitalters.* Zeitschrift für Gletscherkunde, Volume 15, 1926/27.
31. *Wieviel Jahre sind seit dem Hochstand der letzten Vergletscherung verstrichen?* Zeitschrift für Gletscherkunde, Volume 15, 1926/27.
32. *"Die Eiszeit". Zeitschrift für allgemeine Eiszeitforschung usw.* Zeitschrift für Gletscherkunde. Volume 15, 1926/27.
33. *Die geologische und archäologische Stellung des Hochgebirgspaläolithikums in der Schweiz.* Zeitschrift für Gletscherkunde, Volume 15, 1926/27.
34. With G. H. Steinmann: *Über das Diluvium des Niederrheins und die Gliederung des Eiszeitalters.* Zeitschrift für Gletscherkunde, Volume 15, 1926/27.
35. *Die Ostalpen in der Eiszeit.* In Deuticke (Ed): "Die österreichischen Alpen". Wien, 1927.

MORPHOLOGY

1. *Die feste Erdrinde und ihre Formen. Ein Abriß der allgemeinen Geologie und der Morphologie der Erdoberfläche.* (Volume 2 of "Allgemeine Erdkunde" von Hann, Hochstetter und Pokorny) g. Aufl. 1897. (1903 a Russian translation by M.A. Engelhardt was published in Petersburg.)
2. *Notice preliminaire sur la morphologie du Jura Suisse et Français.* Archives des Sciences physiques et naturelles. T. 14, 1902.
3. *Morphologie du Plateau Suisse et du Jura.* Compte rendu des travaux de la Soc. helvetique de sc. nat. 1902.
4. *Die glazialen Züge im Antlitz der Alpen.* Mitteilungen des Vereins für Erdkunde Leipzig 1906 (1907).

5. With W. M. Davis: *Über die glazialen Skulpturformen in Gebirgen.* Zeitschrift für Gletscherkunde, Volume 2, 1907.
6. With W. Kilian: *Über Glazialerosion und Übertiefung.* Zeitschrift für Gletscherkunde, Volume 2, 1907.
7. *Das Alter der alpinen Landschaftsformen.* Jahresbericht der Geographischen Gesellschaft Bern, Volume 21, 1907.
8. *Glazialmorphologische Exkursion in das Chamounixgebiet, ins Wallis und ins Berner Oberland.* Compte rendu des travaux du IX. Congrès international de Géographie Gèneve 1908, T. 1 (1910).
9. *Die glazialen Züge im Antlitz der Alpen.* Naturwissenschaftliche Wochenschrift N.F., Volume 8, No. 50, 1909 und Compte rendu des travaux du IX. Congrès international de Géographie, Gèneve 1908, T. 2 (1910).
10. *Zur Frage der Entwicklung der Rhein-Rhone-Wasserscheide.* Zeitschrift der Gesellschaft für Erdkunde, Berlin 1909.
11. *Entstehung der quartären Schotterterrassen im Umkreis der Alpen.* Zeitschrift für Gletscherkunde, Volume 4, 1909/10.
12. *Bemerkungen zu V. Hilber: Entstehung der quartären Schotterterrassen im Umkreis der Alpen.* Zeitschrift für Gletscherkunde, Volume 4, 1909/10.
13. *Gliederung der diluvialen Schotter in der Umgebung von Basel.* Zeitschrift für Gletscherkunde, Volume 6, 1911/12.
14. *Das Zungenbecken des alten Ennsgletschers als Felsbecken.* Zeitschrift für Gletscherkunde, Volume 7, 1912/13.
15. *Die sogenannten Glazialerscheinungen in der Rhön.* Zeitschrift für Gletscherkunde, Volume 8, 1913/14.
16. *Zur Frage der Entstehung der Sölle.* Zeitschrift für Gletscherkunde, Volume 9, 1914/15.
17. *Zur Morphologie der Ötscherlandschaft.* Mitteilungen der Geographischen Gesellschaft Wien 65, 1922.
18. *Alte Züge im Landschaftsbild der Ostalpen.* Vortrag Berlin. Zeitschrift der Gesellschaft für Erdkunde Berlin, 1922.
19. *Zur Glazialmorphologie von Norwegen.* Zeitschrift für Gletscherkunde, Volume 12, 1921/22.
20. *Über die bodengestaltende Wirkung des vorstoßenden Oberen Grindelwaldgletschers.* Zeitschrift für Gletscherkunde, Volume 12, 1921/22.
21. *Bemerkungen zu Worm: Kare und Schneegrenze.* Zeitschrift für Gletscherkunde, Volume 14, 1925/26.
22. *Glazialmorphologische Forschungen in der Schweiz und in Spitzbergen mittels des Flugzeugs.* Zeitschrift für Gletscherkunde, Volume 14, 1925/26.

23. *Die norwegische Strandplattform und ihre Beziehungen zur Quartärzeit.* Zeitschrift für Gletscherkunde, Volume 15, 1926/27.

HYDROLOGY

1. ***Grundwasser und Typhus.***[464] Mitteilungen der Geographischen Gesellschaft Hamburg, 1887/88.
2. *Über Schwankungen der Seen und Meere.* Verhandlungen des IX. Deutschen Geographentages, Wien, 1891.
3. *Säkulare Schwankungen der Eisverhältnisse des Hudsonflusses.* Meteorologische Zeitschrift, Volume 9, 1892.
4. *Untersuchungen über die tägliche Periode der Wasserführung und die Bewegung der Hochfluten in der oberen Rhone.* Petermann's Mittheilungen, 1895.
5. *Bericht der Flußkommission der Schweizerischen Naturforschenden Gesellschaft für das Jahr 1896/97.* Verhandlungen der Schweizerischen Naturforschenden Gesellschaft 1897.
6. *Bericht der Flußkommission für 1900/1901.* Verhandlungen der Schweizerischen Naturforschenden Gesellschaft, 1901.
7. *Berichte der Flußkommission der Schweizerischen Naturforschenden Gesellschaft pro 1903/04 und 1904/05.* Verhandlungen der Schweizerischen Naturforschenden Gesellschaft 1904 und 1905. (Published 1905 and 1906.)
8. *Die Bilanz des Kreislaufs des Wassers auf der Erde.* Geographische Zeitschrift, Volume 11, 1905.
9. *Bericht der Flußkommission für 1905/06.* Verhandlungen der Schweizerischen Naturforschenden Gesellschaft 1906.
10. *Gutachten betreffend die Folgen, die die Ausführung der Millstätter Kraftanlage für den See voraussichtlich haben wird.* Wien 1908, Ed. Sieger.
11. *Zur Thermik der Alpenseen und einiger Seen Nord-Europas.* Geographische Zeitschrift, Volume 15, 1909.
12. *Über die Temperaturverhältnisse der Flußseen, insbesondere der Alpen.* Compte rendu des travaux du IX. Congrès international de Géographie 1908, T. 2, (1910).
13. *Über das thermische Regime der Seen mit Abfluß. Verhandlungen der K. Russ.* Geographischen Gesellschaft 47, 1911 (Russisch).
14. *Die großen Seen Nordamerikas und ihre Bedeutung für den Verkehr.* Mitteilungen der Geographischen Gesellschaft Wien, Volume 61, 1918.

[464] Chapter 1 in this collection

15. *Bemerkungen zum Aufsatz von K. Fischer: Die Grundgleichungen des Wasserhaushaltes der Flußgebiete.* Meteorologische Zeitschrift, Volume 38, 1921.

OCEANOGRAPHY

1. *Die Schwankungen des Wasserstandes im Schwarzen Meere und ihre Ursachen.* Meteorologische Zeitschrift, Volume 2, 1886.
2. *Meeresspiegel und Klima.* Naturforscher, Volume 20, Tübingen 1887.
3. ***Die Schwankungen des Wasserstandes im Kaspischen Meer, dem Schwarzen Meer und der Ostsee in ihrer Beziehung zur Witterung.***[465] Ann. der Hydrographie und maritime Meteorologie, 1888.
4. *Über Schwankungen der Seen und Meere.* Verhandlungen des IX. Deutschen Geographentages, Wien 1891.
5. *Meer und Regen.* Naturwissenschaftliche Wochenschrift, N.F, Volume 4, No. 26, 1905.
6. *Das italienisch-österreichische Projekt einer gemeinsamen Erforschung des Adriatischen Meeres.* Mitteilungen der Geographischen Gesellschaft Wien, Volume 53, 1910.
7. *Die erste Kreuzungsfahrt S. M. S. "Najade" in der Hochsee der Adria 25. Febr. bis 7. März 1911.* Mitteilungen der Geographischen Gesellschaft Wien, Volume 54, 1911.
8. *Der Zustand des Adriatischen Meeres am Ausgang des Winters 1910/11.* Mitteilungen der Geographischen Gesellschaft Wien, Volume 54, 1911.
9. *Die dritte Terminfahrt S. M. S. "Najade" in der Hochsee der Adria vom 16. Aug. bis 5. Sept. 1911.* Mitteilungen der Geographischen Gesellschaft Wien, Volume 55, 1912.
10. *Das Projekt einer internationalen Erforschung des Mittelmeers.* Mitteilungen der Geographischen Gesellschaft Wien, Volume 57, 1914.

POLAR RESEARCH

1. *Resultate der meteorologischen Beobachtungen der deutschen Polarstationen 1882/83.* Meteorologische Zeitschrift, Volume 4, 1888.
2. *Der Kampf um den Nordpol.* Mitteilungen der Geographischen Gesellschaft Wien, Volume 52, 1909.
3. *Filchners deutsche antarktische Expedition.* Zeitschrift für Gletscherkunde, Volume 5, 1910/11.

[465] Chapter 2 in this collection

4. *Das Eis der Antarktis*. Schriften des Vereins zur Verbreitund naturwissenschaftlicher Kenntnisse. Volume 51, 1910/11.
5. *Die Alaskaexpedition der Amerikanischen Nationalen Geographischen Gesellschaft vom Sommer 1910*. Zeitschrift für Gletscherkunde, Volume 6, 1911.
6. *Neuer Plan einer Durchquerung Grönlands*. Zeitschrift für Gletscherkunde, Volume 6, 1911/12.
7. *Die Schneegrenze in der Antarktis*. Zeitschrift für Gletscherkunde, Volume 7, 1912/13.
8. *Die Ergebnisse der Schweizerischen Grönlandexpedition 1912/*13. Zeitschrift für Gletscherkunde, Volume 12, 1921/22.
9. With E. v. Drygalski: *Über das Eis der Antarktis und der subantarktischen Meere*. Zeitschrift für Gletscherkunde, Volume 13, 1923/24.
10. With Meinardus: *Über die hypsographischen Kurven Grönlands und der Antarktis und die Normalform der Inlandeisoberfläche*. Zeitschrift für Gletscherkunde, Volume 15, 1926/27.

CARTOGRAPHY

1. *Bericht über das Projekt einer Erdkarte im Maßstab 1:1.000.000*. 11. Jahresbericht der Geographischen Gesellschaft Bern, 1891/92.
2. *Schweizerische Reliefkarten*. 12. Jahresbericht Geographischen Gesellschaft Bern, 1893.
3. *Die Frage der Weltkarte 1:1.000.000 vor dem Londoner Geographenkongreß*. 14. Jahresbericht Geographischen Gesellschaft Bern, 1895.
4. *Über Karten der Volksdichte*. Zeitschrift für Schweizerische Statistik, 1903.
5. *Neue Karten des Montblancgebietes*. Zeitschrift für Gletscherkunde, Volume 2, 1908.
6. *Zur Frage der Farbenplastik in der Kartographie*. Mitteilungen der Geographischen Gesellschaft Wien, Volume 52, 1909.
7. *Die internationale Weltkartenkonferenz in London*. Mitteilungen der Geographischen Gesellschaft Wien, Volume 53, 1910.
8. *Oberleutnant E. v. Orels Stereoautograph als Mittel zur automatischen Herstellung von Schichtenplänen und Karten*. Mitteilungen der Geographischen Gesellschaft Wien, Volume 54, 1911.
9. *Die internationale Weltkartenkonferenz in Paris im Dezember 1913*. Mitteilungen der Geographischen Gesellschaft Wien, Volume 57, 1914.
10. *Die Dachstein-Karte*. Mitteilungen des Deutschen und Österreichischen Alpenvereins 1916.

11. *Hundert Jahre Militärgeographisches Institut.* Mitteilungen des Deutschen und Österreichischen Alpenvereins, 1919.
12. *Die Entwicklung des Kartographischen, früher Militärgeographischen Institutes in der Zeit des Umsturzes bis Ende 1923.* Mitteilungen der Geographischen Gesellschaft Wien, Volume 66, 1923.

BIOGRAPHICAL

1. *Nekrolog auf E. Hagenbach-Bischoff.* Zeitschrift für Gletscherkunde, Volume 5, 1910/11.
2. *Dr. Josef Roman Ritter Lorenz von Liburna. Sein Leben und Wirken.* Mitteilungen der Geographischen Gesellschaft Wien, Volume 55, 1912.
3. *F.A. Forel*†. Zeitschrift für Gletscherkunde, Volume 7, 1912/13.
4. *Prof. James Geikie*†. Zeitschrift für Gletscherkunde, Volume 9, 1914/15.
5. *Alfred Grund*†. Mitteilungen der Geographischen Gesellschaft Wien, Volume 58, 1915.
6. *Alexander Woeikof*†. Zeitschrift für Gletscherkunde, Volume 10, 1916/17.
7. *Prof. Dr. F. Mühlberg*†. Zeitschrift für Gletscherkunde, Volume 10, 1916/ 17.
8. *Julius Hann*†. Mitteilungen Geographischen Gesellschaft Wien, Volume 64, 1921.
9. *Sven Hedin zum 60. Geburtstag,* Mitteilungen des Deutschen und Österreichischen Alpenvereins 1925.
10. *Robert Sieger*†. Pet.ermann's Mittheilungen, Volume 72, 1926.
11. *Prof. Olinto Marinelli*†. Zeitschrift für Gletscherkunde, Volume 15, 1926/27.

MISCELLANEOUS

1. *Über die angebliche Änderung der Entfernung zwischen Jura und Alpen.* 11. Jahresbericht der Geographischen Gesellschaft Bern, 1891/92.
2. *Über die Geschwindigkeit der Gebirgsbildung und der Gebirgsabtragung.* Himmel und Erde, Volume 6, 1894.
3. *Die Schweizerische Landschaft einst und jetzt.* Rektoratsrede, given on 18. November 1899, Bern 1900.
4. *Dalmatien und das Österreichische Küstenland.* Talks given on the occasion of the first university roundtrip. Wien und Leipzig, 1911.
5. *Das Pflanzenschaf (Baranetz).* Russische Revue, Volume 21, 1882.
6. *Wanderungen des Elentiers in Rußland.* Kosmos, Volume 1, 1884.

7. *Die Stellung der Geographie auf dem Gymnasium.* Talk given in Bern 1893.
8. *Über die Heranbildung der Geographielehrer an der Universität.* Geographische Zeitschrift, Volume 15, 1905.
9. *Die größte Volkshochschule der Welt.* Internationalen Monatsschrift für Wissenschaft, Kunst und Technik, Volume 8, 1914.
10. *Die Kriegstagung deutscher Hochschullehrer der Geographie Ostern 1916 zu Heidelberg.* Mitteilungen der Geographischen Gesellschaft Wien, Volume 59, 1916.
11. *Rapport sur l'exposition de géographie scolaire.* Compte rendu du V. Congrès International des Sciences Géographiques, 1891.
12. *Bericht über den V. Internationalen Kongreß der geograph. Wissenschaften zu Bern vom 10.–14. Aug. 1891.* 11. Jahresbericht Geographischen Gesellschaft Bern, 1891/92; at the same time published in French
13. *Der VI. Internationalen Geographischen Kongreß zu London 1895.* 14. Jahresbericht der Geographischen Gesellschaft Bern, 1895.
14. *Bericht über den VII. Internationalen Geographenkongreß.* 17. Jahresbericht der Geographischen Gesellschaft Bern, 1899.
15. *Die transkontinentale Exkursion der Amerikanischen Geographischen Gesellschaft durch die Vereinigten Staaten,* August bis Oktober 1912. Mitteilungen der Geographischen Gesellschaft Wien, Volume 56, 1913.
16. *Die k. k. Geographische Gesellschaft und die Entwicklung der Geographie in den letzten zehn Jahren.* Mitteilungen der Geographischen Gesellschaft Wien, Volume 60, 1917.
17. *Die wissenschaftlichen Ergebnisse der Pamirexpedition des* Deutschen und Österreichischen *Alpenvereins 1913.* Mitteilungen des Deutschen und Österreichischen Alpenvereins 1924.
18. *Eine Flaschenpost vom Tegetthoff.* Mitteilungen der Geographischen Gesellschaft Wien, Volume 65, 1922.

Subject Index

Name Index

MIX
Papier aus verantwortungsvollen Quellen
Paper from responsible sources
FSC® C105338

If you have any concerns about our products,
you can contact us on
ProductSafety@springernature.com

In case Publisher is established outside the EU,
the EU authorized representative is:
Springer Nature Customer Service Center GmbH
Europaplatz 3, 69115 Heidelberg, Germany

Printed by Libri Plureos GmbH
in Hamburg, Germany